Navigation: Science and Technology

Volume 6

This series *Navigation: Science and Technology (NST)* presents new developments and advances in various aspects of navigation - from land navigation, marine navigation, aeronautic navigation to space navigation; and from basic theories, mechanisms, to modern techniques. It publishes monographs, edited volumes, lecture notes and professional books on topics relevant to navigation - quickly, up to date and with a high quality. A special focus of the series is the technologies of the Global Navigation Satellite Systems (GNSSs), as well as the latest progress made in the existing systems (GPS, BDS, Galileo, GLONASS, etc.). To help readers keep abreast of the latest advances in the field, the key topics in NST include but are not limited to:

- Satellite Navigation Signal Systems
- GNSS Navigation Applications
- Position Determination
- Navigational instrument
- Atomic Clock Technique and Time-Frequency System
- X-ray pulsar-based navigation and timing
- Test and Evaluation
- User Terminal Technology
- Navigation in Space
- New theories and technologies of navigation
- Policies and Standards

This book series is indexed in **SCOPUS** database.

More information about this series at http://www.springer.com/series/15704

Zheng Yao · Mingquan Lu

Next-Generation GNSS Signal Design

Theories, Principles and Technologies

中国工信出版集团

電子工業出版社
PUBLISHING HOUSE OF ELECTRONICS INDUSTRY
http://www.phei.com.cn

Springer

Zheng Yao
Department of Electronic Engineering
Tsinghua University
Beijing, China

Mingquan Lu
Department of Electronic Engineering
Tsinghua University
Beijing, China

ISSN 2522-0454 ISSN 2522-0462 (electronic)
Navigation: Science and Technology
ISBN 978-981-15-5798-9 ISBN 978-981-15-5799-6 (eBook)
https://doi.org/10.1007/978-981-15-5799-6

Jointly published with Publishing House of Electronics Industry
The print edition is not for sale in China (Mainland). Customers from China (Mainland) please order the print book from: Publishing House of Electronics Industry.
ISBN of the Co-Publisher's edition: 978-7-121-28522-6

Original Chinese edition published by Publishing House of Electronics Industry, China, 2016

This Springer imprint is published by the registered company Springer Nature Singapore Pte Ltd.
The registered company address is: 152 Beach Road, #21-01/04 Gateway East, Singapore 189721, Singapore

Preface

In the past 50 years, navigation and positioning technology has made revolutionary progress in terms of territorial coverage and accuracy. Especially since the emergence of Global Navigation Satellite System (GNSS) such as the United States' Global Position System (GPS) and Russia's GLObal Navigation Satellite System (GLONASS), navigation and positioning technology has significantly changed human enterprise and lifestyles. However, requirements for improvement in navigation and positioning performance are constantly increasing. As demand continues to rise, the performance limitations of traditional satellite navigation systems, which were designed and built half a century ago, have become increasingly limiting. Concurrently, great progress has been made in theoretical research and technological development in physics, aerospace engineering, wireless communications, and in satellite navigation itself. Next-generation GNSSs—such as the modernizing of GPS, the European Union's Galileo system, and China's BeiDou Navigation Satellite System (BDS)—are rapidly developing based on decades of experience and are incorporating new technologies.

Satellite navigation systems are based on efficient and accurate signal design and processing. The inherent performance limitations of satellite navigation signals determine the upper limit of performance for the entire system. The space and the ground segments of the satellite navigation system operate based on the transmission and reception of signals. If the signals broadcast by the satellite have design flaws, even though the design of other components are perfect, the overall system will have significant deficiencies in key functions such as position determination, speed measurement, and timing accuracy, and the application and industrialization of the satellite navigation system will be severely restricted.

With the continuous expansion of satellite navigation applications and the refinement of service requirements, the number of navigation signals broadcast by the next-generation GNSS has increased significantly, and the performance requirements for such signals have risen simultaneously. On the one hand, the radio frequency spectrum resources available for satellite navigation have become increasingly allocated, thus severely limiting improvement in signal performance that might rely on increased bandwidth usage. On the other hand, the power efficiency of existing constant envelope multiplexing techniques has also become lower, while the complexity of the satellite signal has significantly increased. At the same time, the requirements dealing with limitations of signal processing caused by increased signal complexity, the interoperability requirements between systems, and the backward compatibility to legacy systems cause constraint in the achievement of signal design optimization. The lack of design theory, the lag between technological development and future demand, and the contradiction between expanding demand and the scarcity of resources pose a series of challenges to the signal design of the next-generation satellite navigation systems.

In recent years, many novel solutions have been proposed by the academic and industrial community to the problems faced in the design of satellite navigation signals. In terms of spread spectrum modulation, the spectral split modulation schemes represented by binary offset carrier (BOC) and multiplexed BOC (MBOC) have begun to be applied to the next-generation navigation signals. In the realm of constant envelope multiplexing, many advanced techniques have emerged. New achievements have been made in the development of spreading codes, channel structures, and message structures. The emergence of these new techniques has given designers of the next-generation satellite navigation systems wider choices, but at the same time places higher requirements on signal designers to understand the technology and grasp users' needs.

The construction of each satellite navigation system shoulders different historical burdens. The specific mission and application requirements for each system are different as well. Moreover, the design constraints of the various elements of the signal are coupled to each other. Therefore, it is impossible to design a new navigation signal with excellent comprehensive performance by simply piecing together different techniques or mimicking other systems. The signal designer must have a deep understanding of the characteristics and applicable conditions of each technique, of what they are, and how they work in specific situations. This is the only way they can select the most suitable configuration when faced with specific requirements, and finally, design a navigation signal with excellent comprehensive performance. In addition, the signals of the next-generation system are very different from traditional satellite navigation signals in terms of structure and methods of processing. An in-depth and systematic understanding of the characteristics of the next-generation GNSS signals is also very important for GNSS receiver developers.

The authors of this book had been heavily involved in the construction of China's BDS, and have many years of experience in the design of satellite navigation signals and receivers. Academic communication with peers in this field has deepened our understanding of navigation signal design. In the process of our own research, we have also achieved innovative and novel results. We wished to contribute to the construction and application of the next-generation GNSS systems with our knowledge and experience. Therefore, the idea of publishing this book came naturally.

Much academic work has been undertaken on the theories and applications of satellite navigation systems. Many scholars have done excellent research, offering detailed discussions on the working principles, system composition, receiver principles, and application of specific satellite navigation systems. This book focuses on signal design, which complements the existing literature and contributes to the knowledge base related to satellite navigation. With the navigation signals broadcast by satellites as the thread, this book focuses on the structure of navigation signals and the design ideas and theories behind each structure component, as well as the latest research. In preparing the text, we have tried to organize, describe, and evaluate existing work according to our own understanding. This book is different from the conventional "user manual" interpretation, which is based on signal interface control document (ICD) of each satellite navigation system. Instead, it takes a technical angle, incorporating the traditional technology of signal design and newly developed advanced technology into a general framework to be discussed as an organic whole. In our discussion, we tried to avoid simply enumerating various techniques, aiming instead to reveal their essence. This approach requires readers to have a solid mathematical foundation and knowledge of signal processing. There is a tight coupling between chapters as we believe that this style of discussion can help readers view the satellite navigation system from a higher perspective, seeing the overall structure of the navigation signal and the relationship between the various elements more clearly, as well as the nature, origin, and evolution process behind each technique. We hope that this book will help readers learn about relevant research fields, grasp the characteristics of each signal design technique, and understand the scientific problems behind the evolution of this technique in a deeper way. We trust that it will inspire readers to think, and to generate more innovative ideas that will contribute to the research and construction of future satellite navigation systems.

The book is divided into nine chapters. Chapter 1 reviews the development of satellite navigation technology, and then explains the development and current status of GPS, GLONASS, Galileo, and BDS. After a detailed discussion of the importance of navigation signals in GNSS construction, a review is given of the development of the signal structure for the above-mentioned major GNSSs, thus providing a background for the next-generation GNSS signal design discussion.

Chapter 2 begins with explaining the basic principles of satellite navigation so as to clarify the missions and functions to be carried out by a navigation signal. Then, using the structural elements of satellite navigation signals as the main thread, this chapter systematically expounds on the important basic concepts in the design of the signals. We also discuss the effects of various factors on system performance, such as center frequency, transmission power, polarization characteristics, spreading modulation, spreading code, message structure, and the multiplexing of signals. This chapter serves as the basis for subsequent chapters. Since the elements that make up the satellite navigation signal are closely related, the design of many elements of the signal is carried out in the form of trade-offs and compromises. Therefore, it is necessary to have a systematic understanding of the structure of the entire signal before conducting in-depth research on each aspect.

Chapters 3–5 explore the spread spectrum modulation techniques currently used in the field of satellite navigation. Chapter 3 discusses basic concepts and properties of the direct sequence spread spectrum (DSSS) technique, and provides the mathematical model and some important characteristics of the time and frequency domains of the spreading modulated signal. The conclusions of this chapter support the discussions in Chaps. 4 and 5.

Chapters 4 and 5 work in parallel. Chapter 4 details several typical spread spectrum modulation techniques, such as BPSK with rectangular chips (BPSK-R), BOC, and MBOC, while Chap. 5 systematically elaborates on a series of analyses and evaluation methods for signal performance at the design stage. The content of these two chapters is strongly linked. On the one hand, the characteristics of the various modulation methods are reflected by the results of performance analysis. On the other hand, many modulation methods are also designed via optimization which is based on performance evaluation criteria. Therefore, for those who read these two chapters for the first time, we recommend going through the general ideas and basic conclusions of each chapter individually, and then reading the two chapters as a whole in detail.

Chapters 6 and 7 offer an in-depth discussion of the basic theory and implementation methods of on-board constant envelope multiplexing of satellite navigation signals. Chapter 6 mathematically models the constant envelope multiplexing process of general satellite navigation signals, and gives the mathematical and geometric descriptions of the constant envelope on the basis of the baseband complex envelope concept. Further, the concepts of phase mapping tables, constellation diagrams, and multiplexing efficiency are given. Then, starting from the intermodulation component structure, and the phase mapping table, we explore the design equations and analysis methods of general constant envelope multiplexing technology of these two seemingly completely different constant envelope multiplexing design concepts. Finally, by using the concept of the baseband signal to unify the conclusions of the two viewpoints, we give three equivalent representations of a constant envelope multiplexing scheme. After mastering the basic theories presented in this chapter, readers will have a clearer and deeper understanding of the specific techniques of constant envelope multiplexing when reading Chap. 7.

Chapter 7 discusses several typical constant envelope multiplexing techniques in detail. Some of these techniques have already been used in satellite navigation systems in operation, while others have promising prospects for application in future system construction. In the discussion of these implementation techniques, we do not simply enumerate the various technical implementations, but focus on their essence and nature, so that readers can understand the implementation details of each technique while noticing the connections between them, thereby gaining a more systematic understanding of the research area.

Chapter 8 takes a close look at how we can enable future development by implementing excellent signal designs with higher adaptability and flexibility. This chapter can also be regarded as a comprehensive example of applying the theory and techniques provided by this book into the future signal design. In this chapter, the concept of using multicarrier modulation and multiplexing joint design to get out of the cycle of conflicts in future GNSS signal design is proposed. Performance analysis with some typical case studies demonstrates that the proposed multicarrier constant-envelope composite (MCC) signal provides a promising solution for the next-generation GNSS signal design.

Chapter 9 offers a brief summary of the book, and also foresees potential future developments of navigation signals.

The content of this book fully reflects the new ideas and technologies in satellite navigation signal design originating from both academia and industry over recent years. It also incorporates the authors' many years of work in signal design. With a clear theme and structure, it is an academic work that places equal emphasis on theory and practice. It can be used as a reference book for scientific and technological personnel in satellite navigation and related fields, or as a textbook for teachers and graduate students at higher education institutions.

Beijing, China
April 2020

Zheng Yao
Mingquan Lu

Acknowledgements

This work is supported by the National Natural Science Foundation of China (NSFC) under Grant No. 61771272. At the time of publication, we would like to thank students and colleagues in the GNSS Laboratory at the Department of Electronic Engineering, Tsinghua University. Special thanks go to Xiaowei Cui, and Hong Li for their long-term guidance and help. In particular, we would like to thank those current and former graduate students who have worked in this field under our supervision, including Xiaoming Zhang, Fu Guo, Jiayi Zhang, Tengfei Wang, Yang Gao, and Junjie Ma. Some of their research has found their way into this book. Without the dedicated work of these students, this book would not have been possible.

Many thanks to James Torley, who kindly and carefully reviewed the manuscript and, by way of innumerable discussions, helped in our endeavour to polish it. The assistance of Tengfei Wang and Ningyan Guo in rectifying and debugging the manuscript can hardly be overestimated, too. We would also like to thank our peers and colleagues from China and abroad for their help and support.

We bring our gratitude also to Mei Song and her colleagues at Publishing House of Electronics Industry for initiating this project and inspiring us in the course of writing, and our special thanks to Mengchu Huang and his colleagues at Springer Nature for the help at the final stage of preparing the book.

Finally, we thank our family members for making this effort worthwhile and for their patience during the numerous hours that we spent in front of computers.

Contents

1 Introduction 1
- 1.1 Overview 1
- 1.2 The Development and Current Status of GNSSs 4
  - 1.2.1 Predecessors of the Current GNSS Programs 4
  - 1.2.2 GPS and Its Modernization 5
  - 1.2.3 GLONASS and Its Modernization 7
  - 1.2.4 The Galileo Satellite Navigation System 8
  - 1.2.5 The BeiDou Satellite Navigation System 9
- 1.3 The Importance of Navigation Signals 11
- 1.4 Development of the Satellite Navigation Signals 12
  - 1.4.1 Development of the GPS Signals 13
  - 1.4.2 Development of the Galileo Signals 14
  - 1.4.3 Development of the BDS Signals 15
- References 16

2 Structure of Satellite Navigation Signals 19
- 2.1 Introduction 19
- 2.2 Basic Principles of Satellite Navigation 20
  - 2.2.1 Position Determination by Signal Ranging 20
  - 2.2.2 The Linearization Method of Position Estimation 23
  - 2.2.3 Precision of User Position 24
  - 2.2.4 Pseudorange Measurement 24
- 2.3 Key Elements of Satellite Navigation Signals 25
- 2.4 Carrier Frequency 26
- 2.5 Transmission Power 32
- 2.6 Polarization 36
- 2.7 Multiple Access 38
- 2.8 Spreading Modulation 39

2.9 Spreading Sequence and Secondary Code . . . 42
2.9.1 Even and Odd Correlations of Spreading Sequences . . . 43
2.9.2 Construction of Spreading Sequences . . . 45
2.9.3 The Effect of Spreading Modulation on Code Correlation . . . 49
2.9.4 Secondary Code . . . 51
2.10 Pilot Channel and Data Channel . . . 53
2.10.1 The Role of the Pilot Channel . . . 53
2.10.2 Power Allocation of the Data Channel and Pilot Channel . . . 56
2.11 Multiplexing . . . 58
2.12 Message Structure and Channel Coding . . . 59
2.12.1 Message Structure . . . 59
2.12.2 Channel Coding . . . 60
References . . . 63

3 Basic Properties of Direct Sequence Spread Spectrum Signals . . . 65
3.1 Introduction . . . 65
3.2 Spreading Modulated Signal Model . . . 67
3.3 Time Domain Characteristics of Spreading Modulated Signals . . . 70
3.3.1 Measure of Signal Similarity . . . 70
3.3.2 Cross-Correlation and Autocorrelation Functions . . . 70
3.3.3 Periodic Signal Correlation Function . . . 71
3.3.4 Cross-Correlation Function of Spread-Spectrum Signals . . . 72
3.3.5 Cross-Correlation Function of SCS Modulated Signals . . . 74
3.4 Frequency Domain Characteristics of Spreading Modulated Signals . . . 79
3.4.1 Power Spectral Density . . . 79
3.4.2 PSD of Spread-Spectrum Signals Without Message Modulation . . . 80
3.4.3 PSD of Spread-Spectrum Signals with Message Modulation . . . 82
3.4.4 PSD of Aperiodic Spread-Spectrum Signals . . . 85
3.4.5 PSD of SCS Modulated Signals . . . 85
3.4.6 Cross-Power Spectral Density . . . 87
3.4.7 Normalization of PSD . . . 87
References . . . 88

4 Spreading Modulation Techniques in Satellite Navigation 89
4.1 Introduction 89
4.2 BPSK-R Modulation 92
4.3 BOC Modulation 96
4.3.1 Definition of BOC Modulation 96
4.3.2 PSD of BOC Signals 98
4.3.3 The Autocorrelation Function of BOC Signals 101
4.3.4 Characteristic Differences Between Sine-Phase and Cosine-Phase BOC Modulations 103
4.4 BCS Modulation 107
4.5 CBCS Modulation 111
4.5.1 Definition of CBCS Signals 111
4.5.2 The Autocorrelation Function of the CBCS Signal 113
4.5.3 PSD of CBCS Signals 114
4.5.4 Cross-Correlation Deviation of CBCS Signals 117
4.6 TMBCS and QMBCS Modulations 120
4.7 MBOC Modulations 120
4.7.1 TMBOC Modulation 123
4.7.2 CBOC Modulation 124
4.7.3 QMBOC Modulation 127
4.8 The Processing Ambiguity of Split Spectrum Signals 130
4.8.1 Description of the Problem 130
4.8.2 The False Acquisition Probability Under the Serial Acquisition Strategy 131
4.8.3 The False Acquisition Probability Under the Parallel Acquisition Strategy 133
4.8.4 The Ambiguity Thread in Code Tracking 136
4.8.5 Methods to Eliminate Ambiguity 136
4.9 Other Spreading Modulations 147
4.9.1 AltBOC Modulation 147
4.9.2 MSK Modulation 151
4.9.3 GMSK Modulation 153
4.9.4 SRRC Spread Spectrum Waveform 156
4.9.5 PSWF Spreading Waveform 158
4.9.6 Summary of Bandwidth-Limited Spreading Modulation Techniques 159
References 160

5 Performance Evaluation Theory for Satellite Navigation Signals . . . 163
5.1 Introduction 163
5.2 Baseband Equivalent Expressions of Received Signals 167
5.2.1 Complex Envelope Representation of Signals 167
5.2.2 PSD of the Signal and Noise-Plus-Interference 167
5.3 Lower Bound of the Spreading Code Tracking Error 168

5.3.1 Cramér–Rao Lower Bound of the Code Tracking Error . . . 169
5.3.2 CRLB in a Gaussian White Noise Environment . . . 170
5.4 Signal Processing Model for Satellite Navigation Receivers . . . 172
5.4.1 Pre-detection Integration . . . 173
5.4.2 Acquisition . . . 174
5.4.3 Code Tracking . . . 175
5.4.4 Carrier Tracking . . . 177
5.5 Ranging Performance Under Thermal Noise and Interference . . . 178
5.5.1 Spreading Chip Waveform Mismatch . . . 179
5.5.2 Formulation and Statistical Properties of Correlator Output . . . 181
5.5.3 Statistical Characteristics of the Discriminator Output . . . 182
5.5.4 Tracking Error Under Non-coherent Processing . . . 183
5.6 The Performance of Acquisition, Carrier Tracking, and Data Demodulation . . . 187
5.7 Multipath Resistant Performance . . . 191
5.8 Radio Frequency Compatibility . . . 194
5.8.1 The Spectrum Separation Coefficient . . . 195
5.8.2 The Code Tracking Spectral Sensitivity Coefficient . . . 196
5.8.3 Equivalent Carrier-to-Noise Ratio . . . 200
References . . . 202

6 Fundamental Theory of Constant Envelope Multiplexing for Spread-Spectrum Signals . . . 205
6.1 Introduction . . . 205
6.2 High Power Amplifier . . . 207
6.3 Constant Envelope Signal . . . 208
6.4 Constant Envelope Multiplexing . . . 208
6.4.1 Signal Model . . . 209
6.4.2 Inter-modulation Term . . . 212
6.5 Transparency Constraint . . . 214
6.5.1 Orthogonality Constraint . . . 215
6.5.2 Basis Vector Construction in Bipolar Case . . . 217
6.6 Envelope Constancy Constraint . . . 218
6.7 Efficiency of Constant Envelope Multiplexing . . . 219
6.8 Design Methodologies of CEM . . . 220
6.8.1 Waveform Domain Processing . . . 220
6.8.2 Phase Domain Processing . . . 222
6.8.3 Summary . . . 226
6.9 Representation and Implementation of CEM Signals . . . 227
6.9.1 Phase-Mapping-Based Form . . . 228
6.9.2 Phase-Synthesis-Based Form . . . 229

6.9.3 Waveform-Synthesis-Based Form 230
6.9.4 Relationship Between Different Representations 232
6.10 PSD of CEM Signals 233
References 234

7 Constant Envelope Multiplexing Techniques for Spread-Spectrum Signals 237
7.1 Introduction 237
7.2 QPSK Multiplexing 239
7.3 Time Division Multiplexing 241
7.4 POCET Technique 245
7.5 Quadrature Product Subcarrier Modulation 247
7.5.1 QPSM Multiplexing of Three-Signal Case 248
7.5.2 QPSM Multiplexing of Arbitrary Number of Signals . . . 249
7.5.3 QPSM Multiplexing for CBCS Signals 250
7.6 Multiplexing Based on Majority Voting Logic 252
7.6.1 Majority Voting Logic 252
7.6.2 Uniform Weighting MV Multiplexing 252
7.6.3 Non-uniform Weighting MV Multiplexing 255
7.6.4 Phase Mapping Table of Interlaced MV Multiplexing 258
7.7 Constant Envelope Multiplexing via Intermodulation Construction (CEMIC) 261
7.8 Multi-frequency Constant Envelope Multiplexing 263
7.8.1 Sideband Modulation by Using Complex Subcarriers . . . 264
7.8.2 DCEM Based on Square Wave Complex Subcarriers . . . 265
7.9 Rotating POCET Technique 266
7.10 ACE-BOC Modulation/Multiplexing Technique 267
7.10.1 Direct Form of the ACE-BOC Signal 268
7.10.2 Phase Rotation Form of the ACE-BOC Signal 270
7.10.3 ACE-BOC Solutions with Typical Power Allocations . . . 272
7.11 CEM for Multilevel Signals 276
7.12 Cascading of Constant Envelope Multiplexing Techniques 277
7.12.1 InterVote 277
7.12.2 POCET-Vote 277
7.12.3 TD-AltBOC 278
7.13 Remaining Challenges in CEM 279
7.14 Summary 281
References 281

8 Multicarrier Constant Envelope Composite Signal 285
8.1 Introduction 285
8.2 Challenges in Future GNSS Signal Design 286

8.2.1 Carrier Frequency Selection 286
8.2.2 Spreading Modulation Design 287
8.2.3 Multiplexing 287
8.2.4 A Gordian Knot 288
8.3 Multicarrier Constant-Envelope Composite Signal 289
8.4 Case Study of Adding a MCC Signal in L1 Band 291
8.4.1 Signal Description 292
8.4.2 RF Compatibility Analysis 294
8.4.3 Diversified Processing Strategies 295
8.4.4 Processing Mode Switching 298
8.4.5 Selective Availability 299
8.5 Conclusions 300
References 300

9 Conclusion 303
9.1 Performance Evaluation of Satellite Navigation Signals 303
9.2 Possible Future Developments of Satellite Navigation Signals 306

About the Authors

Zheng Yao Dr. Zheng Yao received his B.E. degree in electronics information engineering and his Ph.D. degrees in information and communication engineering from Tsinghua University, Beijing, China, in 2005 and 2010, respectively, where he is currently an Associate Professor with the Department of Electronic Engineering. He is the main inventor of asymmetric constant envelope binary offset carrier and quadrature multiplexed binary offset carrier modulations, which are employed in the BeiDou Navigation Satellite System (BDS) B2 and B1 civil signals, respectively. His primary areas of interest include the next-generation satellite navigation signals design, software-defined GNSS receiver, wireless localization technologies, and personal and vehicular positioning in challenging environments.

Dr. Yao is a Senior Member of Institute of Electrical and Electronic Engineers (IEEE) and Institute of Navigation (ION). He has also been a Technical Program Chair and a Section Chair of several international conferences. He is the recipient of the 2017 ION Early Achievement Award, in recognition to his pioneering contributions in developing new GNSS signals and multiplexing techniques; and advancing the Chinese BDS signal design. He also received the 2008 Student Paper Award from the U.S. Institute of Navigation and the 2010 and 2013 Best Paper Awards from the China Satellite Navigation Conference. e-mail: yaozheng@tsinghua.edu.cn

Mingquan Lu Dr. Mingquan Lu received the M.E. and Ph.D. degrees in electrical engineering from the University of Electronic Science and Technology of China, Chengdu. He is a professor with the Department of Electronic Engineering, Tsinghua University, Beijing, China. He directs the Positioning, Navigation, and Timing (PNT) Research Center, which develops global navigation satellite system (GNSS) and other PNT technologies. His current research interests include GNSS modeling and simulation, signal design and processing, and receiver development. e-mail: lumq@tsinghua.edu.cn

Abbreviations

1-D	One-dimensional
2-D	Two-dimensional
ABCS	Alternative BCS
ACE-BOC	Asymmetric constant envelope BOC
ACF	Autocorrelation function
AEP	Architecture Evolution Plan
A-GPS	Assisted GPS
AltBOC	Alternative BOC
AM/AM	Amplitude-to-amplitude
AM/PM	Amplitude-to-phase
ARNS	Aeronautical Radio Navigation Service
ASPeCT	Autocorrelation side-peak cancellation technique
ASZ	Autocorrelation sidelobe zero
AWGN	Additive white Gaussian noise
BCS	Binary coded symbol
BDS	BeiDou Navigation Satellite System
BOC	Binary offset carrier
BPSK	Binary phase-shift keying
BPSK-R	BPSK with rectangular chips
C/A	Coarse acquisition
CASM	Coherent adaptive subcarrier modulation
CBCS	Composite BCS
CBOC	Composite BOC
CCF	Cross-correlation function
CDMA	Code-division multiple access
CEM	Constant envelope multiplexing
CEMIC	Constant envelope multiplexing via intermodulation construction
CNR	Carrier-to-noise ratio
CPM	Continuous phase-modulation
CRC	Cyclic redundancy check

CRLB	Cramér–Rao lower bound
CS	Commercial service
CSD	Cross-power spectrum density
CT-SSC	Code tracking spectral sensitivity coefficient
DCEM	Dual-frequency constant envelope multiplexing
DET	Double estimate tracking
DLL	Delay locked loop
DME	Distance Measuring Equipment
DOF	Degrees of freedom
DOP	Dilution of precision
DPE	Double phase estimator
DSSS	Direct sequence spread spectrum
DTFT	Discrete-time Fourier transform
EIRP	Equivalent isotropic radiated power
E-L	Early-minus-late
EMLP	Early-minus-late power
EU	European Union
FDMA	Frequency-division multiple access
FFT	Fast Fourier transform
FHSS	Frequency-hopping spread spectrum
FOC	Full operational capability
GBAS	Ground-Based Augmentation System
GCC	Ground control center
GEO	Geosynchronous Earth orbit
GIOVE	Galileo In-Orbit Validation Element
GLONASS	GLObal Navigation Satellite System
GMSDT	GPS Military Signal Design Team
GMSK	Gaussian-filtered MSK
GNSS	Global Navigation Satellite System
GPS	Global Position System
GRASS	General removing ambiguity via side-peak suppression
G-S	Gram–Schmidt
GSS	Galileo sensor station
HPA	High power amplifier
I&D	Integrate and dump
ICD	Interface control document
ICG	International Committee on GNSSs
IF	Intermediate frequency
IGEB	Interagency GPS Executive Board
IGSO	Inclined geosynchronous satellite orbit
IM	Inter-modulation
IMV	Inter-modulation vector
IoT	Internet of Things
IRNSS	Indian Regional Navigational Satellite System
ISI	Intersymbol interference

ITU	International Telecommunication Union
JPO	Joint Program Office
JTIDS	Joint Tactical Info Distribution System
LDPC	Low-density parity check
LEO	Low Earth orbit
LHCP	Left-hand circular polarized
LUT	Look-up table
MAI	Multiple access interference
MBOC	Multiplexed BOC
MCC	Multicarrier constant-envelope composite
MC-CDMA	Multicarrier code-division multiple access
MDR	Multipath-to-direct ratio
MEAG	Mission Evolution Advisory Group
MEO	Medium Earth orbit
MSK	Minimum shift keying
MSPM	Multicomponent signals with phase modulation
MV	Majority voting
NCO	Numerically controlled oscillator
NEML	Narrow early-minus-late
NNSS	Navy Navigation Satellite System
NRL	Naval Research Laboratory
NRZ	Non-return-to-zero
OCX	Operational Control System
OFDM	Orthogonal frequency-division multiplexing
OS	Open service
PAR	Peak-to-average ratio
PDP	Phase domain processing
PLL	Phase lock loop
PMB	Phase-mapping-based
PNT	Position, navigation, and timing
POCET	Phase-optimized constant-envelope transmission
PRN	Pseudo-random noise
PRS	Public regulated service
PSB	Phase-synthesis-based
PSD	Power spectral density
PSWF	Prolate spheroidal wave functions
PUDLL	Pseudo-correlation function based unambiguous delay lock loop
PVT	Position, velocity, and time
QMBCS	Quadrature-multiplexed BCS
QMBOC	Quadrature-multiplexed BOC
QPSK	Quadrature phase-shift keying
QPSM	Quadrature product subcarrier modulation
RDSS	Radio Determination Satellite Service
RF	Radio frequency
RFC	Radio frequency compatibility

RHCP	Right-hand circular polarized
RMS	Root mean square
RNSS	Radio Navigation Satellite Service
RTCA	Radio Technical Commission for Aeronautics
SAR	Search and rescue
SAW	Surface acoustic wave
SBAS	Satellite-based augmentation system
SC	Side-peak cancellation
SCF	Synthesized correlation function
SCS	Step-shape code symbol
SNIR	Signal-to-noise-plus-interference ratio
SNR	Signal-to-noise ratio
SoL	Safety of life survive
SPS	Standard positioning service
SRRC	Square-root raised cosine
SSC	Spectral separation coefficient
SSPA	Solid state power amplifier
TACAN	Tactical Air Navigation System
TD	Time division
TDMA	Time-division multiple access
THSS	Time-hopping spread spectrum
TMBCS	Time-multiplexed BCS
TMBOC	Time-multiplexed BOC
TOA	Time of arrival
TOI	Time of interval
TT&C	Telemetry, track and command
TTFF	Time to first fix
TWTA	Travelling wave tube amplifier
ULS	Up-link station
UWB	Ultra-wide band
VE	Very early
VL	Very late
WDP	Waveform domain processing
WSB	Waveform-synthesis-based

Chapter 1
Introduction

As the Yangtze River forges ahead waves upon waves, the new generation will invariably excel the previous.

— Ancient Chinese proverb

1.1 Overview

Global Navigation Satellite Systems (GNSSs) can provide users with continuous, real-time, and accurate information of position, velocity, and time (PVT). GNSS have become the world's most significant information infrastructure for position, navigation, and timing (PNT) services. Based on GNSS, a whole new industry has emerged which exerts a significant impact on national security and economy as well as social development, and even alters methods of production and peoples' lifestyles. Therefore, the world's major powers attach great importance to the construction and application of GNSS. The United States' Global Position System (GPS) and Russia's GLObal Navigation Satellite System (GLONASS) were put into use in the late 20th century, and both have engendered huge military, economic and social benefits. Since the beginning of the 21st century, there has been a global upsurge in the building of satellite navigation systems.

At present, the United States and Russia are implementing ambitious plans to modernize GPS and GLONASS respectively, while the European Union (EU) is stepping up the building of its own satellite navigation system, the Galileo system. For China, the construction of the BeiDou Navigation Satellite System (BDS) global phase (BDS-3) is ongoing following the first experimental system generation (BDS-1) and the second regional system generation (BDS-2) having been already put into use. It is foreseeable that as soon as 2020, the four major GNSS programs, namely GPS,

Z. Yao and M. Lu, *Next-Generation GNSS Signal Design*, Navigation: Science and Technology 6, https://doi.org/10.1007/978-981-15-5799-6_1

GLONASS, Galileo, and BDS, will create new competition and cooperation around the world, representing in a peak in the development of GNSS.

A typical GNSS consists mainly of three segments: the space segment, the ground segment, and the user segment. These three parts are relatively independent and yet interrelated, and altogether form an organic whole through various radio and optical links. Among them, the satellite navigation signals, which are radio signals broadcast by the space segment, hold a unique and significant position in GNSS implementation. It is the only core link in a GNSS that simultaneously establishes connections between the space segment, the ground segment and the user segment. Its uniqueness and importance are reflected in the following three aspects.

First, through the uplinks and downlinks of the satellite system, the space segment and the ground segment form a closed loop where the navigation signal serves as an indispensable link carrying the feedback information. So the potential performance of the satellite navigation signal largely determines the inherent capability of the whole system.

Second, the satellite navigation signal is the only interface between the space segment and user segment. Hence, the signal quality is directly related to whether the positioning and timing capability of the GNSS can be fully realized by the user receivers, thereby affecting the service quality and the user experience.

Third, since the satellite navigation signal is the only item that receivers process, in a macroscopic view, it also largely determines the prospect of application and industrialization of the total GNSS. Therefore, the design of satellite navigation signals is not only an important aspect in building the GNSS, but also the key to the popularization, application, and industrialization of satellite navigation.

From the perspective of the development history of GNSSs, satellite navigation signals have been an important aspect of each system upgrade. In the first generation of satellite navigation systems represented by the USA Transit system, the signal design directly related to the pulse-Doppler ranging technology of radar, and could only realize discontinuous two-dimensional positioning with a low update rate and poor positioning accuracy.

In the 1970s, in the evolution of the second-generation satellite navigation system represented by GPS, the L1 coarse acquisition (C/A) signal that is widely used nowadays was originally designed as an auxiliary signal to the military encrypted signal. Although the second-generation satellite navigation system used the most advanced modulation technique of its time—the direct sequence spread-spectrum/binary phase shift keying (DSSS/BPSK)—there were many areas to be improved. For instance, this design did not fully consider that a satellite navigation system has the capability of combining distance measurement and data transmission, and underestimated the diverse requirements of future services. In fact, the second-generation satellite navigation system basically copied the signal design of satellite communications. The waveform, message structure, and channel coding were simple, with relatively low ranging code rates. Further evolution of GPS in the 1990s resulted in advanced receiving and processing technology that allowed even further exploitation of the performance potential of the GPS navigation signal. However, limitations of the GPS C/A signal gradually appeared. It was not until then that academia and industry realized that in order to further improve the performance of satellite navigation, it was necessary to update the concept of signal design and introduce new theories and technologies in light of the unique characteristics of satellite navigation systems. Thus, the design of a new navigation signal became a very important part of the modernization of GPS and the construction of Galileo system.

It has been seen that in the new generation of GNSS signal design, many approaches have been proposed that would aim at satellite navigation characteristics in order to satisfy the many diverse application requirements. A current example is that the design of both GPS and Galileo signals involves a new spread-spectrum modulation technique known as binary offset carrier (BOC) [1], which not only enables the systems to largely avoid spectrum interference between existing and new signals, but also improves the inherent tracking accuracy of the signal by providing a wider root mean square (RMS) bandwidth [2]. On this basis, advanced modulation techniques such as multiplexed BOC (MBOC) [3–5] and alternative BOC (AltBOC) [6] have emerged, which further improve the flexibility of design and inherent properties of the navigation signals.

In China, the BDS-2 system has been implemented and is operational, and the follow-on BDS-3 system is currently under construction. During the development and application of BDS-2, the overall BDS program has gained a new understanding of the importance of navigation signals, and has accumulated valuable engineering experience in signal design. However, due to specific objective and subjective conditions, signal design in BDS-2 used relatively conservative techniques, limiting its performance as well as the compatibility and interoperability with other GNSSs.

The construction of BDS-3 is both a precious opportunity and a major challenge for signal design. First, signal design is a significant task in the building of BDS-3. Designers can benefit from the experience and lessons of the construction and application of BDS-2. Meanwhile, they can also learn from other GNSS implementations about new ideas, theories, and techniques of the past 20 years. In particular, the latest achievements of GPS modernization and the building of the Galileo system provide valuable insights for BDS-3 designers. Therefore, in BDS-3 signal design, it is necessary to consider the demands of global users for GNSS services in the coming decades and exploit this trend for GNSS applications. Comprehensive theoretical and technical research based on the latest achievements in signal design should be conducted to achieve greater breakthroughs. On the other hand, BDS-3 faces more challenges. The increasing demand for PNT services has put forward higher requirements for BDS-3 signals in terms of ranging accuracy, service robustness, and security. Furthermore, due to the significant increase in the total number of signals, the improvement of new signal performance in the future is severely limited by a lack of available bandwidth, which brings many constraints to signal design.

It should be noted that the construction of the BDS is facing the general development trend of international GNSS and the competitive nature of the GNSS field. In general, the new generation of GNSS is evolving toward multi-system compatibility and interoperability. Cooperation and exchange are the mainstream of international GNSS construction and applications. One of the key segments of GNSS compatibility and interoperability, advocated by the International Committee on GNSSs

(ICG) and promoted by major countries and relevant International organizations, is GNSS signals. Therefore, the signal design of BDS-3 should be compatible and interoperable with other GNSSs. Especially since GPS and Galileo have reached substantial compatibility and interoperability, the BDS, as a latecomer, should strive for compatibility and interoperability of its signals with other GNSSs. The design of BDS-3 navigation signals is not only to meet China's PNT service needs, but also to determine whether BDS can be successfully integrated into the international GNSS family as an equal partner.

1.2 The Development and Current Status of GNSSs

1.2.1 Predecessors of the Current GNSS Programs

Satellite navigation technology was born in the early 1960s. The Navy Navigation Satellite System (NNSS), also known as the Transit satellite navigation system, built by the USA Navy in 1963 can be regarded as the world's first practical satellite navigation system [7]. It used low Earth orbit (LEO) satellites at about 600 nautical miles above sea level as its constellation and transmitted a precise timestamp with two minute interval in two operating bands of 150 MHz and 400 MHz. Positioning was based on the measurement of the Doppler frequency of satellites' transmitted signal. Since satellites orbit the Earth at a known speed and move relative to user receivers, the relative motion causes Doppler frequency shifts in the received signals. Using the Doppler shifts measured by the ground receiver located at a known elevation, the slant range between the ground receiver and the satellite can be calculated. This is the "measured result". However, another method used to calculate the slant range is based on the assumed position of the ground receiver. This is the "calculated result". If there is an error between the calculated and measured results of the slant range, it indicates that the assumed position is not accurate. If, after several rounds of iterative calculations and comparisons of these two position determining methods, the error falls within an acceptable range, the Transit system developed a final position result.

Due to the low sensitivity to the receiver's motion and the limitation of two-dimensional positioning, the Transit system was not suitable for aviation applications. Moreover, the system suffered discontinuous availability. This was because in order to avoid interference from different signals, there were only five available Transit satellites on orbit. Only when a Transit satellite was located above the same horizontal plane as the receiver could it be used. The elapsed time between two positionings near the Equator was a few hours, and the positioning interval in mid-latitude areas was about one hour. In that era, inertial navigation was generally used to maintain continuous positioning. Even though various defects existed in the Transit system, it was a pioneering satellite navigation system and demonstrated advantages such as wide area coverage and all-weather positioning. It laid an important foundation for the subsequent construction of GPS.

On account of the shortcomings of Transit system in terms of only two dimension positioning, service continuity, and dynamic user support, the US Naval Research Laboratory (NRL) launched the Timation Satellite Program in 1972, led by R. Easton [8]. The main features of Timation included a high-stability Cesium clock, which replaced the crystal oscillator in the satellite payload to achieve high-precision time synchronization, and side-tone ranging as a replacement for Doppler velocimetry. Within the Timation program, there were plans to launch 8 to 12 satellites to form a circular orbit with a height of more than 10,000 kilometers and a period of eight hours, covering the whole globe. Three Timation satellites were launched in medium-high orbit in 1967, 1969, and 1974 respectively, the last of which became the first technical test satellite of the GPS. Many of the technologies first used in the Timation program became an important basis for the development of GPS.

While the USA Navy was developing satellite navigation systems, the USA Air Force was also conducting a satellite navigation system test with the code name 621B [9, 10], with a view to adopting constellations containing high-eccentricity satellite orbits with a period of 24 hours. The most important contribution of the 621B system was to propose the pseudo random noise (PRN) model for radio ranging. The PRN ranging model was tested and verified by broadcasting navigation signals to aircraft using pseudolites.

In 1973, the USA Department of Defense began to develop a new satellite navigation system called Navigation Satellite for Timing and Ranging/Global Position System (NAVSTAR/GPS), known as GPS for short. This new system integrated the requirements of the Navy and Air Force for navigation and positioning and combined the advantages of 621B and Timation.

1.2.2 GPS and Its Modernization

The working principle of the second-generation satellite navigation systems represented by GPS will be elaborated on in Sect. 2.2. In summary, compared with the early generation programs, the second-generation satellite navigation systems have higher satellite orbits, more satellites, higher operating frequency, and their positioning principle is a three-sphere intersection based on estimated signal time of flight. Based on these characteristics, the GPS system offers all-weather, continuous, real-time positioning, navigation, and timing services, covering the whole globe. It can provide precise three-dimensional coordinates, as well as speed and time information for a variety of users on land, at sea, in air and, to a degree, in space.

GPS is the first GNSS to be built to have full operational capability (FOC). In 1995, the USA announced that GPS had entered the FOC phase with a complete constellation of 24 working satellites. The system broadcasts a C/A signal [11] of code division multiple access (CDMA) in the L1 frequency band to provide standard positioning service (SPS) for global civilian users. Simultaneously, P(Y) signals for authorized services were broadcast in the L1 and L2 bands, which allowed military dual-frequency receivers to correct ionospheric errors so as to obtain higher

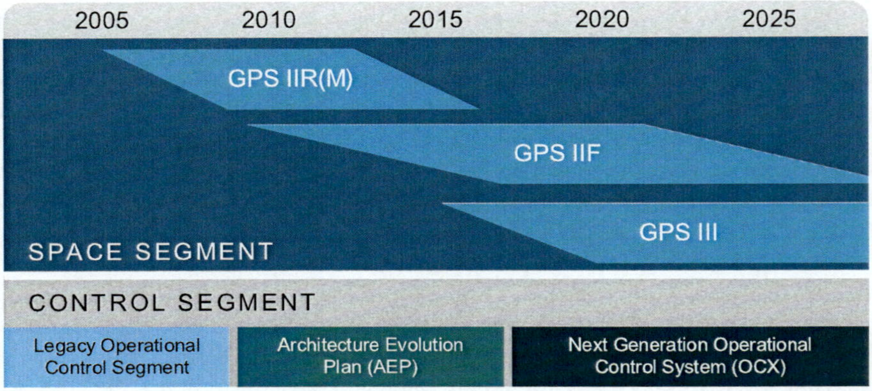

Fig. 1.1 Timetable for GPS modernization

positioning accuracy than SPS. Later, civilian and military users began to make increasing demands for positioning services, and GNSS implementations continued to develop across the world. In this context, in order to strengthen GPS's support to the USA military in modern warfare and maintain its dominance in the global civil navigation field, the USA government approved a multibillion-dollar program for GPS modernization in 1999. The system upgrade mainly focuses on the space segment and the control segment, as shown in Fig. 1.1.

The modernization of GPS enhances the performance potential of the total system. From the perspective of equipment, the GPS Block III satellite has made a major breakthrough in technology compared with the previous two generations of GPS satellites [12]. The Block III satellite is equipped with a high-power multibeam transmitting antenna, which can function in strong interference environments, enhances the signal power in designated areas and ensures the positioning accuracy of military receivers. The Block III satellites can also form inter-satellite links with higher data rate and more accurate pointing to cope with constellation abnormalities.

Another significant mark of GPS modernization is the updating of the signals. To this end, a series of advanced technologies and design concepts were introduced. The design of new signals in modulation, multiplexing, and message structure is more suitable for satellite navigation applications, and the performance of new signals has improved substantially compared with traditional GPS signals. With the addition of new GPS satellites, some of which replaced retired platforms, newly deployed GPS satellites have begun to broadcast a new military signal, M-code [14], and three new civil signals, L2C, L5, and L1C.

The second civil signal of GPS, designated L2C, is broadcast in the L2 band [15]. This signal has been broadcast on the GPS IIR-M satellites since 2005 and is mainly for commercial users. GPS users who require high-precision positioning services can receive and process both the L1 C/A signal and the L2C signal to correct ionospheric errors by combining dual-frequency measurements. The positioning accuracy obtained in this manner is close to or even higher than that of existing GPS military services.

The frequency band for GPS's third civil signal, L5 [16], is located on a band reserved by the International Telecommunication Union (ITU), exclusively for Aeronautical Radio Navigation Service (ARNS). Compared with the L2C and L1 C/A signals, the L5 signal has higher power, a wider bandwidth, and a more advanced signal design concept. Initially, the primary purpose of the L5 signal design was to improve positioning accuracy and robustness for future aviation applications by jointly processing the L1 C/A signal and L5 signal. The L5 signal has been broadcast on the GPS Block IIF satellite since 2010. The number of on-orbit satellites broadcasting L5 signals after 2021 is expected to reach 24.

The fourth GPS civil signal, L1C [17], shares L1 band with the GPS L1 C/A signal. The design of the L1C signal includes many innovative concepts, and fully supports interoperability with other GNSS. The L1C signal was first broadcast on the GPS Block III satellite in 2016, and global coverage of L1C signals should be achieved in 2026.

The control segment is an important part of GPS. The ground control system has been undergoing continuous upgrading and rebuilding in recent years, mainly consisting of the Architecture Evolution Plan (AEP) and the Next Generation Operational Control System (OCX) [18, 19]. AEP has gradually replaced the host-based architecture of legacy master stations with a modern information technology based architecture so as to improve the flexibility and responsiveness of GPS operations. OCX adds various new capabilities to the GPS control segment, including complete control of civil signals L2C, L5, and L1C. These efforts are being carried out step by step with the modernization of GPS satellites.

1.2.3 GLONASS and Its Modernization

GLONASS, Russia's global navigation satellite system, was the second GNSS to be built and put into operation after the USA GPS [20, 21]. Its early design was mainly aimed at military applications. In 1990 and 1991, Russia published the interface control document (ICD) [22] for GLONASS civil signals, marking its offering of free positioning services to civil users all over the world. In January 1996, GLONASS successfully formed a complete constellation of 24 working satellites for the first time, marking its entry into the FOC phase. However, the designed life expectancy of GLONASS's in-orbit working satellites was only two to three years. In the late 1990s, due to Russia's economic recession, new satellites were not launched when replacements were needed. Therefore, GLONASS was faced with a continuous decline in working satellites after FOC. In 2001, the number of in-orbit working satellites dropped to 6. Later, with improvements to Russia's economy since the beginning of the 21st century, GLONASS was restored to FOC in 2011.

In order to serve national defense security and improve the competitiveness of GLONASS, a modernization scheme was launched in 2003 to comprehensively rebuild and upgrade the satellites, signals, and ground control systems [21, 23]. In terms of the signals, traditional GLONASS signals used frequency division multiple access (FDMA) technology. The satellites transmitted signals on different carrier frequencies in the L1 and L2 bands. However, in order to improve signal accuracy, multipath-suppression capability, and interoperability with other GNSS, modern GLONASS broadcasts CDMA signals in addition to existing FDMA signals [24]. On 26 February 2011, the first third-generation GLONASS-K1 satellite was successfully launched, carrying out the testing of the L3C CDMA signal. The designed life expectancy of the GLONASS-K1 satellite is 10 years, and the stability of the satellite clock has been improved over previous versions [25]. The center frequency point of the new L3C signal is located at 1,207.14 MHz, which coincides with the Galileo E5b and BDS B2b. Like the frequency band used by the GPS L5 signal, this band has also been reserved for safety of life survive (SoL) applications by ITU. In 2014, the fourth-generation GLONASS-K2 satellites began to be launched, and the constellation is expected to be completed by 2021. The design life of a GLONASS-K2 satellite is also 10 years. It will broadcast four new CDMA signals and the traditional FDMA signals simultaneously.

1.2.4 The Galileo Satellite Navigation System

The Galileo system is a GNSS built and operated by EU. In 1999, in order to meet the growing demand for PNT services and to lessen dependence on USA GPS and Russian GLONASS, the EU decided to build a system of its own, designated as Galileo. Unlike GPS and GLONASS, Galileo was originally built as an open navigation system for civilian global navigation users [26].

Between 2005 and 2011, by launching the Galileo In-Orbit Validation Element (GIOVE) A and GIOVE B satellites, the Galileo test platform was established, and the verification of ground monitoring equipment was completed [27]. From 2011 to 2013, Galileo entered the stage of in-orbit verification. The system's operational capability was verified by establishing the smallest constellation that could realize positioning with four satellites. The next step was to deploy 18 satellites and more ground monitoring facilities, enabling Galileo to begin initial operation. According to the EU's plan, after the Galileo system entered the FOC phase, the space segment would include 30 medium Earth orbit (MEO) satellites with an orbital altitude of 23,222 km, in three orbital planes with an inclination of 56°. The control segment includes two ground control centers (GCCs) located in Germany and Italy, a series of telemetry, track and command (TT&C) stations, an up-link station (ULS), and a series of Galileo sensor stations (GSSs) [28–30]. Based on these facilities, the Galileo system broadcasts CDMA signals on the E1, E5, and E6 frequency bands [31]. Its satellite navigation signal design is service-oriented. Open service (OS), safety of life survive (SoL), commercial service (CS), public regulated service (PRS), and search and rescue (SAR) are provided to users worldwide [32].

In order to meet the needs of system services, Galileo's designers used more open design concepts and advanced technologies based on the design and construction of GPS and GLONASS. Compared with the traditional signals of these systems, the service-oriented Galileo signals show significant latecomer advantages that can help the receiving equipment designers achieve higher performance.

It usually takes a long time to construct satellite navigation systems. In order to meet the demand for positioning service in the future, Galileo's partner countries made significant efforts to adapt the advanced technology for the system. In 2007, the unfinished Galileo system began to plan its modernization, and set up the Mission Evolution Advisory Group (MEAG) led by GNSS experts to critically evaluate the modernization plan from a variety of perspectives. According to public reports, Galileo will introduce a series of modern navigation, communication, and aerospace application technologies such as inter-satellite links and high-precision atomic clocks in its subsequent phases.

1.2.5 The BeiDou Satellite Navigation System

The BDS is China's independently self-developed GNSS. Its development goals are building a navigation satellite system to meet the needs of the country's national security as well as economic and social development, and to provide continuous, stable and reliable services for global users; developing BDS-related industries to support China's economic and social development, as well as to improve people's living standards; enhancing international cooperation to share the fruits of development in the field of satellite navigation, and to increase the comprehensive application benefits of GNSSs. Since satellite navigation systems are an important element of the spatial information infrastructure, BDS has an important role in consolidating national defense, developing the economy, and even improving daily life.

China's BDS program began in the 1990s. In order to overcome various difficulties such as limited funds, insufficient technology resources and lack of construction and management experience in large scale space-based information systems, China formulated the following three-step development plan for BDS [33, 34].

The first step of BDS is to construct the BDS Experimental System, also known as BDS-1. The project started in 1994, and in 2000, with the launch of two geosynchronous Earth orbit (GEO) satellites, the system was completed and put into operation. With an active-positioning scheme, BDS-1 provided users in China with positioning, timing, wide-area differential and short message communication services. The third GEO satellite was launched in 2003, which further enhanced the system performance. With BDS-1, China became the third country in the world to have an independent satellite navigation system after the USA and Russia.

The BDS-1 system uses two satellites to provide users with the Radio Determination Satellite Service (RDSS) [35], including basic two-dimensional positioning and short message services. The BDS-1 system consists of three parts: the space segment consisting of three geosynchronous satellites, the control segment comprising a ground master station and a number of calibration stations, and the user segment containing user equipment.

The positioning principle of BDS-1 is as follows: By using the known position coordinates of the two GEO satellites as two different sphere centers and using the measured distances from the satellites to the user terminal as radiuses, respectively, two spheres can be formed. The user terminal must be on the intersection of the two spheres. Using an elevation map provided by the ground control segment, a non-homogeneous Earth-centered curved surface can be established with the Earth center as the origin and the distance from the Earth center to the user terminal as its radius. The exact position of the user terminal is the intersection of the spherical arc and the non-homogeneous curved surface. The ground master control station calculated the user terminal's position and sent the position to the user terminal via the outbound signal.

The second step of BDS was to construct BDS-2. The project began in 2004, and by the end of 2012, a total of 14 satellites, including 5 GEO satellites, 5 inclined geosynchronous satellite orbit (IGSO) satellites and 4 MEO satellites, had been launched to complete the space constellation deployment. In addition to be compatible with BDS-1, BDS-2 integrated a passive-positioning design, and provided users in the Asia-Pacific region with positioning, velocity measurement, and timing as well as short message communication services.

The launch time and track type of each BDS-2 satellite are shown in Table 1.1 [36].

At the end of 2012, the constellation composed of 14 satellites were completed, forming a network with a service capability covering China and surrounding areas.

The third step of BDS is to construct BDS-3. The project began in 2009, and by the end of 2018, a total of 19 satellites had been launched to complete a preliminary

Table 1.1 BDS-2 satellites launch record [36]

Satellite	Launch date	Launch vehicle	Orbit
First BDS navigation satellite	2007-4-14	CZ-3A	MEO
Second BDS navigation satellite	2009-4-15	CZ-3C	GEO
Third BDS navigation satellite	2010-1-17	CZ-3C	GEO
Fourth BDS navigation satellite	2010-6-2	CZ-3C	GEO
Fifth BDS navigation satellite	2010-8-1	CZ-3A	IGSO
Sixth BDS navigation satellite	2010-11-1	CZ-3C	GEO
Seventh BDS navigation satellite	2010-12-18	CZ-3A	IGSO
Eighth BDS navigation satellite	2011-4-10	CZ-3A	IGSO
Ninth BDS navigation satellite	2011-7-27	CZ-3A	IGSO
10th BDS navigation satellite	2011-12-2	CZ-3A	IGSO
11th BDS navigation satellite	2012-2-25	CZ-3C	GEO
12th and 13th BDS navigation satellites	2012-4-30	CZ-3B	MEO
14th and 15th BDS navigation satellites	2012-9-19	CZ-3B	MEO

system for global services. By launching additional 11 satellites, the deployment of BDS-3 was planned to be completed with the 30-satellite constellation by the end of 2020. BDS-3 has inherited the technical designs of both active and passive services, and can provide positioning, velocity measurement, timing, global short message communication, and international search and rescue services to global users. Users in China and surrounding areas can also enjoy regional short message communication, satellite-based augmentation, and precise point positioning services, etc.

Compared with other GNSS, BDS features a hybrid constellation of GEO satellites, IGSO satellites, and MEO satellites. The sub-satellite point trajectory of the IGSO satellite is a symmetrical "8"-shape of the same size across the northern and southern hemispheres, with the intersection at the Equator. Monitoring most of the IGSO satellites can be achieved by establishing monitoring stations in China. The coverage of the IGSO satellites is as high as 80% for China, so the IGSO constellation can achieve good usage with this fewer number of satellites [37]. The MEO satellites are characterized by the ability to form a navigation constellation with a small number of satellites to achieve global service. In addition, the GEO satellites in BDS are stationary relative to the ground, which means that they can meet the requirements of continuous coverage in a specific area. The mixed constellation of satellites in these three separate orbital planes can improve the coverage in the Asia-Pacific region.

The integration of navigation and communication capabilities is another major feature of BDS. The GEO satellites in the BDS constellation provide passive navigation and positioning services, and also inherit RDSS from BDS-1, with active navigation positioning as well as short message communication, and position reporting. In summary, BDS-3 now has multiple service functions including basic navigation, short message communication, satellite-based augmentation, international search and rescue, and precise point positioning, etc.

1.3 The Importance of Navigation Signals

Within a GNSS, the basic principle of positioning and navigation is trilateration. The key in the positioning process based on this principle is to obtain the current position of the satellite and the distance measurement between the satellite and the observation point.

The receiver calculates the instantaneous position and velocity of the satellite with the information constantly broadcast in navigation signals, such as satellite ephemeris, almanac data, satellite clock correction, ionospheric delay correction model parameters, and satellite status. The distance from the satellite to the observation point is obtained by measuring the time difference between the navigation signal leaving satellite and arriving at the receiver. Observations of the arrival time of satellite signals at the receiver are the basis for the positioning, navigation, and timing of the entire system. In the receiver, the satellite signal's arrival time is extracted by tracking the code and carrier phases. The code and carrier tracking accuracy largely determine the measurement accuracy of the signal arrival time. Factors affecting the accuracy of the code and carrier tracking include the navigation signal structure, receiver technology, and transmission environment. Among these factors, the structure of the navigation signal and the quality of the transmission determine the performance limit that can be achieved for ranging. No matter how advanced the receiving technology is, the best performance that can be achieved will be limited by the navigation signal itself. With state-of-the-art receiver technology, the navigation signal characteristics will become a bottleneck for performance improvement.

The navigation signal can be considered as the core of the satellite navigation system. In fact, the navigation signal is the sole means for realizing the basic functions of the navigation system, and plays a decisive role in key performance such as positioning, speed measurement, timing accuracy, compatibility and interoperability, and anti-interference ability. The ground control segment, the on-board payload, and the user terminal all work around the precise generation, transmission, reception, and processing of navigation signals. If the signal design is flawed, the performance of the system will be inherently insufficient, even if the equipment of the ground segment, space segment, and user segment is perfect. This can greatly hinder the future promotion and application of the system.

The importance of navigation signal design is also reflected in the fact that a complete signal scheme is the basis for the development of other subsystem, such as the on-board payload or the user terminal equipment, and for the compatibility with other systems. For interoperability in international negotiations and collaborations, a complete signal scheme is the primary input. More importantly, as a global coverage broadcasting system, GNSS's signal design has a wide range of influence. Once the signal structure is determined and published, any subsequent modifications could cause service interruptions and necessitate equipment modifications for extremely large user groups, causing incalculable economical losses. Therefore, the signal design for a satellite navigation system cannot simply follow the "release first, update later" route. It must be forward-looking and undergo extensive demonstration in the early stages of design.

1.4 Development of the Satellite Navigation Signals

Because the design of the signal plays an important role in the construction of satellite navigation systems, a vast amount of international research has been undertaken on the topic. In the 1990s, the USA and Europe established signal design teams to design and verify GPS and Galileo signals. In 1997, the USA's GPS Joint Program Office (JPO) set up the GPS Modernized Signal Design Group. In 2002, the design of L2C, L5C, and military signals on the L1 and L2 bands were basically completed.

The EU had been conducting preliminary research on the design of signals for Galileo system since 1997. In 2001, the Galileo Signal Design Task Force was formally established with this purpose. Due to the possibility of spectrum overlap between GPS and the Galileo system, the EU and the USA held more than 30 technical and political negotiations between 2004 and 2007, and finally reached an

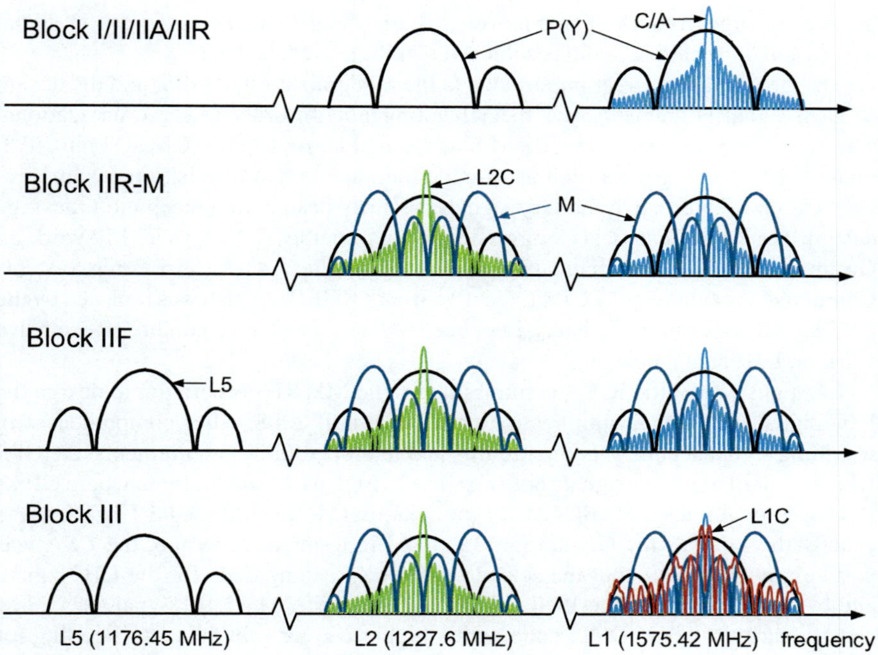

Fig. 1.2 Evolution of GPS signal modernization [38]

agreement in 2007 by adjusting the signals of the initial plan, and strengthening GPS and Galileo collaboration on compatibility and interoperability. At this point, the signal design for GPS and Galileo was basically completed. A large number of experimental verifications were carried out, and the on-orbit verification of the new signals for both GPS and Galileo were undertaken using the GPS Block IIR(M) satellites and the Galileo GIOVE satellites respectively.

1.4.1 Development of the GPS Signals

Figure 1.2 shows the evolution of GPS signal modernization.

The signals used in early GPS implementations were direct sequence spread spectrum (DSSS) with traditional rectangular chip waveforms. As technology improved, in order to face the higher performance requirements from military and civilian users and the competition brought by the emergence of the Galileo system and the BDS, USA's JPO established the GPS Modernization Signal Design Group in 1997, and officially proposed the GPS Modernization Plan in 1999. In 2002, the design of several new signals such as L2C, L5C, and the military signal on the L1 and L2 bands were basically completed. In the process of GPS modernization, the USA has invested a large amount of manpower, material, and financial resources in signal design and verification, and has achieved significant results [39].

The USA attaches great importance to the modernization of GPS, and the design of each signal is handled by a specialized agency. In order to serve navigational warfare, GPS JPO set up the GPS Military Signal Design Team (GMSDT) in 1998, including multiple groups such as M-code modulation and acquisition design, pre-verification of alternative military signals, security design, and navigation message design. The Radio Technical Commission for Aeronautics (RTCA)'s SC159 Working Group defines the L5 civilian signal requirements. The JPO, the Air Force Systems Command, the Interagency GPS Executive Board (IGEB) (which was replaced by the PNT Executive Committee later), Lockheed Martin, and Boeing conducted extensive work on L2C signal design.

In August 2003, the IGEB commissioned The MITRE corporation to design the L1C signal, and signal optimization continued until 2008. After comprehensively weighing signal accuracy, compatibility, and interoperability requirements, the GPS L1C and Galileo E1 OS signals both now use MBOC(6, 1) modulation, where GPS's L1C specifically uses TMBOC(6, 1) modulation [5]. Since the GPS L1C signal is exactly the same as the Galileo E1 OS signal in terms of spectrum, the USA was particularly cautious about the selection of the spreading code for the L1C signal, and eventually selected the Weil code for this purpose. The L1C signal uses a low density parity check (LDPC) code with a higher coding gain than BCH coding for forward error correction coding, and is optimized in terms of message structure, which greatly improved the signal robustness and flexibility.

It is worth noting that in the design process of the modernized GPS signals, several ideas proposed in the design of the 621B project were revisited and implemented, such as high-performance error correcting code which can increase the robustness of navigation message demodulation, and the additional pilot channel which can enhance the robustness of signal acquisition and tracking.

1.4.2 Development of the Galileo Signals

The EU began conducting preliminary research on the signal design for Galileo in 1997. In 2001, the Galileo Signal Design Task Force was formally established for this purpose. Many aerospace institutes, research institutes, and universities in the EU participated in the design and verification of Galileo's signal system, and undertook a significant amount of in-depth research.

Since Galileo system was a latecomer to the field, it references GPS signal design in many aspects. At the beginning, the Galileo signal was similar to that of GPS. After consistent progress and development, Galileo proposed its own improved BOC signals as well as several new technologies. For example, in the MBOC signal jointly proposed with GPS, Galileo uses a self-developed composite BOC (CBOC) modulation [4]. In the E5 band, it also uses in-house developed AltBOC modulation/multiplexing and memory codes. In the design process for the Galileo signal, researchers also analyzed and tested a large number of candidate signals (such as binary coded symbol (BCS) and composite BCS (CBCS) [40–42]). In 2004, Europe and the USA agreed to cooperate on the performance, interoperability, and compatibility of the GPS and Galileo signals. In 2007, the first-generation Galileo signal design was completed.

Since Galileo's signal design and the modernization of the GPS signal are concurrent in time, their design ideas were shared with each other. Galileo's signal and the modernized GPS signals have many common aspects such as: both employ BOC modulation or other split-spectrum modulation methods derived from BOC modulation; both use additional pilot channels on the basis of data channels; both use the hierarchical structure of the spreading code; and both use high-performance channel coding for the correction of errors in the navigation message. Of course, there are also significant differences between the two, for example: Galileo uses cosine phase BOC modulation and applies more flexible AltBOC modulation, while the GPS M-code signal uses a sinusoidal phase BOC modulation; the Galileo signal generally has a higher information rate than the GPS signal; some of Galileo's signals use a so called random-code as a spreading code, which can only be stored and read, while GPS employs the Weil code, which can be generated using shift registers; and in the MBOC implementation, Galileo uses CBOC modulation, while GPS uses time-multiplexed BOC (TMBOC) modulation.

In addition, an important part of the signal design is performance analysis and verification. In the process of signal design, Galileo follows a four-step process which includes theory-based performance analysis, simulation verification based on a software defined transceiver system, hardware-based simulation verification, and on-orbit verification based on test satellites. Different stages of experimental evaluations have enabled the Galileo signal itself and the signal-related equipment to be fully validated, laying a solid foundation for the Galileo system to provide high-performance services.

1.4.3 Development of the BDS Signals

Due to national conditions in the 1980s, China began to build its first generation of satellite navigation systems. The BDS-1 satellite navigation signal adopted the code division multiple access RDSS structure, and the signal operating frequencies were selected in the S and L frequency bands [46].

In 2004, China began to build the BDS-2. The State and corresponding military departments demonstrated and gradually solidified the overall system plan through three phases, namely in-depth investigation, top-level design, and construction program demonstration. BDS-2 drew on the experience of GPS signal design, and broadcast navigation signals using quadrature phase shift keying (QPSK) modulation in B1, B2 and B3 frequency bands. The subsequent research and application of various types of terminals applied to BDS-2 enhanced the research results for the regional signal implementation and accumulated experience for global system signal design.

BDS-2 has been completed and put into use, and a global system is currently under construction. In the building and application of the BDS-2 system, a new understanding of the importance of navigation signals has been gained. Valuable practical engineering experience has also been accumulated in the design of satellite navigation signals.

In view of the important position of satellite navigation signals in GNSS construction, application promotion, and industrialization, and in order to gain a favorable position in the field by sharing BDS's technical competence with users all over the world, signal design has been given high priority in China. In the past few years, many Chinese research institutes have undertaken extensive research into signal design, and phased results have been achieved. This includes aspects such as frequency resources, system compatibility and interoperability, intellectual property rights and other strict constraints, as well as increasingly refined and higher-performance service requirements. A variety of innovative technologies such as quadrature multiplexed BOC (QMBOC) [47, 48], asymmetric constant envelope BOC (ACE-BOC) [49–52], and double QPSK [53] have been proposed.

Of course, the completion of the BDS-3 construction does not mean the end of satellite navigation design. For BDS and other GNSS systems, the day when the system's construction is completed and launched is the beginning of the system renewal phase. In the context of the development of China's PNT system with BDS as the core, the signal design for the next-generation BDS will face fresh challenges. Therefore, the design of satellite navigation signals will be a long-term and continuing project.

References

1. Betz JW (2001) Binary offset carrier modulations for radio navigation. Navigation 48(4):227–246
2. Betz JW, Kolodziejski KR (2009) Generalized theory of code tracking with an early-late discriminator part I: lower bound and coherent processing. IEEE Trans Aerosp Electron Syst 45(4):1538–1556
3. Avila-Rodriguez J-A, Hein GW, Wallner S, Issler J-L, Ries L, Lestarquit L, de Latour A, Godet J, Bastide F, Pratt T et al (2008) The MBOC modulation: the final touch to the Galileo frequency and signal plan. Navigation 55(1):15–28
4. Hein GW, Avila-Rodriguez J-A, Wallner S, Pratt AR, Owen J, Issler J-L, Betz JW, Hegarty CJ, Lenahan LS, Rushanan JJ et al (2006) MBOC: the new optimized spreading modulation recommended for Galileo L1 OS and GPS L1C. In: 2006 IEEE/ION PLANS, 2006, vol 2006, pp 884–892. Coronado, CA, USA
5. Betz JW, Blanco MA, Cahn CR, Dafesh PA, Hegarty CJ, Hudnut KW, Kasemsri V, Keegan R, Kovach K, Lenahan CL et al (2006) Description of the L1C signal. In: Proceedings of ION GNSS 2006, vol 2006, pp 2080–2091
6. Lestarquit L, Artaud G, Issler J-L (2008) AltBOC for dummies or everything you always wanted to know about AltBOC. In: Proceedings of ION GNSS 2008, vol 2008. Savannah, GA, USA, pp 961–970
7. Parkinson BW, Stansell T, Beard R, Gromov K (1995) A history of satellite navigation. Navigation 42(1):109–164

8. Easton RL (1978) The navigation technology program. Navigation 25(2):107–112
9. Spilker JJ, Axelrad P, Parkinson BW, Enge P (1996) Global positioning system: theory and applications, volume I. American Institute of Aeronautics and Astronautics
10. Betz JW (2013) Something old, something new: signal structures for satellite-based navigation: past, present, and future. Inside GNSS, pp 34–42
11. NavSTAR GPS (2000) Space segment/navigation user interfaces. Arinc Research Corporation
12. Gruber B (2010) GPS modernization and program update. In: Stanford's 2010 PNT challenges and opportunities symposium. Stanford, CA
13. National Coordination Office for Space-Based Positioning (2014) GPS modernization. http://www.gps.gov/systems/gps/modernization/
14. Barker BC, Betz JW, Clark JE, Correia JT, Gillis JT, Lazar S, Rehborn KA, Straton III JR (2006) Overview of the GPS M-code signal. DTIC document
15. Fontana RD, Cheung W, Stansell T (2001) The modernized L2 civil signal. GPS World 12(9):28–35
16. Tran M (2004) Performance evaluations of the new GPS L5 and L2 civil (L2C) signals. Navigation 51(3):199–212
17. Stansell T, Hudnut KW, Keegan RG (2001) GPS L1C: enhanced performance, receiver design suggestions, and key contributions. In: Proceedings of ION GNSS 2010, pp 2860–2873
18. Martin H, Lead PC (2011) GPS status and modernization, in the ninth meeting of the national space-based positioning, navigation, and timing advisory board
19. Crews M (2008) Long-term future of GPS. In: Proceedings of the 2008 national technical meeting of the institute of navigation (NTM 2008), pp 1–26
20. Langley RB (1997) GLONASS: review and update. GPS World 8(7):46–51
21. Alkan RM, Karaman H, Sahin M (2005) GPS, Galileo and GLONASS satellite navigation systems & GPS modernization. In: Proceedings of the 2nd international conference recent advances in space technologies, RAST 2005, vol 2005, pp 390–394
22. GLONASS ICD (2002) "GLONASS interface control document" version 5.0, Moscow
23. Revnivykh S (2011) GLONASS status and modernization. In: Proceedings of ION GNSS. Portland, OR, USA 2011:839–854
24. Gibbons G (2008) Russia approves CDMA signals for GLONASS, discussing common signal design. Inside GNSS
25. Thoelert S, Erker S, Furthner J, Meurer M, Gao GX, Heng L, Walter T, Enge P (2011) First signal in space analysis of GLONASS K-1. In: Proceedings of the 24th international technical meeting of the sateuite division of the institute of navigation. The Institute of Navigation, Portland OR, USA, pp 3076–3082
26. Benedicto J, Dinwiddy S, Gatti G, Lucas R, Lugert M (2000) Galileo: satellite system design. European Space Agency
27. Montenbruck O, Gunther C, Graf S, Garcia-Fernandez M, Furthner J, Kuhlen H (2006) GIOVE-A initial signal analysis. GPS Solut 10(2):146–153
28. Falcone M (2008) "Galileo programme status. In: Proceedings of ION GNSS 2008, pp 16–19
29. Lugert M, Falcone M, Claes P, Castro A, Matussi S, Martinez-Rosique J-A, Manteiga M, Blonski D (2002) Galileo ground segment for navigation and integrity. In: Proceedings of ION GPS 2002, pp 2135–2147
30. Enderle W, Weber T, Kuhlen H (2003) The architecture of the European global navigation satellite system-Galileo
31. Galileo open service signal in space interface control document, (OS SIS ICD) Draft 0, 19th May 2006
32. Trautenberg HL, Weber T, Schafer C (2004) Galileo system overview. Acta Astronaut 55(3):643–647
33. Han C, Yang Y, Cai Z (2011) The BeiDou navigation satellite system and its time scales. Metrologia 48(4):S213
34. Ran C (2012) Development of the BeiDou navigation satellite system. In: Global navigation satellite systems. Report from the joint workshop of the national academy of engineering and the chinese academy of engineering. Washington, DC, USA

35. Tan S (2009) Theory and application of comprehensive RDSS positioning and reporting. Acta Geodaetica et Cartographica Sinica 38(1):1–5
36. Official BDS website, launch record. http://www.beidou.gov.cn/attach/fsjl.jpg, 2015-10-1
37. Xu Q (2001) Satellite constellation of regional satellite navigation system. Surv Mapp Eng 03:1–5
38. Hegarty C, Chatre E (2008) Evolution of the global navigation satellite system (GNSS). Proc IEEE 96(12):1902–1917
39. Zhu X, Wang F (2005) The signal design of present satellite navigation system and its inspiration. In: International conference on space information technology
40. Hegarty C, Betz JW, Saidi A (2001) Binary coded symbol modulations for GNSS. In: Proceedings of the 60th annual meeting of the institute of navigation, pp 56–64
41. Hein GW, Avila-Rodriguez J-A, Ries L, Lestarquit L, Issler J-L, Godet J, Pratt T (2005) A candidate for the Galileo L1 OS optimized signal. In: Proceedings of ION GNSS 2005, pp 833–845
42. Ries L, Issler J-L, Lestarquit L, Avila-Rodriguez J-A, Hein GW (2012) Spread spectrum signal
43. Bian S, Jin J, Fang Z (2005) The BeiDou satellite positioning system and its positioning accuracy. Navigation 52(3):123–129
44. China Satellite Navigation Office (2013) "BeiDou navigation satellite system signal in space interface control document open service signal (Version 2.0)
45. Cao C (2009) Status of COMPASS/BeiDou development. In: 2009 PNT challenges and opportunities symposium
46. Tan S (2007) Satellite navigation and positioning engineering. National Defence Industry Press, Beijing
47. Yao Z, Lu M, Feng ZM (2010) Quadrature multiplexed BOC modulation for interoperable GNSS signals. Electron Lett 46(17):1234–1236
48. Yao Z, Lu M (2011) Optimized modulation for compass B1-C signal with multiple processing modes. In: ION GNSS 24th international technical meeting of the satellite division. OR, USA, Portland, pp 1234–1242
49. Yao Z, Zhang J, Lu M (2016) ACE-BOC: dual-frequency constant envelope multiplexing for satellite navigation. IEEE Trans Aerosp Electron Syst 52(2)
50. Yao Z, Lu M (2013) Constant envelope combination for components on different carrier frequencies with unequal power allocation. In: Proceedings of ION ITM. San Diego, CA, USA, pp 629–637
51. Yao Z, Lu M (2013) Design, implementation, and performance analysis of ACE-BOC modulation. In: Proceedings of ION GNSS+. Nashville, TN, USA, pp 361–368
52. Zhang J, Yao Z, Lu M (2014) Applications and low-complexity implementation of ACE-BOC multiplexing. In: Proceedings of ION ITM. San Diego, CA, USA, pp 781–791
53. Zhang K (2013) Generalized constant-envelope DualQPSK and AltBOC modulations for modern GNSS signals. Electron Lett 49(21):1335–1337

Chapter 2
Structure of Satellite Navigation Signals

Hardly any information technology goes obsolete. Each new one throws its predecessors into relief.

— James Gleick

2.1 Introduction

A satellite navigation system is a broadcasting system, and the signal it broadcasts is the only interface connecting the system and the receiver. Hence, for technicians involved in the signal design and developers of satellite navigation receivers, it is beneficial to acquire a deep understanding of the design of the navigation signal.

In the past few decades, satellite navigation systems have undergone major evolution. Since the construction of the first-generation GPS, new global navigation satellite systems, regional navigation satellite systems, and satellite-based augmentation systems (SBASs) have been developed. Meanwhile, the design of navigation signals has been going on for more than fifty years. During this period, with the construction of a series of satellite navigation systems, new signal structures have been proposed, many of which have been put into use. Although the new signals introducing novel technologies in recent years differ greatly from the legacy GPS and GLONASS signals in many ways, when comparing them with the legacy signals designed half a century ago, we find that major features such as design concepts and principles are not fundamentally different.

The elements that make up a satellite navigation signal are closely related, and there will sometimes be conflicts between the independent optimization of multiple elements. Actually, an efficient signal design is a joint optimization of multiple elements and is carried out in a process of trade-off and compromise. Therefore, it is important to gain a systematic understanding of the whole signal structure before

Z. Yao and M. Lu, *Next-Generation GNSS Signal Design*, Navigation: Science and Technology 6, https://doi.org/10.1007/978-981-15-5799-6_2

going into details and in-depth discussions about signal elements in the following chapters. In this chapter, the basic principles of satellite navigation are first introduced to clarify the mission of a navigation signal. Following this, we take the structural elements of satellite navigation signals as the main topic, then systematically sort out the important basic concepts in the signal design, as well as the impact of each signal element on the system performance. Many of the concepts and topics will appear frequently in subsequent chapters, so it is necessary to understand the relationships and interactions among them.

2.2 Basic Principles of Satellite Navigation

2.2.1 Position Determination by Signal Ranging

The four major GNSSs currently in operation are all based on the Radio Navigation Satellite Service (RNSS) system. This system is essentially a radio positioning system based on the principle of time of arrival (TOA). For TOA positioning, the user receiver obtains multiple measurements of the signal propagation time from several known sources to the receiver, and then converts them into distances. The position of the receiver can be obtained by finding the intersection of multiple spheres.

A basic TOA positioning process includes the following steps:

Step 1: Measure the distances between the user receiver and several satellites;

Step 2: Calculate the positions of the satellites;

Step 3: Obtain the user's position from the satellites' positions and the user's distances to the satellites, by solving geometric equations as described below.

We use Fig. 2.1 to illustrate this process in detail.

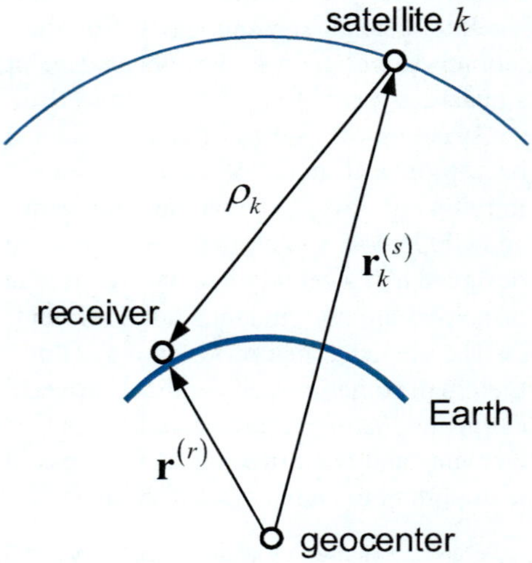

Fig. 2.1 Basic principle of satellite positioning

Here, it is assumed that in the epoch t, the relative position between the receiver and the center of the Earth is defined by a spatial vector $\boldsymbol{r}^{(r)} = (x_u, y_u, z_u)^{\mathrm{T}}$. Meanwhile, a number of satellites above the receiver are broadcasting ranging signals with modulated message data. The clock on each satellite is strictly synchronized with the time reference of the total system as well as the phase of the ranging signal they broadcast.

For simplicity, let's assume that the satellite is stationary in space. After receiving the ranging signal broadcast by satellite k, the receiver can use the ephemeris parameters in the signal message to calculate satellite k's space vector $\boldsymbol{r}_k^{(s)}$ relative to the center of the Earth.

From the ranging signal, the receiver can also recover the system time when the satellite k transmitted the signal. The user receiver has a clock which is also synchronized with the system time. When the receiver determines that the signal transmitted from satellite k at system time t_s arrives at the receiver at a local (receiver) time t_r, it can then determine that the propagation time from satellite k to the receiver is $t_r - t_s$. Multiplying this time difference by the speed of light c, the receiver can obtain the distance ρ_k from the satellite k to the receiver. Then we can obatin an equation which is

$$\rho_k = \left\| \boldsymbol{r}_k^{(s)} - \boldsymbol{r}^{(r)} \right\| . \tag{2.1}$$

In (2.1), ρ_k and $\boldsymbol{r}_k^{(s)}$ are known and $\boldsymbol{r}^{(r)}$ is unknown.

Every equation in the form of (2.1) limits the user's position to a sphere centered at a satellite, as shown in Fig. 2.2. It is obvious that if we conduct this measurement on three satellites at the same time, we will have three equations, and the user's position $\boldsymbol{r}^{(r)}$ can be determined by the intersection of three spheres. However, the prerequisite for the above solution is that the user receiver's clock should be perfectly synchronized with the time reference, which is difficult to achieve in most cases.

Considering the issues of cost and portability, the receiver clock usually uses an inexpensive crystal oscillator, which results in a constantly changing bias b_u of local time at the receiver from the "official" system time. With this bias, the measured distance from the satellite to the receiver is not equal to the real geometric distance. In order to distinguish it from the real distance, we usually call this calculated measurement *pseudorange*. However, the influence of the receiver clock's bias is the same when measuring the signal transmission time of each satellite. In the actual estimation of user position, in addition to the three unknowns in $\boldsymbol{r}^{(r)}$, b_u becomes the fourth unknown item to be estimated.

Other error sources are also taken into consideration in actual measurements, including the influence of random noise, propagation errors, and relativistic effects. For a detailed analysis of various error sources, see [1]. Some of these errors can be monitored by the system's operation control station and transmitted to the user via navigation messages broadcast by satellites for the correction of pseudorange. If user equipment and mission requirements permit, users can further correct these errors. For example, one can use carrier phase measurements to smooth code pseudorange; conduct ionospheric correction by simultaneously receiving dual-frequency signals;

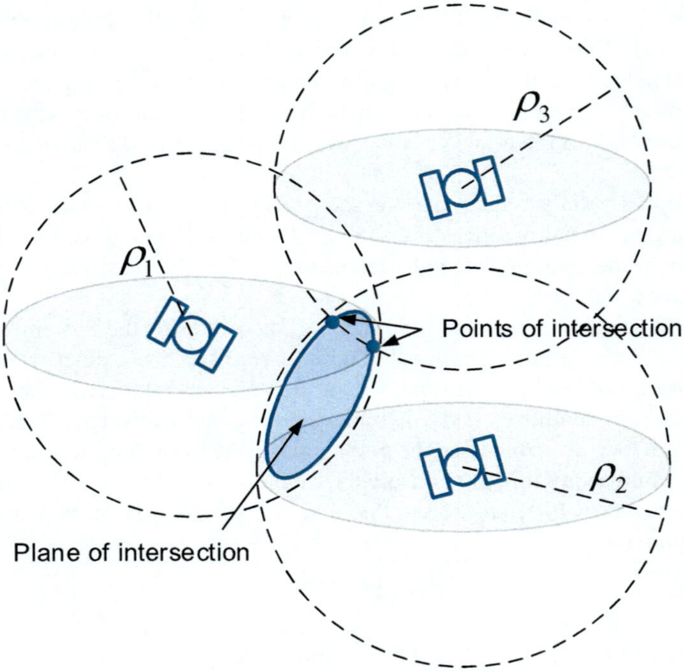

Fig. 2.2 Determining a user's position by finding the intersection of three spheres

use more precise ephemeris; and use correction values provided through wide-area or local-area differential methods.

Let $\tilde{\rho}_k$ be the pseudorange after error elimination. It can be written as

$$\tilde{\rho}_k(\boldsymbol{x}) = \left\| \boldsymbol{r}_k^{(s)} - \boldsymbol{r}^{(r)} \right\| + cb_u + n_k\,, \tag{2.2}$$

where

$$\boldsymbol{x} = [x_u, y_u, z_u, cb_u]^{\mathrm{T}} \tag{2.3}$$

is unknown and to be solved, n_k represents the residual error after applying various error elimination methods. In order to estimate $\boldsymbol{x}$, at least four equations are needed, i.e. we need to simultaneously measure the pseudoranges from at least four satellites. If the user is in an open area and there are no obstructions limiting sky visibility, it is easy to observe four satellites at the same time. In many cases, the receiver can even obtain pseudorange observations from many more satellites. If we assume that the pseudorange values of K satellites are obtained at the same time, the simplest way to solve a total of K nonlinear equations with the form of (2.2) is to linearize them at the user's approximate position and then solve it by the iterative method.

2.2.2 The Linearization Method of Position Estimation

First, a joint vector composed of a user's approximate position $\boldsymbol{r}_0^{(r)}$ and the user's approximate clock bias is given as

$$\boldsymbol{x}_0 = [x_0, y_0, z_0, cb_0]^{\mathrm{T}} . \tag{2.4}$$

The corresponding rough pseudorange of the kth satellite is

$$\tilde{\rho}_k(\boldsymbol{x}_0) = \left\| \boldsymbol{r}_k^{(s)} - \boldsymbol{r}_0^{(r)} \right\| + cb_0 . \tag{2.5}$$

Then, we expand (2.2) at $\boldsymbol{x}_0$ by first-order Taylor series, which can be written as

$$\tilde{\rho}_k(\boldsymbol{x}) = \tilde{\rho}_k(\boldsymbol{x}_0) + \delta\rho_k , \tag{2.6}$$

where

$$\begin{aligned} \delta\rho_k &\approx \left.\frac{\partial \tilde{\rho}_k(\boldsymbol{x})}{\partial \boldsymbol{x}}\right|_{\boldsymbol{x}=\boldsymbol{x}_0} (\boldsymbol{x} - \boldsymbol{x}_0) + n_k \\ &= \left[\left.\frac{\partial \rho_k}{\partial x_u}\right|_{x_u=x_0}, \left.\frac{\partial \rho_k}{\partial y_u}\right|_{y_u=y_0}, \left.\frac{\partial \rho_k}{\partial z_u}\right|_{z_u=z_0}, 1 \right] \delta\boldsymbol{x} + n_k \\ &\triangleq \boldsymbol{h}_k \delta\boldsymbol{x} + n_k . \end{aligned} \tag{2.7}$$

The receiver linearizes the pseudorange observations of each satellite. Then, a total of K linear equations of the above form can be listed. We can then combine them using matrix symbology, which can be written as

$$\begin{aligned} \delta\boldsymbol{\rho} &= \begin{bmatrix} \delta\rho_1 \\ \delta\rho_2 \\ \vdots \\ \delta\rho_K \end{bmatrix} = \begin{bmatrix} \boldsymbol{h}_1 \\ \boldsymbol{h}_2 \\ \vdots \\ \boldsymbol{h}_K \end{bmatrix} \delta\boldsymbol{x} + \begin{bmatrix} n_1 \\ n_2 \\ \vdots \\ n_K \end{bmatrix} \\ &= \boldsymbol{H}\delta\boldsymbol{x} + \boldsymbol{v} . \end{aligned} \tag{2.8}$$

In (2.8), $\boldsymbol{H}$ is referred to as a geometric matrix, which is a $K \times 4$ matrix. The value of its elements is related to the geometric relationship between the user and satellites. When $K > 4$, the least squares solution of (2.8) is

$$\delta\hat{\boldsymbol{x}} = \left(\boldsymbol{H}^{\mathrm{T}}\boldsymbol{H}\right)^{-1} \boldsymbol{H}^{\mathrm{T}} \delta\boldsymbol{\rho} . \tag{2.9}$$

The result is used to correct the initial rough estimation of user's position and the receiver clock bias. Therefore, we have $\hat{\boldsymbol{x}} = \boldsymbol{x}_0 + \delta\hat{\boldsymbol{x}}$. Then, (2.2) can be linearized on the corrected position and clock bias estimates. The solution of the equations can be iterated continually until the difference between the two successive estimates is small enough.

2.2.3 *Precision of User Position*

In order to have a rough estimate of the precision of the user's position, we assume that the pseudorange measurement errors of each satellite are zero-mean, independent and identically distributed, with a standard deviation of σ_{URE}. Then, the covariance of the positioning errors can be written as

$$\operatorname{cov}\left\{\hat{\boldsymbol{x}} - \boldsymbol{x}\right\} = \sigma_{\text{URE}}^2 \left(\boldsymbol{H}^{\text{T}}\boldsymbol{H}\right)^{-1} . \tag{2.10}$$

From (2.10), we can tell that both the satellite geometry distribution and the precision of pseudorange measurement affect precision when solving the user's position from the equations. The partial influence of satellite geometry is shown by $\left(\boldsymbol{H}^{\text{T}}\boldsymbol{H}\right)^{-1}$ in (2.10), and its quality can be characterized as the dilution of precision (DOP). Another major factor that affects the precision of the user's position is the pseudorange measurement accuracy.

2.2.4 *Pseudorange Measurement*

The pseudorange measurement value is obtained by signal tracking loops within the receiver. The loops track and measure the code phase of the received ranging signal or the phase of the radio-frequency carrier.

Satellite navigation systems use PRN code signals as ranging signals. This signal is modulated on the carrier frequency of the satellite. PRN codes appear like random binary sequences, but in fact they have specific generating schemes, and at every moment their values are pre-determined. Moreover, as mentioned earlier, the signal generation device on the satellite is strictly synchronized with the system time, i.e. there exists a unique and definite relationship between the phase of the PRN code signal and the system time.

It is known to the receiver how the PRN code of each satellite is generated, so the receiver can reproduce the PRN code of the satellite. By continually adjusting the initial phase of the locally reproduced PRN code signal through a feedback loop, synchronization can be achieved with the PRN code received from the satellite signal. Once the code phase synchronization is achieved, the receiver can recover the satellite transmission time t_s corresponding to the signal received at the receiver's local time t_r. This can be achieved by reading the time information in the message modulated on the PRN code signal, and a series of parameters such as the internal state of the local PRN code generator and the state of the numerically controlled oscillator

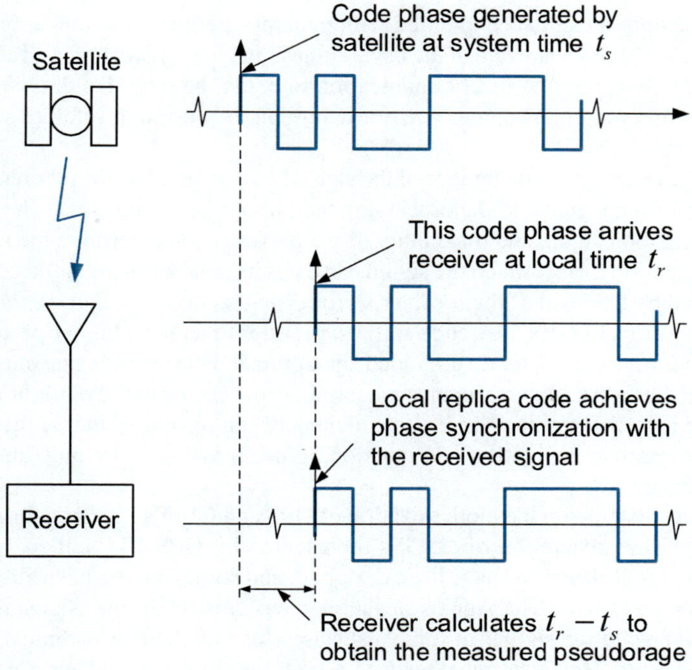

Fig. 2.3 Pseudorange measurement by locally reproduced signals

(NCO) which drives the internal code generator. Then, as mentioned in Sect. 2.2.1, the receiver subtracts this transmission time from the corresponding signal receiving time t_r to obtain the measured pseudorange value for this satellite. This process is shown in Fig. 2.3.

In addition to obtaining pseudorange estimates by measuring PRN code phase, pseudorange estimates can also be obtained by measuring the phase of the radio-frequency carrier, the precision of which is much higher than measuring the code phase. However, the receiver needs to resolve the difficulty of carrier integer cycle ambiguities. For more information on carrier phase measurement techniques, see [1, 2].

2.3 Key Elements of Satellite Navigation Signals

As discussed in Sect. 2.2, satellite navigation signals have two important roles in the positioning solution: (1) Measure the distance between the user receiver and satellites; (2) Transfer data including ephemeris and atmospheric model correction parameters. Regarding data transferring ability, there is no special requirement for satellite navigation signals compared with general satellite communication signals. This is because the update rate of the navigation signal message is low, and a less strict requirement is applied to its communication rate. On the other hand, the pursuit of high-precision ranging becomes a major feature in the design of satellite navigation signals.

The actual ranging performance of the signal is determined by the inherent ranging capability of the signal and the processing technology of the receiver. The structure of the navigation signal and the quality of the transmission determine the maximum ranging capability. How much the signal can exert its maximum potential accuracy is determined by the level of the receiver's processing technology. During the decades of using traditional GPS C/A code signals, in order to achieve higher performance, receiver designers consistently upgraded and optimized the receiving technology, and exhausted the GPS C/A performance potential. As the performance of traditional GPS signals reaches its physical limits, technicians began to realize that if they wanted to further improve navigation and positioning performance, they must design new navigation signals.

In recent years, with the modernization of GPS and GLONASS, and the construction of the new generation of GNSSs represented by GPS III, Galileo, and BDS, more new signal structures have been designed, and some of them have already been put into practical use. The signals of the new systems retain the characteristics of traditional navigation signals in some respects, while adding new technical elements in other aspects. However, whether it is a traditional GPS signal, or a new signal designed for future systems, the essential features can be described as a series of key elements [3]. These elements are carrier frequency, signal power, polarization mode, multiple-access mode, spreading modulation, spreading chip waveform, message structure, channel coding, data modulation, data and pilot components, secondary coding, and multiplexing. These elements are all related to one another. Sometimes there will be conflicts between two or more key elements during optimization. In the following sections of this chapter, the above-mentioned elements and their impacts on the receiver's performance will be discussed and analyzed. After reading this chapter, we will see that a well-established signal design is achieved through the joint optimization of multiple objectives, after a series of trade-offs and compromises.

2.4 Carrier Frequency

The selection of the carrier frequency of the navigation signal is one of the most important issues in signal design. The carrier frequency determines many properties of the signal, including the spatial propagation characteristics, the implementation cost of the hardware for transmission and receiving, the degree of Doppler shift, and the extent of interference with other radio systems.

From the perspective of transmission path loss, when the distance R from the user receiver to the satellite is known, transmission loss in free space

$$L_s = (4\pi R f_{\mathrm{RF}}/\mathrm{c})^2 \tag{2.11}$$

is proportional to the square of the signal frequency f_{RF}. Therefore, theoretically, the lower the carrier frequency of the signal, the smaller the path loss. However, the group delay when the signal passes through the ionosphere is inversely proportional to the square of f_{RF} [4]. Therefore, the higher the carrier frequency of the signal, the less is it affected by the ionosphere. Moreover, in order to create a dual-frequency combination to eliminate the ionospheric error, the satellite navigation system needs to broadcast two sets of signals at two frequency points distant to each other. In addition, when antenna gain is constant, the higher the carrier frequency of the signal, the smaller the size of the antenna (See (2.22) in Sect. 2.5).

In the initial design phase of the first-generation GPS, various possible frequency bands were considered, including the frequency band around 400 MHz, the L-band (1–2 GHz), and the C-band (4–6 GHz). Even though the free space loss in the band around 400 MHz is the lowest, the ionospheric delay and the magnitude of delay fluctuations in this band are relatively large. It is also difficult to place the dual-frequency signal at two points distant enough from each other for the band near 400 MHz. For the C-band, because of the high frequency, its free space loss is about 10 dB higher than the that of the L-band and is more affected by rainfall and atmospheric attenuation. Therefore, the final selection of the L-band for GPS was a compromise, taking multiple factors into account.

The first-generation GPS uses two segments of the L-band: the L1 band with a center frequency of 1,575.42 MHz, and the L2 band with a center frequency of 1,227.6 MHz. It can be noted that the center frequency of both bands is an integer multiple of 10.23 MHz with $L_1 = 1575.42$ MHz $= 154 \times 10.23$ MHz and $L_2 = 1227.6$ MHz $= 120 \times 10.23$ MHz, where 10.23 MHz is the clock reference frequency of the basic GPS satellite signal.[1] In order to ensure the coherence between the components of the signal, all the ranging codes, navigation messages, and basic clocks of the radio-frequency carrier in the GPS signal are obtained by dividing or multiplying by the reference frequency of the same atomic clock.

However, why would a frequency of 10.23 MHz be chosen as the reference frequency for the atomic clock on the satellite? This is because the cycle duration of the C/A code in the GPS signal is designed to be 1 ms, and the PRN sequence used is a Gold code sequence with a period of 1,023, so the basic clock frequency of the C/A code is 1.023 MHz. The rate of the GPS P code is designed to be 10 times that of the C/A code, which results in 10.23 MHz becoming the reference frequency. Interestingly, although the initial choice of this value is determined by the period of the GPS C/A code, this frequency has become the on-board clock reference for most of satellite navigation systems. Moreover, not only has it had an impact on the choice of central frequency of the subsequent systems, but it also affects the choice of the period length of the PRN code for these systems. It is not difficult to verify that the

[1]To compensate for the relativistic effect, the actual frequency on the satellite is 10.22999999543 MHz.

center frequency of many new systems is an integer multiple of 10.23 MHz or 1.023 MHz, and the period of the spreading code is usually an integer multiple of 1,023.

As mentioned earlier, the ionospheric group delay is inversely proportional to the square of the carrier frequency. Therefore, if the satellite simultaneously broadcasts the navigation signals at two different frequencies, the receiver can derive the ionospheric delay at each frequency by calculating the difference in the signal's pseudorange delay. The pseudorange measured $\mathring{\rho}$ that eliminates the ionospheric delay can be obtained by

$$\mathring{\rho} = \frac{(f_1/f_2)^2 \rho_1 - \rho_2}{(f_1/f_2)^2 - 1}, \tag{2.12}$$

where ρ_1 and ρ_2 are the pseudorange measurements of the signal broadcasted on f_1 and f_2. Without losing generality, we assume that $f_1 > f_2$. Equation (2.12) is actually a weighted linear combination of the pseudorange measurements on two frequencies. In the combination, the effect of ionospheric delay is eliminated, but at the same time, other errors in the pseudorange, such as thermal noise and multipath error, are inevitably amplified. The farther apart f_1 and f_2 are, the smaller the amplification factor is, and the smaller the amplification effect of the combination on other errors. The frequency gap between L1 and L2 is 347.82 MHz, which is sufficient for estimating the ionospheric delay using the dual-frequency pseudorange measurements combination.

In addition to L1 and L2, the first-generation GPS has two other center frequencies: L3 at 1,381.05 MHz and L4 at 1,379.913 MHz. The L3 frequency has been used to broadcast time-gated signals, but these signals are not used for navigation services. L4 was considered to provide triple-frequency services, but so far it has not been put into use. Therefore, these two frequencies are not well known to navigation users. With these in mind, it is not surprising that a third civilian frequency of 1,176.45 MHz was added in the modernized GPS to the ARNS band and is named L5. L5 is also an integer multiple of 10.23 MHz with $L_5 = 1176.45$ MHz $= 115 \times 10.23$ MHz. L5 is further from L1 than L2. Therefore, from the perspective of error amplification, a better pseudorange result can be obtained if we combine the signals on L1 and L5 to correct the error caused by the ionospheric layer.

The BDS-2 system uses both the L-band and the S-band. The three frequency bands of service in the L-band are the B1 band with a center frequency of 1,561.098 MHz, the B2 band with a center frequency of 1,207.14 MHz, and the B3 band with a center frequency of 1,268.52 MHz. In the subsequent construction of BDS, in order to provide better multi-system interoperability for civil users, a new B1 center frequency was added at 1,575.42 MHz and a new B2 center frequency, named B2a, was added at 1,176.45 MHz. The original B2 center frequency was thereby renamed B2b.

For the Galileo system, several rounds of adjustments for band selection have been carried out since 2000. The Galileo system was originally designed to use six frequency bands: E1 had a center frequency of 1,598.742 MHz with a bandwidth of 4 MHz (4×1.023 MHz more precisely, but often abbreviated when indicating bandwidth); the center frequency of E2 coincided with BDS-2 B1, and the bandwidth was 4 MHz; E4 had a center frequency of 1,256.244 MHz, with a bandwidth of 4 MHz; E5 was located on 1,202.025 MHz, with a bandwidth of 20 MHz; and the E6 center frequency was 1,278.75 MHz, with a bandwidth of 40 MHz. Galileo also planned to broadcast a 20-MHz-wide navigation signal at 5,014.746 MHz in the C-band. In addition, this system had an E3 band between E5 and GPS L2, but there is no specific plan for how to use it.

However, according to the ICD [5] released by Galileo system in 2010, plans for the initial frequency band allocation and bandwidth are quite different from the current actual frequency allocation scheme. In fact, the subsequent Galileo frequency band design has undergone several rounds of adjustments, mainly considering compatibility with GPS and several other radio system signals, and interoperability with GPS signals. For example, in order to avoid the frequency bands used by the US Army's Joint Tactical Info Distribution System (JTIDS), Distance Measuring Equipment (DME), and Tactical Air Navigation System (TACAN), the Galileo E5's center frequency shifted to 1,191.795 MHz, and its band was further divided into two subbands, E5a and E5b. The center frequency of E5a coincides with the GPS L5, and the center frequency of E5b coincides with BDS-2 B2. The original E1 and E2 bands along with the L1 band in the middle were combined into a new E1 band with the center frequency adjusted to 1,575.42 MHz, which coincides with GPS L1. E3 and E4 did not reappear in subsequent signal planning.

As a FDMA system (see Sect. 2.7), GLONASS has a carrier frequency setting that is quite different from other satellite navigation systems. It also uses three different frequency bands, which are located near 1.6 GHz, 1.25 GHz, and 1.2 GHz respectively. In the GLONASS ICD, the first two frequency bands are recorded as L1 and L2. However, in order to distinguish them from GPS's L1 and L2, G1 and G2 are usually used in the literature to refer to these two frequency bands, and G3 is used to refer to the third frequency band.

Different GLONASS satellites transmit signals at different carrier frequencies, allowing the receiver to distinguish the source of the signal while avoiding interference between the signals. The carrier frequency used by each GLONASS satellite is specified according to the following rules:

$$f_{G1,k} = f_{G1} + \Delta f_1 k = 1602\ \text{MHz} + 0.5625k\ \text{MHz}\,, \tag{2.13}$$

$$f_{G2,k} = f_{G2} + \Delta f_2 k = 1246\ \text{MHz} + 0.4375k\ \text{MHz}\,, \tag{2.14}$$

$$f_{G3,k} = f_{G3} + \Delta f_3 k = 1204.704\ \text{MHz} + 0.4230k\ \text{MHz}\,, \tag{2.15}$$

where k is used to distinguish between different channels. The factors Δf_1, Δf_2, and Δf_3 are the increments of the satellite signal frequency between two adjacent channels. Numerically, there is a proportional relationship between the three frequencies assigned to the same satellite and between the frequency intervals

$$\frac{f_{G1,k}}{f_{G2,k}} = \frac{f_{G1}}{f_{G2}} = \frac{\Delta f_1}{\Delta f_2} = \frac{9}{7}, \tag{2.16}$$

$$\frac{f_{G1,k}}{f_{G3,k}} = \frac{f_{G1}}{f_{G3}} = \frac{\Delta f_1}{\Delta f_3} = \frac{94}{125}, \tag{2.17}$$

and there exists the following relationship between the reference frequency and the frequency increment of each frequency band:

$$\frac{f_{G1}}{\Delta f_1} = \frac{f_{G2}}{\Delta f_2} = \frac{f_{G3}}{\Delta f_3} = 2848\,. \tag{2.18}$$

Since the PRN codes used by the signals of each GLONASS satellite are the same, in order for the receiver to distinguish the source of the signals and avoid interference between the signals, the k values used by different satellites should be different in principle. GLONASS did allocate 25 different channels to the satellites at the beginning of system construction, i.e. $k = 0, 2, \ldots, 24$. However, the frequencies of the navigation signal interfere with the frequencies of the radio astronomy and the satellite communication service. For example, the frequency of 1,612 MHz is the radio frequency used for cosmic background radiation detection. Therefore, at the request of the ITU, GLONASS agreed to change the way channels were allocated in several stages. From 1998 to 2005, the number of channels was reduced to 12, and the same channel was assigned to two satellites 180° apart on the same orbital plane. After 2005, the entire G1 band was adjusted to a lower frequency, and it started to be used to transmit signals when $k = -7, -6, \ldots, +5, +6$. Satellites launched after 2005 use filters to limit their out-of-band emissions and interference. The above adjustments have had some impact on the performance of the system, especially the halving of the number of channels. Although the ground receiver never simultaneously observes two satellites 180° apart on the same orbital plane, for aerospace receivers with altitudes higher than 200 km, there is a risk of receiving signals from these two satellites at the same time. In this case, the aerospace receiver cannot determine which satellite the signal came from, which may result in a positioning error. In order to avoid this problem, the receiver at high altitude must consider other methods to distinguish these two satellites other than the channel number [6].

The above discussion shows that the frequency band selection of each satellite navigation system is the result of multiple-factor trade-offs. Various constraints are imposed on the frequency selection, especially for later systems. Among all available frequency resources, the L-band has many advantages for satellite navigation applications, such as better spatial propagation characteristics, moderate antenna size, and relatively small atmospheric impact. Therefore, it is favored by many major satellite navigation systems. Figure 2.4 shows that at present, the signal frequencies of the four satellite navigation systems are basically located in the upper L-band (1,559–1,610 MHz) and the lower L-band (1,164–1,300 MHz).

Moreover, for the geolocation surveying receivers, if the carrier phase measurements from three frequencies can be used simultaneously, the receiver can obtain an observation combination with long wavelength, small ionospheric interference, and

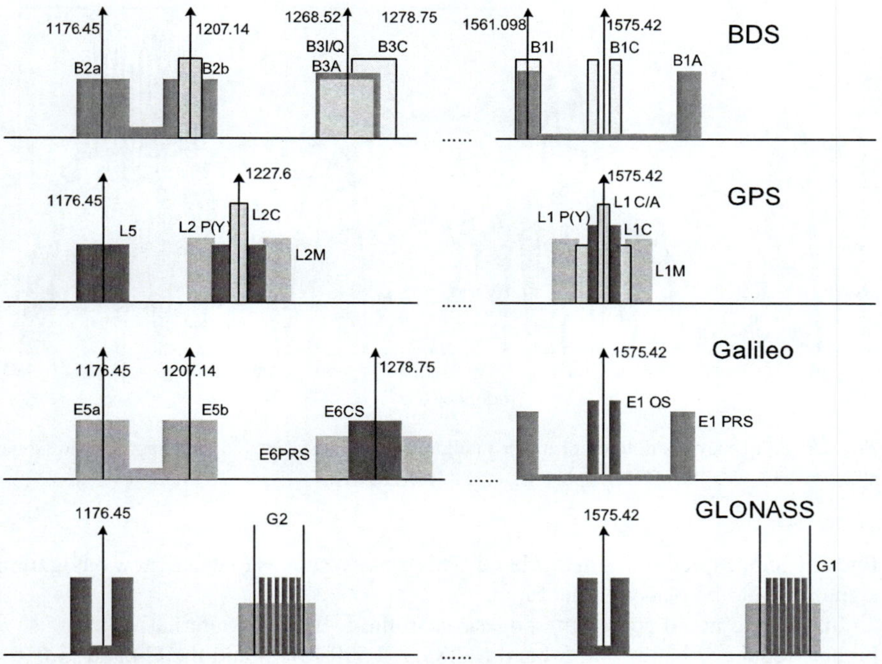

Fig. 2.4 Spectrum allocation of conventional and new-generation satellite navigation systems (some spectra may be further adjusted)

low measurement noise. By using this combination for carrier phase positioning, the carrier's integer ambiguity can be quickly fixed with high accuracy. Therefore, almost all next-generation satellite navigation systems provide signals at three frequencies.

In order to reduce the complexity of receivers and improve the performance of multi-system users, several new systems also incorporate interoperability in the frequency selection. The overlapping of the signal center frequencies of different systems can reduce the complexity of the antenna and the radio frequency (RF) front-end of the multi-system receiver. This advantage is especially important for civil signals. However, it may also exacerbate the interference between signals. Compatibility, which means to ensure that new signals and systems do not interfere with the existing signals and systems, has become a major consideration in signal design for new generation satellite navigation systems. The compatibility between signals with the same or similar center frequency is mainly achieved by adjusting the spectrum shape of the signal (see Sect. 2.8). Even so, finding a suitable center frequency in the L-band for a new navigation signal is becoming increasingly difficult.

Figure 2.5 shows the signal spectrum allocation of systems that are already built or are going to be built in the upper L-band (1.56–1.61 GHz). Obviously, it is difficult to

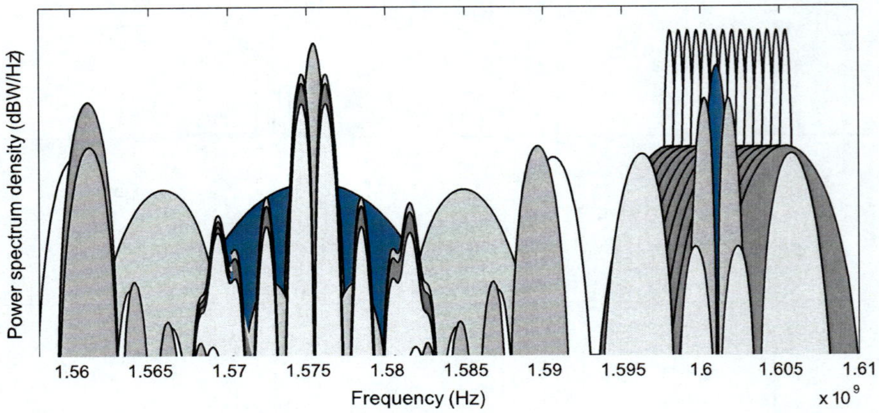

Fig. 2.5 Frequency distribution of major satellite navigation systems in the upper-L band (some signal spectra may be further adjusted)

find any unoccupied space in this band. This raises issues as to where new navigation signals should be placed in the future.

In order to avoid spectrum congestion in the L-band, an alternative is to use a higher frequency band. The S-band (2,483.5–2,500 MHz) and the C-band (5,010–5,030 MHz) have both been considered. The S-band is already used in the BDS, and the IRNSS also plans to use it. However, compared with the upper L-band, the frequency ranges allocated to the navigation service in the S-band and C-band are relatively narrow, which may limit the performance of receiving, and cannot accommodate too many signals. In addition, as mentioned at the beginning of this section, these higher frequencies will result in greater free space loss and phase noise. The Doppler shift caused by the relative movement of the receiver and the satellite is also larger. Given that the transmission loss of the C-band is the greatest, in order to obtain the same receiving performance as the current L-band signal, the satellite needs greater transmission power when broadcasting the signal in the C-band, and the antenna and the RF link of the receiver also need larger gain, which results in complexity increasing of both transmission and reception [7].

2.5 Transmission Power

When a suitable frequency band has been selected to transmit signals, the next thing to consider is the power of the transmitted signal. A variety of problems associated with the transmission and interference of the signal, receiving antenna, and receiver design can be addressed by changing the transmission power of the signal. Single-system users would certainly want the power of the signal it receives to be as high as possible, without sacrificing performance. Greater power means higher signal-to-noise ratio (SNR) and will enable the signal to gain greater penetration into buildings and occlusions and therefore improve navigation and positioning performance. However, on the one hand, to obtain greater received signal strength, the satellite needs higher transmitting power. This requires an increase in the satellite's RF power or in the gain of the transmitting antenna, and these measures will increase the cost and complexity of implementation of the satellite. Furthermore, higher signal power can cause greater interference with other radio services in the same frequency band.

In fact, the power of the transmitted signal is limited by the cost, weight, and power consumption of the transmitting related devices on the satellite. When the satellite is exposed to the Sun, the energy for the transmitted signal comes from solar panels. When the satellite is not exposed to the Sun, the energy of the transmitted signal comes from internal batteries. To obtain a stronger signal, more solar panels and batteries are needed, which will increase the weight and launch cost of the satellite. In addition to the limitations imposed by on-board equipment, greater transmitting power can also cause interference with non-satellite navigation systems using the same or adjacent frequency bands, and can also introduce greater multiple access interference (MAI) in the receiver. As will be seen in the analysis in Sect. 2.9, although current satellite navigation systems use CDMA to avoid mutual interference between signals, the actual PRN codes are not ideally orthogonal. Therefore, when a specific satellite's signal power is much stronger than other satellites, the multiple access performance of the receiver will be affected.

Faced with constraints from both on-board transmission power and ground receiver signal strength, we must establish a relationship between the two. Suppose that the power of the signal transmitted by the navigation satellite on one carrier is P_T, and the distance between the satellite and the user is R. If the strength of the signal transmitted by the satellite antenna in all directions is the same, the power density received on the surface of the Earth is

$$P = \frac{P_T}{4\pi R^2}(\mathrm{W/m^2})\,. \tag{2.19}$$

Figure 2.6 shows the geometric relationship between the satellite and a receiver on the surface of the Earth. If the earth is approximately considered to be a sphere of radius R_E, R can be calculated as a function of the satellite's elevation θ_e, as follows:

$$R = -R_E \sin(\theta_e) + \sqrt{R_E^2\left(\sin^2(\theta_e) - 1\right) + R_{\mathrm{SV}}^2}\,, \tag{2.20}$$

where R_{SV} is the orbital radius of the satellite.

In the actual design, in order to improve the efficiency of signal transmission, the satellite's transmitting antenna is usually not omnidirectional, but has a certain directivity, which can radiate the signal power in a specific direction. In order to compensate for the greater path loss of users with low elevation, navigation satellites typically use directional pattern shaped antenna to transmit signals. The vast majority of satellite navigation users are near the surface of the Earth. As shown in Fig. 2.7, for a navigation satellite about 20,000 km from the ground, the field angle α is about 28°.

Therefore, in order to uniformly apply the transmitted signal to the surface of the Earth, an ideal antenna pattern would be one in which the energy emitted by the antenna is near zero when the field of view angle α is greater than 28°, and when the angle is less than 28° the radiant energy received on the Earth surface is uniformly distributed. However, the actual antenna pattern is smooth, so the range of radiant energy exceeds the Earth's apparent subtended angle to some extent. This also allows receivers at high altitude, such as spaceborne receivers, to receive a limited number of navigation signals for positioning. When the satellite's elevation angle to the receiver is 90° the path loss is minimal, and when the satellite is on the line of sight horizon, the path loss is maximum. For MEO satellites, the path length difference between the two cases is about 5,000 km, so the path loss caused by this path length variation is about 2.1 dB. The satellite's transmission antenna is designed with some compensation for the difference in path loss at this altered elevation angle.

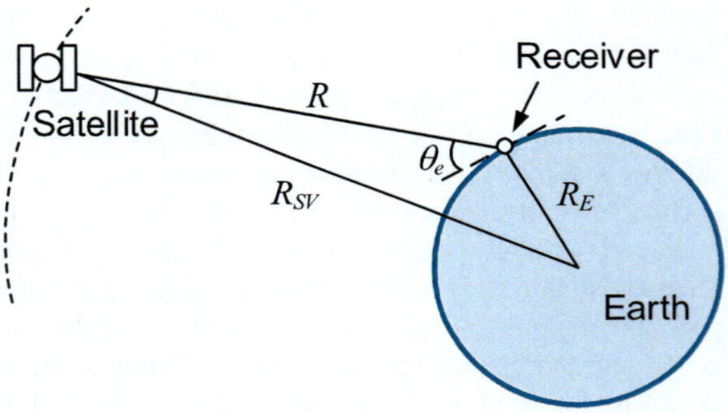

Fig. 2.6 The geometric relationship between the receiver and the satellite

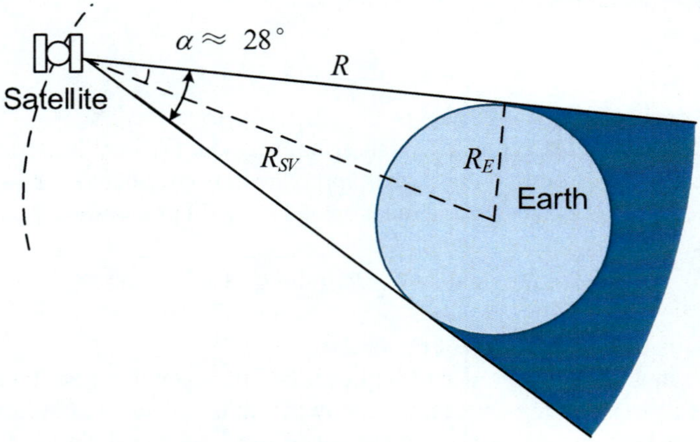

Fig. 2.7 The geometric relationship between a satellite and Earth

For traditional GPS satellites, the antenna gain is approximately 12 dB at low elevation angles. The antenna has a maximum gain of about 13 dB at an elevation angle of 40°. When the satellite is at its zenith, the gain is approximately 10.2 dB. For the new generation of navigation satellites, the antenna gain in all elevations is higher. Assuming that the gain of the satellite antenna in a certain direction is G_T, and considering the antenna polarization mismatch loss L_i and atmospheric loss L_A, then the signal power density that can be received at R distance from the satellite in this direction is

$$P' = \frac{P_T G_T}{4\pi R^2 L_A L_i} (\mathrm{W/m^2}) , \tag{2.21}$$

where $P_T G_T$ is referred to as equivalent isotropic radiated power (EIRP), which characterizes the transmission power in the direction of the maximum antenna gain that can be obtained by the transmitter compared to an omnidirectional transmitting antenna.

Similar to satellite transmission antennas, receiving antennas are also generally directional. If the gain of the receiving antenna in a certain direction is G_R, the corresponding effective receiving area is

$$A_E = \frac{G_R c^2}{4\pi f_{\mathrm{RF}}^2} . \tag{2.22}$$

Thus, at a R distance from the satellite, the satellite signal power that the receiving antenna can receive is

$$P_R = P' A_E = \frac{P_T G_T G_R}{L_s L_i L_A} , \tag{2.23}$$

where the transmission path loss L_s is defined by (2.11). Expressed in terms of decibels, the above formula becomes

$$P_{R,\mathrm{dB}} = \underbrace{P_{T,\mathrm{dB}} + G_{T,\mathrm{dB}}}_{\mathrm{EIRP_{dB}}} + G_{R,\mathrm{dB}} - L_{s,\mathrm{dB}} - L_{i,\mathrm{dB}} - L_{A,\mathrm{dB}} . \tag{2.24}$$

Example 2.1 (*Link budget for satellite navigation signals*)
Consider a MEO navigation satellite with an orbital height of $R_{\mathrm{SV}} \approx 26{,}560$ km. If the Earth's radius is approximately 6,371 km, using (2.20), it is not difficult to obtain that, when the satellite's elevation angle is 5°, the satellite-to-user distance R is about 25,235 km. For the signal on the L1 carrier frequency, the corresponding transmission path loss is

$$L_{s-L1,\mathrm{dB}} = 20 \log \left(4\pi R f_{\mathrm{L1}}/\mathrm{c}\right) = 184.4 \text{ dB} . \tag{2.25}$$

If the transmission power of this satellite on the L1 carrier frequency is 14.3 dBW, and at a 5° elevation angle the corresponding satellite antenna gain is 12.3 dB, then the corresponding EIRP is

$$\mathrm{EIRP}_{\mathrm{dB}} = P_{T,\mathrm{dB}} + G_{T,\mathrm{dB}} = 26.6\ \mathrm{dBW}\,. \tag{2.26}$$

For an omnidirectional receiving antenna on the ground, assuming that $G_{R,\mathrm{dB}} = 0$ dBi and assuming an atmospheric loss of 0.5 dB and a polarization mismatch loss of 1.5 dB, the received power is approximately −159.8 dBW according to (2.24).

When the elevation angle of the satellite is 90°, the satellite-to-user distance R becomes 20,189 km. At this time, for a signal on the L1 carrier frequency, the corresponding transmission path loss becomes

$$L_{s-L1,\mathrm{dB}} = 20\log\left(4\pi R f_{\mathrm{L1}}/\mathrm{c}\right) = 182.5\ \mathrm{dB}\,. \tag{2.27}$$

If the satellite antenna gain at this elevation angle is 10.7 dB and the other conditions are unchanged, the received power is approximately −159.5 dBW.

Keeping all other parameters unchanged, if we change the carrier frequency from L1 to L5, we will find that the corresponding transmission path loss is reduced to

$$L_{s-L5,\mathrm{dB}} = 20\log\left(4\pi R f_{\mathrm{L5}}/\mathrm{c}\right) = 179.9\ \ \mathrm{dB}\,. \tag{2.28}$$

From the ground receiving power of each signal specified in the GPS ICD, we can also find that the L5 signal is the civil GPS signal with the strongest signal receiving power [8]. ■

2.6 Polarization

Satellite navigation signals are transmitted in the form of electromagnetic waves. The orientation of the electromagnetic wave in space is based on the spatial orientation of the electric wave vector $\boldsymbol{E} = \left(E_x, E_y\right)^{\mathrm{T}}$. The polarization of electromagnetic waves can be divided into three types: linear, circular, and elliptical, in which circular polarization can be regarded as a special case of elliptical polarization.

In an isotropic uniformly distributed static medium, the electric field vector $\boldsymbol{E}$ and the magnetic field vector $\boldsymbol{B}$ are orthogonal to each other and perpendicular to the direction of propagation $\boldsymbol{P}$ of the electromagnetic wave. The spatial relationship of the three meets the right-hand rule. In the electric wave vector $\boldsymbol{E}$, components E_x and E_y vibrate at the signal frequency. The electromagnetic wave is said to be linearly polarized if the electric field vector oscillates along a straight line viewed in the direction of electromagnetic wave propagation. If the electric field vector rotates clockwise, the electromagnetic wave is said to be right-hand polarized. If the electric field vector rotates anticlockwise, the electromagnetic wave is said to be left-hand polarized. Furthermore, when the amplitudes of E_x and E_y are equal and their phase difference is 90°, the elliptical polarization becomes circular polarization, as shown in Fig. 2.8.

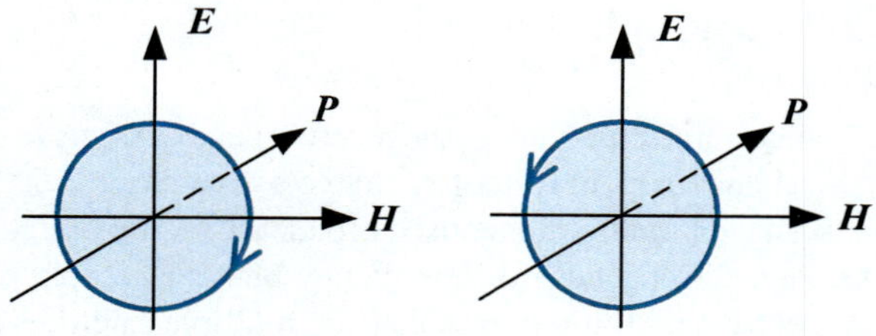

Fig. 2.8 Difference between circularly polarized right-handed and left-handed modes

In order to minimize the polarization loss of the received signal, the polarization form of the receiving antenna should be the same as the polarization of the received signal. Although the generation and reception of linearly polarized waves are relatively simple, receiving the linearly polarized signals requires that the antenna polarization direction of the receiver be the same as that of the transmitted signal. This is very difficult in practical engineering and requires precise pointing adjustment of the receiving antenna. This requirement can only be achieved if there is no relative motion of the transmitting and receiving antennas. For the satellite navigation system, the use of circular polarization can avoid the continuous adjustment in the direction of the receiving antenna.

In all satellite navigation systems currently in use, the polarization form of the signal is right hand circular polarized (RHCP), which moderates the mismatch between the incident electromagnetic field and the receiving antenna, thus reducing the resulting loss. Moreover, the use of circular polarization also has a certain effect on the suppression of multipath signals. If the signal is reflected by the electromagnetic interface, the polarization state of the signal may change. It can be roughly assumed that the satellite navigation signal using RHCP will change to left hand circular polarized (LHCP) after an odd number of reflections. In this way, an antenna with strong resistance to LHCP signals can effectively suppress odd-numbered reflections, especially multipath signals after one reflection.

In fact, it is also possible to use LHCP for navigation signals. Because antennas with different polarization modes will generate large polarization loss when receiving, if some signals are changed to LHCP mode in future navigation signal design, the RF interference between these signals and those RHCP signals in the same frequency band can be reduced [3]. However, the effect of this reduction may be diminished by the actual antenna's polarization out-of-roundness. Moreover, when a multi-system receiver is designed to receive both LHCP signals and RHCP signals, the antenna becomes more complicated. The use of the same polarization for civil signals from different systems is critical to interoperability. Therefore, a difference in polarization modes is not the fundamental way to reduce interference between signals.

2.7 Multiple Access

To determine positioning, the receiver needs to receive signals from multiple satellites simultaneously. In order to achieve a diversity of services, each navigation satellite also needs to simultaneously broadcast multiple signal components. As explained in Sect. 2.4, because the frequency bands available for satellite navigation signals are extremely limited, signals from multiple satellites must share a relatively narrow frequency band. Therefore, the signal design must consider how to minimize mutual interference between multiple signals from the same satellite and different satellites, when simultaneously transmitting these signals in a bandwidth-limited channel.

Conceptually, there is a subtle difference between the techniques used to prevent mutual interference of signals broadcast from multiple satellites and to prevent interference from multiple signals on the same satellite. The former is often referred to as *multiple access* technology and the latter is referred to as *multiplexing* technology. Although both of them can be implemented by code division, frequency division, and time division, in multiplexing the parameter relationships such as the number of signals, relative phase, relative delay, and power ratio between multiple signals are usually fixed, while in multiple access, the parameter relationships between the individual signals are more random. For satellite navigation, multiplexing has a more stringent constraint, that is, when multiple signals are multiplexed on a carrier, the envelope of their composite signal must be constant (see Sect. 2.11 and Chap. 6). Given the differences between the two technologies, we will explain them separately. In this section, we focus on multiple access, and will discuss multiplexing in Sect. 2.11.

Typical multiple access methods include code division multiple access (CDMA), frequency division multiple access (FDMA), and time division multiple access (TDMA).

CDMA systems use PRN codes having good cross-correlation properties to identify signals from different satellites. Although the CDMA method has a so-called near-far effect problem, i.e. when there is a large difference in signal power received from different transmitters, the receiver may be mistakenly locked to the cross-correlation sidelobes of the strong signal. However, for satellite navigation systems, almost all satellites are roughly equal in distance to the user, so unless there is occlusion, the differences between received signal levels are generally small. The PRN code design for satellite navigation systems will be discussed in the next section.

Another possible multiple access scheme for satellite navigation systems is FDMA, in which different satellites use different carrier center frequencies. The GLONASS satellite navigation system uses this multiple access method (see also Sect. 2.4). Strictly speaking, the GLONASS signal is a combination of FDMA and PRN code spreading. The advantage of this access method is that the signal carrier frequencies of each satellite are offset from each other, so the correlation between the signals of each satellite is much smaller than that of the code division. Moreover, FDMA has better resistance to narrowband interference than CDMA. Narrowband interference can affect all CDMA signals on the same carrier frequency, but only affects FDMA signals on one of the multiple carrier frequencies. In addition, FDMA has the advantage that each satellite can use the same ranging code, which means that the receiving channels of different satellites in the receiver can be identical. This feature had considerable advantages in the early development of GLONASS. Although the processing cost of FDMA signals was greater for digital receivers, at that time in the Soviet Union, the design level of analog circuits and the manufacturing process of analog devices were relatively higher than those of digital circuits and digital devices. For a receiver baseband channel implemented with an analog device, multiple structurally identical surface acoustic wave (SAW) filters can be used instead of digital correlators [9].

With the advancement of digital processing technology over time and the reduction in the cost of broadband interference sources, these characteristics of FDMA are no longer an advantage. Compared to the CDMA system, the shortcomings of the FDMA system become more obvious. For a given code rate, the FDMA architecture occupies a wider total bandwidth. In order to reduce this shortcoming, the designers of GLONASS had to set the ranging code rate of its civil signals at only 1/2 the rate of GPS's. In the following chapters, we will see that with equal receiving power and the same chip waveform, the reduction of the ranging code rate will reduce the ranging accuracy. As we explained in Sect. 2.4, GLONASS's overly wide band occupancy has forced it to reduce the number of channels to prevent interference with other nearby radio services. In addition, from the point of view of current digital receiver development, the requirements for processing FDMA signals on RF channels are more complicated. Signals at different frequencies are subject to different device group delay fluctuations, which brings significant difficulty to precise positioning by using carrier phase measurements of GLONASS signals. Therefore, except for the legacy GLONASS, almost all other GNSSs adopt CDMA system. GLONASS has even turned to the CDMA system in the modernization of its signal.

TDMA is rarely used in satellite navigation systems. For a TDMA system, different satellites transmit signals in different time slots. The receiver distinguishes different satellites according to the time slot in which the received signal is located. In general, since satellite navigation requires continuous phase measurements of the signal, it is not a good choice to broadcast intermittent signals in many cases. However, for some positioning systems that seriously suffer from near-far effects, such as ground-based pseudolite navigation systems or ground based augmentation systems, the hybrid mode of TDMA and CDMA can effectively reduce the interference of over-strong signals to weak signals [10].

2.8 Spreading Modulation

CDMA for satellite navigation signals is implemented by DSSS technology. The basic form of the DSSS signal is multiplying the data signal $D(t)$ with bandwidth B_d by a wide band spreading waveform $g(t)$ and then modulating it onto a sinusoidal carrier. The code rate f_c of the spreading waveform $g(t)$ is much higher than the rate of the data signal $D(t)$, thereby significantly broadening the spectrum of the entire signal. The mathematical expression of the spread spectrum signal can be written as

$$s(t) = \sqrt{2P} D(t) g(t) \cos \omega_0 t , \tag{2.29}$$

where P is the power of the spread spectrum signal, and ω_0 is the angular frequency of the sinusoidal carrier. The content to be considered in the design of these two key elements has been discussed in Sect. 2.5 and Sect. 2.4 respectively. This section and the following Sect. 2.9 will focus on the spread spectrum waveform $g(t)$.

The spreading waveform $g(t)$ is composed of a PRN sequence which is usually binary, and a chip waveform with which the bit values of each PRN sequence entry are modulated. It can be expressed as (See also Sect. 3.2)

$$g(t) = \sum_{n=-\infty}^{+\infty} (-1)^{c_n} p(t - nT_c) , \tag{2.30}$$

where $\{c_n\}$ is the PRN sequence, $p(t)$ is the corresponding chip waveform, and the duration of each chip waveform is T_c, which is the reciprocal of f_c. For an ideal PRN sequence, the value of each bit in the sequence is statistically independent of the other bits, and the probability of taking 0 or 1 for each bit is $1/2$. Each satellite broadcasts a spread-spectrum signal using a different PRN sequence, which are approximately orthogonal to each other (see Sect. 2.9). With this orthogonality, signals from different satellites can be distinguished in the receiver by correlating the locally reproduced spread-spectrum signal with the received signal.

In addition to achieving CDMA by using good orthogonality between PRN sequences, there is another important reason why satellite navigation signals use DSSS technology. The PRN sequence in DSSS signals is broadcast in a very high rate, and the values change frequently, which makes the polarity of chip waveforms constantly flip. Moreover, each PRN signal has a unique phase pattern. These features make this signal desirable for ranging.

Since the accurate phase measurement of the spreading chip waveform is one of the core tasks of the receiver, the design of the spreading chip waveform $p(t)$ plays a very important role in the design of the navigation signal. In the signal design, there is a big difference between the optimization goals of the chip waveform $p(t)$ and those of the PRN sequence $\{c_n\}$, although they are related to some extent. Chip waveform design focuses on the potential ranging accuracy of the signal, while the design of the PRN sequence pays more attention to the orthogonality between sequences. Therefore, in the field of navigation signal design, the design and optimization of these two elements are usually carried out separately. We are accustomed to calling the former spreading modulation design and the latter spreading sequence design. In this section, we will focus on spreading modulation. A study of spreading sequence design can be found in Sect. 2.9.

As we have mentioned above, a major function of satellite navigation signals is to measure the phase of the spreading chip waveform, thus achieving the purpose

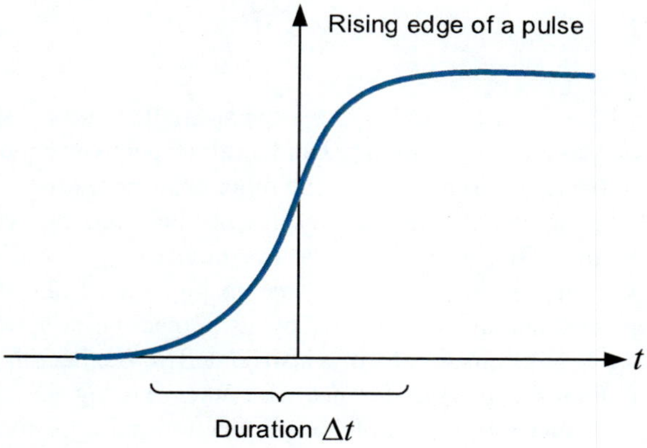

Fig. 2.9 Delay measurement uncertainty

of measuring TOA. This measurement is accomplished by estimating the time when the rising edge of the spread chip waveform of the received signal arrives at the receiver. In a noisy environment, a one-time estimate of the rise edge position of a single chip is unreliable. The DSSS technology provides a long sequence of signals with constantly changing polarities instead of a single pulse. The arrival time of the signal can be estimated by using multiple spreading chip waveforms received in a period of time, to obtain a better capability to counteract noise and interference.

As shown in Fig. 2.9, for the TOA estimate of a single chip waveform pulse, the uncertainty is proportional to the duration Δt of the rising edge of this signal waveform, which is inversely proportional to the bandwidth of the signal. Therefore, the steeper the rising edge of the chip waveform, the larger the corresponding signal bandwidth, and the higher the theoretical resolution of TOA estimation that can be achieved. For the multipath resistant performance of the signal, since the multipath signal always arrives at the receiver later than the direct signal, the key to anti-multipath capability is finding the earliest chip waveform arriving at the receiver and minimizing the impact of its multipath replica. Obviously, a longer-lasting chip pulse is more likely to alias with its multipath replica, while a narrower chip pulse is more easily separated from its multipath replica. Therefore, the narrower the chip pulse used by the ranging signal, the higher the corresponding multipath resolution. The above qualitative analysis shows that the different shapes of the chip waveform affect the measurement accuracy of the signal in thermal noise and its multipath resistance capability. In Chap. 5, we will offer a more in-depth analysis of this issue.

In addition, the analysis in Sect. 2.9 tells us that the actual PRN sequences cannot be completely orthogonal to each other. When multiple navigation signals coexist in the same frequency band, the signal that the receiver wants to receive will be interfered by other signals, resulting a decrease in its SNR. This mutual interference exists between different signals in the same system and signals of different systems. Spectral separation coefficient (SSC) is an indicator that can qualitatively reflect the degree of mutual interference between two signals (see also Sect. 5.8) and is defined as

$$\chi = \int_{-\beta_r/2}^{+\beta_r/2} G_1(f)\, G_2(f)\, \mathrm{d}f\,, \tag{2.31}$$

where $G_1(f)$ and $G_2(f)$ are the normalized power spectral densitys (PSDs) of the two signals on the infinite bandwidth respectively, and β_r is the receiving bandwidth of the receiver. In cases where other conditions are the same, the greater the value of χ, the greater the mutual interference between the two signals.

In (2.31), χ indicates the magnitude of the overlap of the two signals' PSDs. As will be seen in Chap. 3, the shape of the PSD of a spreading spectrum signal is substantially determined by its spreading chip waveform $p(t)$. Therefore, the spreading chip waveform is also related to whether this signal will cause unacceptable mutual spectral interference with other existing signals.

In fact, the purpose of the optimization and adjustment of the spreading modulation in the signal design is to find the spreading chip waveform with the best performance in terms of spectrum occupancy, ranging accuracy, and receiving complexity. All of the traditional satellite navigation signals use a rectangular non-return-to-zero (NRZ) waveform. However, in general, the chip waveform of the DSSS signal is not necessarily rectangular. Essentially, any shape is usable and chips of different time slots can take different shapes.

In recent years, through the process of upgrading satellite navigation systems, variants of several basic DSSS signals using non-rectangular spreading chip waveforms have been studied. For example, the spreading chip waveform of a BOC signal (see Sect. 4.3) uses a square wave. The BCS signal (see Sect. 4.4) further expands the selection of binary chip waveforms. In addition, spreading modulation techniques such as CBCS and MBOC (see Sects. 4.5 and 4.7 respectively) are also emerging.

Since the design of the spreading modulation of the navigation signal is the most important in the signal design, in Chaps. 3–5 we will explain these principles in greater detail along with the application and performance analysis of the spreading modulation.

2.9 Spreading Sequence and Secondary Code

As mentioned previously, good multiple access characteristics of CDMA are achieved based on good autocorrelation and cross-correlation properties between spreading sequences of different signals. In order to minimize the interference from other signals, the spreading sequence $\{c_n\}$ should be completely random. However, as discussed in Sect. 2.2.4, in code synchronization process the receiver must be able to independently generate the spreading sequence that is identical to the signal transmitted by the satellite. Therefore, a truly random sequence cannot be used. Instead, PRN sequences are used.

The difference between a PRN sequence and a random sequence is that the random sequence is unpredictable, and its value in the future can only be described statistically. The pseudo-random sequence is not random at all but has certain predetermined generating rules that both the transmitter and the receiver know. At the same time, it can also exhibit good statistical properties close to a random sequence. It looks like a random sequence to a bystander who does not know how it is generated. This is why it is called a pseudo random noise sequence.

2.9.1 Even and Odd Correlations of Spreading Sequences

In general, a family of PRN sequences $\mathcal{M}$, consists of a set of sequences that have both good autocorrelation and cross-correlation properties. The PRN sequence we are discussing here is binary. For two periodic sequences of length N, the value in one cycle is represented by vectors as

$$\boldsymbol{a} = (a_0, a_1, \ldots, a_{N-1})^{\mathrm{T}} , \tag{2.32}$$

$$\boldsymbol{b} = (b_0, b_1, \ldots, b_{N-1})^{\mathrm{T}} . \tag{2.33}$$

For the convenience of later discussion, we map the binary sequence to a bipolar sequence with the following rules

$$0 \Longrightarrow 1,\ 1 \Longrightarrow -1 \tag{2.34}$$

and in the latter, we always limit the value a_i and b_i to ± 1. Then, the discrete periodic even autocorrelation function of the sequence $\boldsymbol{a}$ is defined as

$$r(\boldsymbol{a}; \tau) = \sum_{i=0}^{N-1} a_i a_{i+\tau} , \tag{2.35}$$

where the subscript of a_i are modulo N, i.e. the above formula is actually a periodic autocorrelation. When τ modulo N is zero, the value of (2.35) is called the autocorrelation main-peak, while when τ modulo N is nonzero, the value of (2.35) is called the autocorrelation sidelobe. Similarly, $\boldsymbol{a}$ and $\boldsymbol{b}$ have their discrete periodic even cross-correlation function, which is

$$r_c(\boldsymbol{a}, \boldsymbol{b}; \tau) = \sum_{i=0}^{N-1} a_i b_{i+\tau} , \tag{2.36}$$

where the subscript of a_i and b_i are modulo N. For cross-correlation, all values of (2.36) are called cross-correlation sidelobes.

Since the data message or the secondary code may also be modulated on the spread spectrum signal, the PRN sequence may have a polarity inversion before and after the data bit transition. In the design of the PRN sequence family, in order to ensure that there is still good cross-correlation between the two sequences when a sequence has a polarity reversal, the odd correlation properties of two PRN sequences must also be considered. Respectively, the definitions of the discrete periodic odd autocorrelation and discrete periodic odd cross-correlation are

$$r_{\text{odd}}(\boldsymbol{a}; \tau) = \sum_{i=0}^{N-\tau-1} a_i a_{i+\tau} - \sum_{i=N-\tau}^{N-1} a_i a_{i+\tau}, \tag{2.37}$$

$$r_{\text{odd},c}(\boldsymbol{a}, \boldsymbol{b}; \tau) = \sum_{i=0}^{N-\tau-1} a_i b_{i+\tau} - \sum_{i=N-\tau}^{N-1} a_i b_{i+\tau}, \tag{2.38}$$

in which the subscript of a_i and b_i are modulo N. It is not difficult to see from (2.37) that when the τ modulo N is zero, the odd autocorrelation value is exactly the same as the even autocorrelation value, and the value is N. That is to say, the autocorrelation main-peak is not divided into parity. The odd correlation and the even correlation only have differences in their autocorrelation or cross-correlation sidelobe values.

Whether even or odd, the autocorrelation and cross-correlation sidelobes of PRN sequences in $\mathcal{M}$ should be as small as possible. The ratio of the autocorrelation main-peak of the signal to the highest sidelobe of the cross-correlation is called the acquisition margin. For satellite navigation, it is necessary for a signal to have a large acquisition margin for better acquisition capability of weak signals. In an environment where the sky is occluded, such as indoors, the receiver may simultaneously obtain a less attenuated satellite signal coming in through the window, and a decaying satellite signal passing through the wall. If the signal strength difference between the two exceeds the acquisition margin and the receiver reproduces the weaker signal and correlates it with the received signal, the cross-correlation sidelobe may appear higher than the auto-correlation main-peak of the weaker signal. At this point, the receiver may mistakenly capture the cross-correlation sidelobes, which will introduce a very large error in the measurement of the pseudorange. The smaller the autocorrelation/cross-correlation sidelobes of the signal, the greater the power gap between stronger and weaker signals that can be tolerated.

In addition to the good autocorrelation/cross-correlation properties of PRN sequences in $\mathcal{M}$, the number of PRN sequences within a PRN sequence family must be sufficient. It is often necessary to assign sequences in the same family to the same type of service signal in signal design. The PRN sequences of the same family have the same period length and similar generation approach, so the code generators of different satellites have the same structure, and the code generators of the respective channels of the receiver can also be identical. For example, legacy GPS defines 37 different spreading codes for its C/A code signals, all of which belong to the Gold code family with a period of 1,023. These spreading codes can be generated using the same generator structure, and different Gold codes can be generated by changing the position of the code taps. With the further division of the new generation of GNSS services and the use of data channel and pilot channel separation structures, the new signals are placing increasing demands on the number of sequences in the code family. For example, Galileo's E1 OS signal design requires 100 code sequences, 50 of which are for the data channels and 50 for the pilot channels. GPS III requires 420 code sequences at the L1 frequency, of which 63 are allocated to the L1C data channels, 63 to the L1C pilot channels, 112 to the Ground-Based Augmentation System (GBAS) data and pilot channels, and 78 to the SBAS data and pilot channels. In addition, 104 are used for future services.

Although the PRN sequences in $\mathcal{M}$ should be as numerous as possible and the correlation property between the sequences should be as good as possible, unfortunately, according to the theoretical constraints, it is very difficult to construct a good PRN sequence family with many sequences that are well correlated to each other.

The normalized Welch bound [11]

$$\theta_{\text{Welch}}^{(\text{norm})} = \sqrt{\frac{M-1}{MN-1}} \tag{2.39}$$

gives the lower bound of the ratio of the largest sidelobe to the autocorrelation main-peak between M sequences of length N. It can be seen that when the sequence length N is given, the greater the total number M is in a sequence family, the higher the worst-case cross-correlation sidelobes between these sequences. When the size M of the sequence family is given, the longer the sequence length N, the better the cross-correlation performance between the signals. When the number of sequences in a sequence family is great enough, the bound can be approximated to

$$\theta_{\text{Welch}}^{(\text{norm})} \approx \frac{1}{\sqrt{N}}. \tag{2.40}$$

At this point, the Welch bound is only related to the period length of the sequence. The longer the sequence period, the lower the Welch bound.

2.9.2 Construction of Spreading Sequences

The Welch bound is simply a lower bound. How far the actual correlation performance of PRN sequences is from the Welch bound depends on the level of sequence design.

There are several ways to construct a family of PRN sequences of a given period length, the most direct of which is the exhaustive method. If we want to construct a code family consisting of M sequences, each of which has a length of N, since the bits of the sequence have 0 and 1 values, the values of all the bits of the entire code family has 2^{MN} possible permutations. In theory, with given constraints and optimization goals, all types of 2^{MN} arrangements can be searched to find the optimal M sequences that meet the design requirements.

This method is used in the design of the PRN sequence of the E1 OS signal and the E6 signal of the Galileo system. The sequences generated by this method are called *random code* or storage code [12]. In the process of finding the optimal sequence family for Galileo system, the optimization goal is set to have the best possible odd/even correlation performance. A constraint on the sequence is that it has an equilibrium property, i.e. the number of 0 and 1 in each sequence should be approximately equal. Another constraint on the sequence is the autocorrelation sidelobe zero (ASZ) property, i.e. when $\tau = \pm 1$, the autocorrelation sidelobe is zero at the time. This requirement is critical to the unbiased nature of the receiver code tracking loop. In the following sections, when we discuss the chip waveform optimization of CBCS signals and the tracking error of time-division multiplexed signals, there will be a more detailed analysis of the problem of the inherent deviation of the tracking loop caused by the shape of the correlation function. Although random codes can theoretically be found by exhausting all permutations, the burden of exhaustive searching is too large to be achieved in a limited amount of time. In real implementation, some heuristic algorithms such as genetic algorithms and simulated annealing algorithms can be used for acceleration.

Since the sequences are obtained by searching, the random codes have no clear real-time generation mode, and can only be generated in advance, stored in the satellite and receiver, and reading and playing back when in use. Hence this type of code is also known as storage code. Although random codes can guarantee good correlation between sequences, when N is large, storing the entire code family for each signal of each satellite in the receiver is still excessively resource-intensive. Therefore, in some systems, such as GPS III, random code is not considered due to the required code generator complexity [13]. Designers are more inclined to find code families with real-time generation mode.

Table 2.1 lists some of the good real-time generated sequence family structures, with their period length, family size, and maximum even correlated sidelobe amplitude. From Table 2.1 we can see that the choice of sequence length is very limited for most PRN sequence families. Among them, the most flexible choice of length is the Weil sequence, of which the period length can be any value as long as it is a prime number, which provides great flexibility in the choice of code length.

The length p of the Weil sequence can take any prime number. In order to explain the method of generating such sequence, it is necessary to briefly introduce the concepts of the quadratic residue, Legendre symbol, and Legendre sequence. When we say that an integer x is a quadratic residue of p, it means that there is an integer y to make $x \equiv y^2 \pmod{p}$. For example, 2 can be called a quadratic residue of 7, because $2 \equiv 3^2 \pmod{7}$. The Legendre symbol is defined as

$$\left(\frac{x}{p}\right) = \begin{cases} 0, & p \text{divides } x\,, \\ 1, & x \text{is quadratic residue of } p\,, \\ -1, & x \text{ is not quadratic residue of } p\,. \end{cases} \tag{2.41}$$

The Legendre sequence is denoted by $\mathrm{Leg}_p\,[n]$. At the time $n = 0$, there is $\mathrm{Leg}_p\,[0] = -1$, and when n takes other integers, there is $\mathrm{Leg}_p\,[n] = (n/p)$.

The Weil sequence was derived from the Legendre sequence. The Weil sequence $\mathrm{Weil}_p^k\,[n]$ with length p is obtained by shifting and adding the Legendre sequence of the same length, i.e.“‘WA

$$\mathrm{Weil}_p^k[n] = \mathrm{Leg}_p[n]\,\mathrm{Leg}_p[n+k]\ , \tag{2.42}$$

where k indicates different Weil sequences, which can take any integer between 1 and $(p-1)/2$.

The spreading sequence used by the GPS III L1C signal is based on the Weil sequence. The L1C spreading sequence design is taken as an example to analyze the design process of the spreading sequence of satellite navigation signals.

Example 2.2 (*Design of L1C spreading sequence structure*) The design constraints of the spreading sequence used for GPS L1C are as follows [13]:

1. First, as mentioned earlier, the code family has a minimum size of 420, i.e. 210 pairs of sequences are prepared for the L1C data channels and the pilot channels;
2. The length of the sequence is 10,230;
3. In the L1C signal, the allocated power of the pilot channel is greater than the data channel, so attention must be paid to the corresponding characteristics of the sequences allocated to the pilot;
4. Both the even and the odd correlation characteristics of the sequences should be considered in the design;
5. The sequence should be approximately zero-balanced, i.e. the probability of 0 and 1 in the sequence should be approximately equal;
6. The data channel and the pilot channel of each L1C composite signal are in-phase superimposed, so the cross-correlation of each pair of data/pilot spreading sequences must be almost zero when $\tau = 0$;

Table 2.1 Typical PRN sequence family

Name	Length N	Code family size	Maximum even correlation sidelobe amplitude[a]
Odd order Gold	2^n-1, n is odd	$N+2$	$1+\sqrt{2}\sqrt{N+1}$
Even order Gold	2^n-1, $n=4k+2$	$N+2$	$1+2\sqrt{N+1}$
Small Kasami set	2^n-1, n is even	$\sqrt{N+1}$	$1+\sqrt{N+1}$
Large Kasami set	2^n-1, $n=4k+2$	$(N+2)\sqrt{N+1}$	$1+2\sqrt{N+1}$
Bent	2^n-1, $n=4k$	$\sqrt{N+1}$	$1+\sqrt{N+1}$
No	2^n-1, $n=2k$	$\sqrt{N+1}$	$1+\sqrt{N+1}$
Gong	$2\times(2^n-1)$	$\sqrt{N}$	$3+2\sqrt{N+1}$
Paterson-Gong	p^2, p prime, $p\equiv 3 (\mathrm{mod}\ 4)$	$\sqrt{N}+1$	$3+2\sqrt{N+1}$
Paterson	p^2, p prime, $p\equiv 3 (\mathrm{mod}\ 4)$	N	$5+4\sqrt{N+1}$
Z_4 linear family I	$2\times(2^n-1)$, n odd	$N/2+1$	$2+\sqrt{N+2}$
Z_4 linear family II	$2\times(2^n-1)$, n even	$(N+2)^2/4$	$2+2\sqrt{N+2}$
Weil	p prime	$(N-1)/2$	$5+2\sqrt{N}$

[a]The corresponding autocorrelation main-peak is N

7. In addition to the good correlation characteristics within the sequence family, the correlation characteristics with other code families of already used signals should be acceptable;
8. Each sequence should be easy to generate.

By the last constraint, random codes are obviously excluded. As seen in Table 2.1, the correlation performance of the Gold sequence is good. However, the difference between its length and 10,230 is too large. Designers have tried a variety of ways, including truncating the long-period Gold sequence [14] and extending the short-period Gold sequence, but the results were not satisfactory until the final selection of the Weil sequence.

Since the length of the Weil sequence can take any prime number, this flexibility makes it very convenient to design the L1C spreading sequence based on the Weil sequence family. The L1C spreading sequence period is 10,230. The two prime numbers closest to it are 10,223 and 10,243. For the former, in order to obtain the L1C sequence, seven bits need to be inserted in the Weil sequence. In the latter case, 13 bits need to be truncated from the Weil sequence. Some studies have shown that inserting bits into a short sequence has less impact on the correlation sidelobes than truncating bits from long sequences. Therefore, the final Weil sequence is chosen to be $p = 10223$.

Figure 2.10 shows how the Weil sequence is generated and how the extra seven bits are inserted. This generation method is easily implemented by logic circuits. The receiver can store the Legendre sequence, export it when needed, and XOR by its shift to obtain the Weil sequence of each satellite. Although the Legendre sequence can be generated in real time, the length of this sequence is only 10,223, and the Weil sequences of all satellites share the same Legendre sequence. Therefore, in many cases, the storage scheme is simpler.

The insertion position of the extra seven bits in the L1C sequence is found by exhausting all possibilities. The criterion for testing is the odd and even autocorrelation and cross-correlation result. However, the search space at this time is much smaller than the search space of the random code. When $p = 10{,}223$, there is a total of $M = (p - 1)/2 = 5{,}111$ Weil sequences in the sequence family. The seven bits have 10,223 insertion methods. At the same time, there are $2^7 = 128$ permutations of the seven bits. However, if we further restrict the number of 0s and 1s in the sequence, then the pattern of these seven bits can be limited. After searching, the inserted seven bits are selected as $(+1, -1, -1, +1, -1, +1, +1)$, and 739 sets of sequences with a length of 10,230 and good correlation characteristics are constructed. Among them, there are 109 sequences with even correlations better than -28 dB and odd correlations better than -26.5 dB. There are 150 sequences with even correlations better than -27.5 dB and odd correlations better than -26.5 dB. There are 480 sequences with even correlations better than -27.2 dB and simultaneous odd correlations better than -26.2 dB.

After finding the 739 sequences, it is also necessary to pair the pilot and data channels of each satellite, and finally construct 210 pairs of pilot/data sequence combinations. Since the transmission power of the pilot channel is higher, the sequence

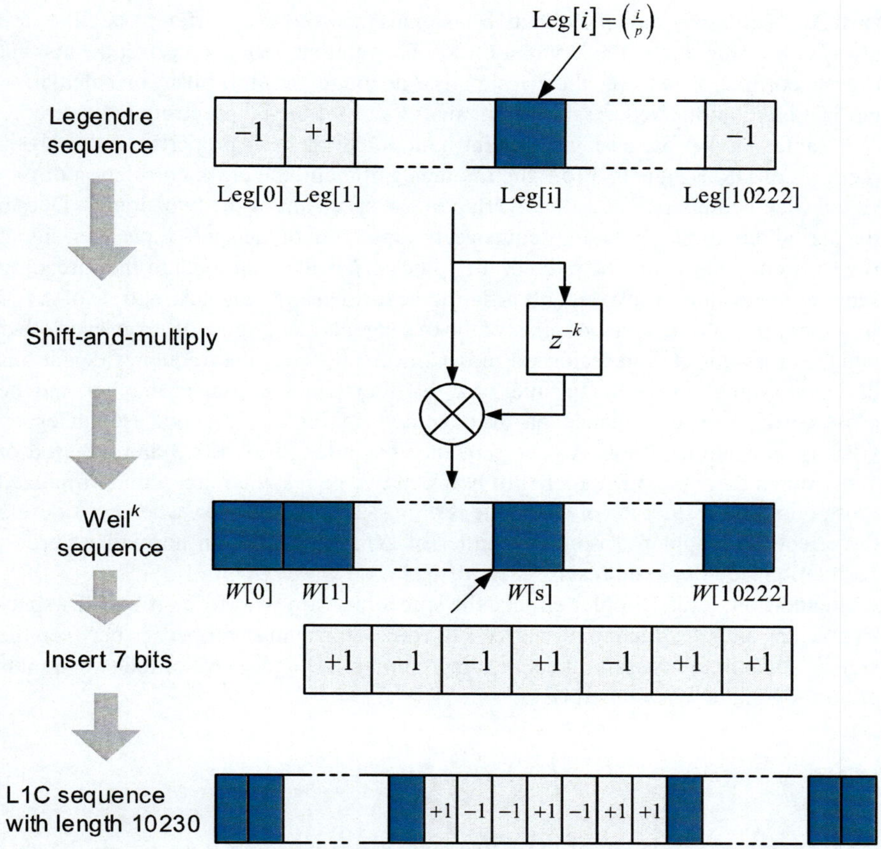

Fig. 2.10 L1C code construction method based on the Weil sequence

with the best correlation in each pair of sequences is assigned to the pilot channel. The top 63 pairs from the best performance of the 210 pairs of sequences are the 63 groups currently used in the GPS L1C signal [15]. ■

2.9.3 The Effect of Spreading Modulation on Code Correlation

In the above example, our analysis of signal correlation characteristic is focused on the sequence level. In practice, it is also necessary to consider the effect of Doppler frequency difference between the two signals and the influence of the spreading chip waveform. Navigation satellites are constantly moving, so for receivers on the Earth surface, the relative frequencies between signals received from different satellites are always changing due to the Doppler effect. The relative frequency offset means that the real correlation between the two signals is no longer a simple bit-to-bit calculation but is more complicated. At this point, analysis of the signal spectrum is required.

Chapter 3 will provide an in-depth introduction to the basic properties and analysis methods of DSSS signals. From the frequency domain, the cross-correlation of two signals can be understood as the overlap of the spectrum of the two signals. Due to the periodicity of the spreading sequence, the spectrum of the spread spectrum signal is a series of spikes, and the pitch of the spikes Δf is the reciprocal of the spreading sequence period. If the two signals arrive at the receiver with an odd multiple of $\Delta f/2$ frequency deviation, then the spikes of the two signals can stagger off from each other and the cross-correlation is close to zero. Conversely, when the frequency deviations of the two signals are even multiples of $\Delta f/2$, their peaks overlap each other, and the cross-correlation performance may be degraded. Taking the C/A code used in legacy GPS as an example, the C/A code sequence has a length of 1,023 and a period of 1 ms. When there is no Doppler shift between two codes, the maximum normalized cross-correlation sidelobe of C/A code is only -24 dB. However, when considering the effects of Doppler, at some multiples of Δf, the maximum normalized cross-correlation sidelobes can reach -21.1 dB [2].

In addition to the Doppler effects, the spreading chip waveform affects the spectrum of the signal, which in turn affects the cross-correlation properties between the signals. In extreme cases, if the spreading chips $p_1(t)$ and $p_2(t)$ of the two signals are orthogonal to each other, i.e.

$$\int_0^{T_c} p_1(t)\, p_2(t)\, \mathrm{d}t = 0\,, \tag{2.43}$$

then, the cross-correlation of these two signals can be completely zero if the chip transition bits are aligned. Even when there is no alignment at the edge of the chip, the orthogonality between such chips contributes to suppressing the correlation between the two.

The civil signals of the next-generation satellite navigation system at the L1-E1-B1 frequency point generally use BOC(1,1) and its derived MBOC(6, 1) spreading modulation. One of the reasons for this is to reduce the interference with the original C/A signal in the same center frequency as much as possible. The C/A code signal uses BPSK-R(1) modulation (see Sect. 4.2), the chip waveform of which is

$$p_{\text{BPSK-R}}(t) = \begin{cases} 1/\sqrt{T_c}, & 0 \le t < T_c \\ 0, & \text{otherwise} \end{cases} \tag{2.44}$$

while the chip waveform of the BOC(1,1) signal is

$$p_{\text{BOC}(1,1)}(t) = \begin{cases} 1/\sqrt{T_c}, & 0 \le t < T_c/2\,, \\ -1/\sqrt{T_c}, & T_c/2 \le t < T_c\,, \\ 0, & \text{otherwise}\,. \end{cases} \tag{2.45}$$

It is easy to verify that these two waveforms satisfy the orthogonality defined by (2.43). This orthogonality can also be understood in the frequency domain. The highest main lobe of the BOC(1,1) signal's power spectrum happens to be at the zero position of the BPSK-R(1) signal's power spectrum.

Using a similar principle, in the design of the GPS L1C signal, the data channel and the pilot channel of the same signal use different modulation modes, wherein the data channel uses BOC(1,1) and the pilot channel uses TMBOC(6, 1, 4/33). The BOC(1,1) chip in the TMBOC signal is orthogonal to the BOC(6, 1) chip. This feature further improves the actual correlation performance of the L1C signal on the basis of the sequence correlation performance.

2.9.4 Secondary Code

The civil signal of the legacy GLONASS uses the Meander sequence as the basic secondary code to aid in data bit synchronization. Next-generation satellite navigation signals typically employ a structured tiered code that combines a primary code and a secondary code on the pilot channel. Secondary code, also known as overlay code, is essentially a PRN sequence, except that the rate is much slower than the rate of the spreading code sequence. The length of one secondary code chip is usually equal to the length of one period of the primary spreading code. By using a secondary code with a low code rate and long period to modulate a primary spreading code with a high code rate and short period, a new spreading code with a longer equivalent period but a high code rate can be constructed. This process is shown in Fig. 2.11.

As for the use of secondary coding in satellite navigation signals, its first role is to improve the cross-correlation properties between satellite signals. As mentioned earlier, the most direct and effective way to improve the correlation of signals is

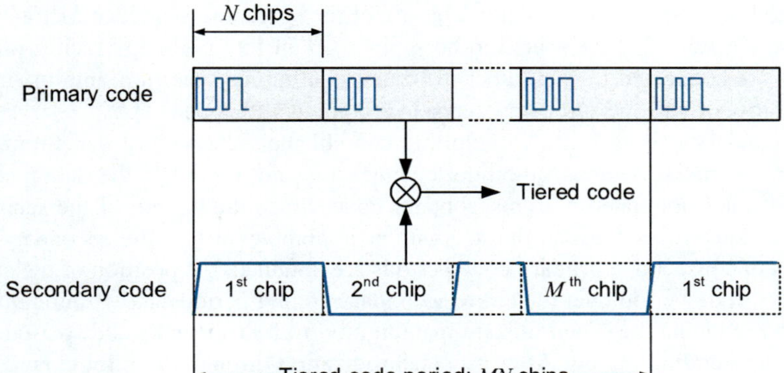

Fig. 2.11 Constructing a long period tiered spreading code using primary and secondary codes

to increase the period length of the PRN sequence. However, an excessively long PRN sequence can significantly increase the cost of signal acquisition. If a new PRN sequence with a period of M times the original PRN sequence is used to replace the original primary code, in theory, the correlation sidelobes can be reduced to $1/\sqrt{M}$ of the original, but at the same time, the code phase uncertainty in signal acquisition stage is increased M times. Moreover, the complexity of the corresponding code generator or storage circuit is also multiplied.

The structured layered code comprised of primary code and secondary code introduces a certain pseudo-random aspect between different periods of the primary code, so that the correlation sidelobes of the combined PRN sequence have a certain reduction. At the same time, chips values in different primary code periods as a whole differ by only one sign at best. Therefore, the uncertainty in signal acquisition in the receiver is not significantly increased.

For applications with nominal received signal strength, the receiver can treat the secondary codes as data bits during the cold start acquisition phase. When the time uncertainty is sufficiently reduced, the current symbol of the secondary code can be completely predicted. When a high-sensitivity receiver needs to use a long coherent integration across multiple primary code periods to improve the SNR, the secondary code will be processed more easily than the completely random data bits since its change pattern is known to the receiver and some strategies can be taken to reduce the uncertainty of the search [16, 17].

However, the improvement of cross-correlation performance by using secondary coding is not as significant as is the direct increasing of the primary code period length. After all, the value of the chip in different primary code periods is not random.

Another function of the secondary code is to reduce the line spectrum width of the power spectrum and improve the signal's ability to resist narrowband interference. As will be explained in Sect. 3.4.2, due to the periodicity of the primary code, its spectrum is a series of spikes, and the spacing of the peaks Δf from each other is the reciprocal of the spreading sequence period. If the narrowband interference just overlaps with a spike, the effect of the interference can be significant. The secondary code with a period of M makes the repetition period of the spreading sequence M times, and the peak distance of the spectrum becomes $1/M$ of the original, which means the peaks are connected to each other to form a continuous spectrum, minimizing the possibility of narrowband interference overlapping with a peak [1].

Secondary coding can also help the data channel achieve fast synchronization of data symbols. Although the pilot channel does not modulate the data message, it is usually modulated with a secondary code. Since the pattern of the secondary code is known, as long as the acquisition is completed and the secondary code values in consecutive primary code periods are obtained, the position of the current primary code period over the entire secondary coding period can be found quickly, thereby reducing the synchronization ambiguity from the primary code period to the secondary coding period. After bit synchronization through the pilot channel, this time information can assist in the data channel processing to demodulate the data message. Moreover, after the secondary code bit synchronization is completed, the inversion of the secondary code does not affect the long-term coherent integration of the pilot channel.

2.10 Pilot Channel and Data Channel

As previously noted, satellite navigation signals play two important roles, namely ranging and transmitting navigation related messages. However, in the optimization of the key elements of the signal, the ranging performance of a signal and the performance of transmitting data are often in conflict. For example, in order to achieve good ranging performance in thermal noise and multipath environments, there should be as many high frequency components in the signal as possible, with the main energy concentrated toward the edge of the band (see Sect. 5.3). However, for transmitting data, the frequency component of a good signal should be as concentrated as possible in the center of the carrier frequency, as we will see more clearly in Chap. 5. In addition, whether the signal is modulated by a data message and the message rate itself also affect the ranging performance of the signal.

2.10.1 The Role of the Pilot Channel

In the original design of the legacy GPS signals, consideration was given to implementing two components that would be provided simultaneously in the transmitted signal – a pilot component for ranging and a data component for transmitting messages. The main reason for this separation is to avoid some of the inevitable performance conflicts that occur when both functions are implemented within a single signal component. However, due to the limitations of satellite payload and receiver technology at that time, especially the limitations of on-board constant envelope multiplexing (CEM) technology (see Sect. 2.11), legacy GPS used a single component to perform the functions of both ranging and message transmitting.

Nevertheless, in new generation GNSSs, almost all signals use the structure of a data channel plus a pilot channel. These two channels use different spreading sequences. For some signals, the spreading modulation waveforms used by the two channels are also different. The data message is modulated on the data channel, and the pilot channel has no message modulated. The two channels are combined by multiplexing techniques. For example, in the GPS L2C signal, the two channels are interleaved by means of time division multiplexing [18], while in the GPS L5 signal and the Galileo E5 signal, the data channel and the pilot channel are modulated in phase quadrature with each other [5, 8]. For the GPS L1C signal and the Galileo E1 OS signal, the two channels are superimposed in phase [15].

For legacy GPS signals, a single signal must not only carry data, but also be used for phase tracking. When data bits are modulated on the carrier, the carrier phase produces a 180° flipping at the bit transition point, which has two main effects:

1. The coherent integration time is limited by the bit length. Because the flipping position of the data bits is usually random, the energy of the coherent integration result across the bit edges may be lost due to bit flipping, unless additional source of information is used, such as assisted GPS (A-GPS), to predict the message information. Otherwise, the coherent integration time cannot exceed the length of a data bit. This imposes a significant limitation on acquisition sensitivity.

2. Carrier phase tracking must use discriminators which are not sensitive to data modulation, i.e. the phase discriminator that can eliminate the influence of the carrier phase 180° flipping. This type of phase discriminator is generally referred to as a Costas discriminator. Costas phase discriminators come in several forms, and there are some differences in the quasi-linear range, thermal noise characteristics, and robustness of the different discriminator forms. References [19, 20] include detailed analysis and summaries of the phase error and characteristics of various phase discriminators.

Here, we use the Costas discriminators in the form of a dot product

$$\Delta\hat{\varphi} = \frac{I_P Q_P}{2T_{\text{coh}} C/N_0} \tag{2.46}$$

as an example, where I_P and Q_P are the correlation values of the receiver's in-phase branch and that of the quadrature branch respectively, T_{coh} is the length of the coherent integration, and C/N_0 is the carrier-to-noise ratio (CNR). When the data bit sign of the signal changes, I_P and Q_P change the sign at the same time, so the product operation in (2.46) eliminates the sensitivity of the discriminator to the data bit signs. Based on ideal code tracking assumptions, when the frequency error and phase error are small, the expectation of the discriminator output can be approximated as

$$\mathbb{E}\left[\Delta\hat{\varphi}\right] = \frac{1}{2}\sin(2\Delta\varphi) \approx \Delta\varphi \tag{2.47}$$

and the closed loop thermal noise variance is

$$\sigma^2_{\text{Costas}} = \frac{B_L}{C/N_0}\left(1 + \frac{1}{2T_{\text{coh}} C/N_0}\right). \tag{2.48}$$

In (2.48), the term in parentheses is called the *squared loss*, and is a number greater than one. Increasing the coherent integration length can reduce the squared loss.

Both the acquisition performance and the tracking performance can be improved in the new GNSS signals by using the pilot channel. Since the signal of the pilot channel has no data modulation, the coherent integration length is not subjected to the data bit length. Therefore, for the acquisition and tracking of the spreading code, a longer T_{coh} can be used to mitigate the squared loss.

For carrier tracking, the benefits are more than that. Since there is no bit flip on the carrier, a pure phase lock loop (PLL) can be used instead of the Costas loop to track it. The phase discriminator used at this time is called the coherent discriminator

[21], the expression of which is

$$\Delta\hat{\varphi} = \frac{Q_P}{\sqrt{2T_{\text{coh}}C/N_0}}, \tag{2.49}$$

When the frequency error and phase error are small, the expectation of the output can be approximated as

$$\mathbb{E}\left[\Delta\hat{\varphi}\right] = \sin\left(\Delta\varphi\right) \approx \Delta\varphi, \tag{2.50}$$

and the closed-loop thermal noise variance is [22]

$$\sigma_{\text{PLL}}^2 = \frac{B_L}{C/N_0}. \tag{2.51}$$

Comparing (2.51) with the closed-loop thermal noise variance expression (2.48) of the Costas loop, it can be seen that since the multiplication operation is avoided, the closed-loop tracking error of the pure PLL has no squared loss, and its performance is only related to the loop bandwidth, but is not sensitive to the length of the coherent integration.

In addition, although the expectations of the Costas discriminator output and the coherent discriminator output are the same if they are linearized near $\Delta\varphi = 0$, the pull-in range and the quasi-linear range of the discriminator curves are different. Figure 2.12 shows the discriminator curves for these two loops.

The pull-in range of the coherent discriminator used in the pure PLL can be $(-\pi, +\pi)$ and the quasi-linear range is about $(-\pi/4, \pi/4)$, while the pull-in range and the quasi-linear range of the Costas discriminator are only half those of the coherent discriminator.

A wider pull-in range allows the tracking loop to withstand greater dynamic stresses. Moreover, the thermal noise tracking threshold[2] of the pure PLL is improved by 6 dB compared to the Costas loop [19]. Therefore, in next-generation GNSS receivers, the pilot channel is recommended to be used for carrier synchronization, and the data channel is only used to complete the demodulation of the message. This not only improves the tracking robustness, but also avoids the square loss, thus increasing the tracking accuracy of the carrier accordingly.

[2]Thermal noise tracking threshold is the minimum CNR to ensure that the 3-σ jitter of the loop tracking error does not exceed the quasi-linear range of the phase detector. It is an empirical indicator of the robustness of the loop under thermal noise.

2.10.2 Power Allocation of the Data Channel and Pilot Channel

In the design of the signal, there is also a problem of power allocation between the data channel and the pilot channel due to the constraint of total signal power, which requires design compromises and trade-offs.

The signal component of the data channel has the disadvantages of limited coherent integration time, narrow phase discriminator range, and squared loss of tracking error. Therefore, the data channel is not recommended to be used in the signal acquisition stage. Moreover, in the signal tracking phase, especially with high dynamics as well as with low SNR, there is also no performance advantage to using the data channel for tracking. Therefore, in most application scenarios, the receiver only needs to capture and track the pilot channel signal component to obtain the ranging measurement. Since the carrier phase and code phase of the data channel are strictly synchronized with those of the pilot channel, the receiver can demodulate the data channel using the code phase and carrier phase information recovered by the pilot channel.

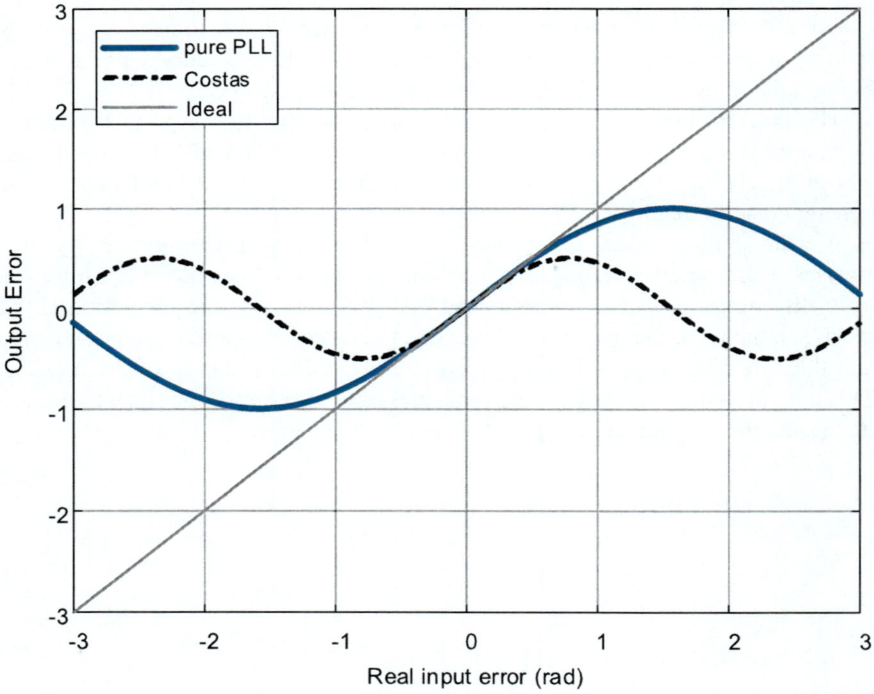

Fig. 2.12 Discriminator curves of the coherent discriminator and Costas discriminator

Increasing the power ratio of the pilot channel can improve the acquisition sensitivity, the tracking accuracy and the robustness of the signal. However, this comes at the cost of a reduction in the power of the data channel, which may affect the bit error rate of demodulation. Conversely, increasing the power ratio of the data channel will lose performance in acquisition and tracking. If we look at the conflict between demodulation, acquisition, and tracking performances out of actual demand, it seems that it is difficult to make efficient trade-offs. However, when we look at the performance of the receiver more comprehensively from the actual requirements of positioning, we can see that is reasonable to allocate more power to the pilot channel. The main reasons are:

1. Degradation of data channel power does not necessarily equate to a reduction in demodulation performance. By using advanced channel coding techniques, the data demodulation threshold can be greatly improved, which can compensate for the impact on signal demodulation from power reduction to some extent. Moreover, the update rate of the navigation data message is very slow, and it is not necessary to continually perform demodulation. It is often used over the course of a few minutes to collect ephemeris data for at least a one- or two-hour time span. By utilizing the slow update characteristic of the navigation data message, the sensitivity of the message demodulation can be improved by the appropriate design of the message structure [23, 24] (see Sect. 2.12).

2. For GNSS receivers, acquisition and tracking performance have much greater impact on usage than demodulation performance. The weakest sensitivity of the navigation receiver is the acquisition sensitivity, and the positioning accuracy is largely determined by the tracking accuracy. For positioning applications, the performance of demodulation just needs to be "good enough", while the accuracy of the tracking and the stability of the loop should be as high as possible. After all, there is almost no difference perceived by a user for a bit error rate of either 10^{-5} or 10^{-12}, but the accuracy of the ranging is 0.2 m or 0.1 m, which is very clearly reflected in the positioning result.

3. Maintaining stable tracking of the carrier loop is a prerequisite for demodulation. In applications such as vehicle navigation and surveying, there are many scenarios that only need to maintain stable tracking without demodulation in harsh environments. One example is, when a vehicle-mounted receiver is blocked by an overpass or buildings for a short time, or when the surveying receiver in the long-term tracks a signal of the satellite whose elevation gradually decreases and the SNR keeps falling. However, there is almost no possibility that only demodulation needs to be maintained but no tracking is required. Moreover, carrier tracking stability is also a prerequisite for data demodulation. Therefore, it is reasonable that the tracking sensitivity is higher than the demodulation sensitivity.

4. In the future development of integrating various means of navigation and communication, message information may be obtained not only from the navigation signal itself, but also from other communication networks and enhanced systems. All of these reduce the pressure of data demodulation in the future.

Based on the above reasons, it is the development trend of the new generation of signal design that the power of pilot channel is higher than that of data channel so that the signal acquisition sensitivity, tracking accuracy and robustness can be improved to a greater extent.

The GPS L1C signal already allocates 75% of the total signal power to the pilot channel and only 25% of the total power to the data channel. However, some readers may wonder why the power ratio of the pilot channel to the data channel is 1 : 1 in the Galileo E1 OS signal, the E5 signal, and the GPS L2C signal. In the design of the signal, the key elements are related to each other. In fact, the design of the channel power allocation is not only related to the performance requirements of each component, but also the constraints of spreading modulation and multiplexing. For some signals with special spreading modulations, not all power ratios are achievable at the transmitter. As we will see in Chap. 7, the selection of equal power ratios for these signals does not indicate that the signal designer is reluctant to increase the pilot power. For some signals, equal power of the two channels was the easiest choice given the multiplexing technology level at the time, under the constraints of the specific spreading modulation used. This also shows that from the research perspective, one of the key tasks in the design of navigation signal is to develop spreading modulation and multiplexing techniques with excellent performance and greater flexibility, so as to provide more options for other elements of the design. This can be seen in recent research trends in spreading modulation and multiplexing.

2.11 Multiplexing

A satellite navigation system usually serves the requirements of many different users at the same time, providing differentiated services. For example, legacy GPS and GLONASS both provide signals for military and civil users. In some new satellite navigation systems, the classification of services is more detailed. For example, the Galileo system provides open service, commercial service, public regulated service, and integrity service.[3] When a transmitter on the satellite has to broadcast multiple signals on a single carrier, it is desirable to combine these signals into a constant envelope composite signal in order to maximize the efficiency of the transmitter (see Sect. 6.2).

When the legacy GPS was originally designed, the problem of multiplexing was easy to solve. Two bipolar DSSS signals for military and civilian users can be combined using QPSK technology (see Sect. 7.2). In QPSK multiplexing, two signals are modulated on a common RF carrier with phases in quadrature. The two components of the QPSK signal are referred to as the in-phase components and the quadrature-phase components. However, in the construction of new generation satellite navigation systems, when it is desirable to combine more than two signals on one carrier, more complex multiplexing techniques are needed, such as Interplexing (see Sect. 7.5), majority voting (see Sect. 7.6), constant envelope multiplexing via intermodulation construction (CEMIC) (see Sect. 7.7), and phase-optimized constant-envelope transmission (POCET) (see Sect. 7.4).

[3]The service was previously called the safety of life survive (SoL) service by the Galileo system.

With continuous refinements of the services of the new generation satellite navigation systems, the number of signals onboard is increasing. Each signal is further subdivided into data channel and pilot channel. More and more signal components need to be multiplexed. Finding effective and flexible CEM techniques has become crucial in the field of navigation signal design. New CEM techniques are emerging, and in recent years, joint constant-envelope multiplexing design on two or more different subcarrier frequencies [25, 26] has become an active and important topic. Chapters 6 and 7 will include more in-depth and complete discussions of multiplexing techniques.

2.12 Message Structure and Channel Coding

2.12.1 Message Structure

The signal modulated by a data message contains a variety of information such as satellite clock correction coefficients, estimated group delay differential, ionospheric delay correction parameters, ephemeris data, and almanac data. This information is used by the receiver to calculate the position and speed of the satellite, as well as making other necessary corrections.

In legacy GPS and GLONASS, navigation messages are arranged in frames and broadcast in order of frame number [18, 27]. In the next-generation satellite navigation systems represented by modernized GPS and Galileo, the structure of the message has changed significantly. Compared with the legacy GPS and GLONASS messages, the next-generation message structure has been greatly improved in terms of the effectiveness, transmission rate, reliability of data, utilization of communication resources, and scalability.

One of the significant changes in the modern message structure compared to the legacy message structure is the type-based block format. The frame structure used in legacy navigation messages is nested by word, subframe, frame and superframe. Different contents are placed in the superframe in a fixed order, and the broadcast time of each word in the superframe is strictly synchronized with the system time. This fixed format superframe structure is convenient for the user to receive all the data sequentially but has restrictions on the flexibility of the broadcasting and the receiving. In order to leave room for the system upgrade, in the frame structure design, it is necessary to reserve some data fields in the message for incorporating new contents in the future. In the frame-based message structure, although it has been indicated that some bit values are meaningless to the user, the satellite still broadcasts them, and the receiver also demodulates them, which wastes channel resources to some extent. However, in the message structure of the new generation GNSSs, the huge superframe structure is broken up into a series of data blocks, each of which corresponds to different content. When the satellite broadcasts, the order of data blocks is not fixed. The obvious benefits of doing this are:

1. The design of the message has room for expansion. The system can reserve a variety of different types of data blocks. When the system is initially built, the specific content of a few data blocks is defined according to the application requirements at that time, and only these data blocks are broadcast. Once the function is extended and new data types need to be added, this can be resolved by defining a new data block type.

2. Both the data transmission sequence at the transmitter and the data processing strategy at the receiver can be more flexible. Some data blocks with faster content updates can be broadcast at a higher update rate, while those with slower content changes can be broadcast at a lower update rate. Moreover, the order of message broadcasting in different signals of the same satellite and different satellites can also be optimized [28], so that users can collect all necessary data in a shorter time.

2.12.2 Channel Coding

After message information is converted into a bit stream, channel coding is carried out to make the signal better resist the influence of noise, interference and fading in the transmission channel. The essence of channel coding is adding redundancy in a controllable manner to the transmitted information sequence so that the receiver can exploit this redundancy to detect or correct the errors that occurred during transmission.

The channel coding of legacy GPS and GLONASS navigation messages often uses linear block codes such as Hamming codes and Bose–Chaudhuri–Hocquenghem (BCH) codes. In recent years, with the development of computer science and communication technology, more advanced channel coding techniques have emerged. Furthermore, the computing power of the receiver has been significantly improved. In the new generation satellite navigation systems, the channel coding of navigation messages has begun to use convolutional codes and LDPC, while also adding cyclic redundancy check (CRC) code and interleaving for combating burst errors.

Example 2.3 *(CNAV-2 message structure of GPS L1C signal)* In recent years, the most promising achievement in the design of a new generation of navigation message and channel coding is the CNAV-2 message structure used by GPS L1C signals [15]. CNAV-2 message structure combines legacy data frame structure with modern data block structure and employees several different channel coding techniques. Figure 2.13 shows the structure of the CNAV-2 message.

One frame of the CNAV-2 message is divided into three subframes of unequal lengths. The clock and ephemeris parameters broadcast by the navigation satellite are updated in a certain period, and within a few hours, these parameters may remain unchanged. Other content in the message, such as the information used to identify the seconds of the week corresponding to the current frame, and some auxiliary information that requires more timeliness, must be frequently updated. Thus, fast-changing information and slow-changing information are treated separately in the

CNAV-2 message. The update period of the clock and ephemeris information is two hours, which is called an interval of the CNAV-2 message. All contents that do not change in one interval are placed in the second subframe, such as satellite orbit parameters, satellite clock parameters, and group delay correction values. The position index marking which two hours of the week the current interval corresponds to is also placed in the second subframe.

When the receiver calculates the time of transmission, the time information obtained from the tracking loop, the PRN code counter, and the secondary coding counter have a temporal ambiguity of some seconds. Therefore, the time of week information is needed to eliminate this ambiguity. The two-hour scale time information in the second subframe of the CNAV-2 message is somewhat rough. Each frame needs to provide its position index within these two hours. The first subframe of the CNAV-2 message is used to undertake this task. The content is called time of interval (TOI), and nine bits are used to identify the current frame position within the entire two-hour period. The resolution is 18 s. When the receiver obtains the TOI, it can be combined with the interval number in the second subframe into the intra-week second information. If the clock uncertainty of the receiver is less than the resolution of TOI, even without demodulating the time information of the second subframe, the intra-week second information can be directly obtained from the TOI. For most navigation receivers with time keeping circuits or mobile communication links, a time uncertainty of less than 18 s is very easy to satisfy.

In addition to the ephemeris and time of week which are the two most basic pieces of information required for positions, the message also needs to provide some auxiliary information that may not be required by all users, such as almanac, single-frequency ionospheric correction, and interoperability deviation parameters. These contents are placed in the third subframe of the CNAV-2 message. The third subframe is designed according to the type-based data block format which we explained earlier.

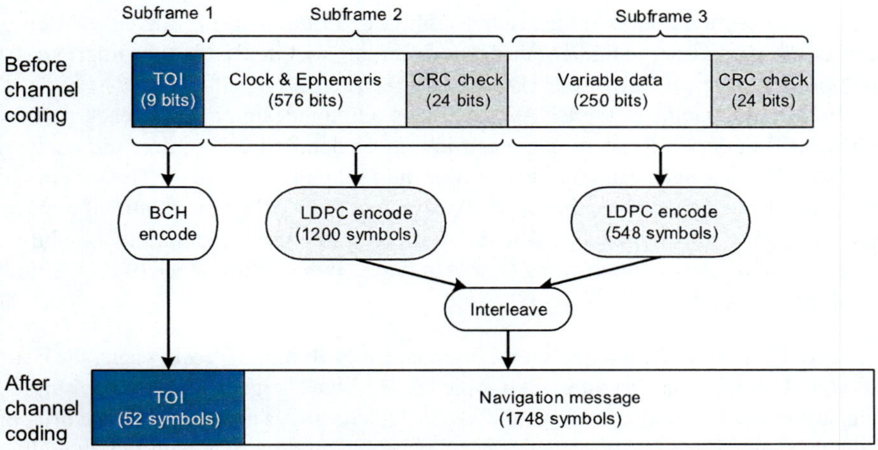

Fig. 2.13 Structure of the CNAV-2 message

The content of the block will be different depending on the page number. The third subframe of the CNAV-2 message currently defines only a few pages of content, leaving many available page numbers, which provides sufficient space for future supplements and updates.

The second subframe and the third subframe are first added with the CRC check code, and then the LDPC code. In addition, in order to resist channel fading, these two subframes are also interleaved after encoding, and finally contain 1,748 symbols. LDPC encoding has already achieved a fairly high channel gain, but demodulation at a lower SNR is further considered in the design of the CNAV-2 message. Since the content of the second subframe is completely unchanged within two hours, every time this subframe is broadcast, the bit sign in the same position is identical. When the SNR is too low that the bit error rate of a single symbol cannot meet the demodulation requirements, the receiver can also superimpose the same symbols of the second subframe of successive frames to further improve the SNR.

Since the content of the first subframe changes quickly, there is no way to perform multi-frame superposition for its demodulation in the case of low SNR. However, the content of the first subframe is the most important of the entire message. As mentioned above, for receivers with time keeping circuits and additional auxiliary data sources, the contents of the second and third subframes may not need to be demodulated at all. However, the TOI contents of the first subframe are directly related to the elimination of the 18s-time uncertainty in the time of transmission calculation. Therefore, in the channel coding design of the first subframe, a highly redundant BCH coding is used. The nine-bit TOI information is encoded by BCH(51, 8) and becomes 52 code symbols. It can be verified that the demodulation sensitivity of this part is much higher than the rest of the message. Even so, CNAV-2 also provides a technique to further improve the sensitivity of TOI demodulation. Since the time difference between signals from different satellites reaching the receiver is only tens of milliseconds, the value of the TOI in the message frames of different satellites obtained simultaneously by the receiver is the same in most cases. Therefore, similar to the principle of multi-frame joint demodulation of the second subframe, for the first subframe, the receiver can combine and demodulate data from different satellites to further improve the demodulation sensitivity of the TOI.

In [29], the details of the CNAV-2 message and the design ideas of each part are elaborated in depth. Readers interested in a more detailed description and analysis of CNAV-2 messages can also read further literatures [15, 23, 24]. The success of the message design enables the GPS L1C signal to reduce the power ratio of the data channel in the transmitted signal without affecting the user's demodulation ability, and puts more power on the pilot channel to improve the performance of acquisition, tracking, and ranging. ■

The structure of the navigation message is still evolving. In recent years, the issue of signal security has become a hot topic. Studies have suggested the use of digital signatures or other authentication features in data messages to increase the security of signal usage. However, this requires a considerable amount of evaluation, including whether it will affect other performance aspects of the signal, and whether there are other ways to enhance signal security.

References

1. Misra P, Enge P (2006) Global positioning system: signals, measurements and performance, 2nd edn. Ganga-Jamuna Press, Kathmandu
2. Kaplan ED, Hegarty CJ (2006) Understanding GPS: principles and applications. Norwood, Artech House, MA
3. JW Betz (2013) Something old, something new: signal structures for satellite-based navigation: past, present, and future. Inside GNSS, pp 34–42
4. Klobuchar J (1996) Ionospheric effects on GPS. In: Parkinson B, James J, Spilker J (eds) Global positioning system: theory and applications, vol I. American Institute of Aeronautics and Astronautics Inc, Washington
5. EU (2010) European GNSS (Galileo) Open service signal in space interface control document, Issue 1
6. Branets V, Mikhailov M, Stishov Y et al (1999) The Soyuz-Mir orbital flight GPS/GLONASS experiment. In: Proceedings of the 12th international technical meeting of the satellite division of the institute of navigation, TN, Nashville, pp 2303–2312
7. Irsigler M, Hein GW, Schmitz-Peiffer A (2004) Use of C-band frequencies for satellite navigation: benefits and drawbacks. GPS Sol 8(3):119–139
8. IS-GPS-705 (2005) NAVSTAR GPS space segment/navigation L5 User interfaces
9. Campbell C (1998) Surface acoustic wave devices for mobile and wireless communications. Academic, New York
10. Cobb H (1997) GPS pseudolites: theory, design, and applications, PhD Dissertation, Stanford University
11. Welch LR (1974) Lower bounds on the minimum correlation of signals. IEEE Trans Inf Theory IT-20
12. Winkel JO (2006) Spreading codes for a satellite navigation system, Patent number WO2006063613 A1. Accessed 22 Jun 2006
13. Rushanan JJ (2007) The spreading and overlay codes for the L1C signal. ION NTM. San Diego, CA, pp 539–547
14. Pratt R, Powe M (2005) Concatenated truncated codes: structure performance and mechanization. In: ION-NTM-2005, Institute of Navigation
15. IS-GPS-800 (2006) Navstar GPS space segment/user L1C interfaces
16. Borio D (2011) M-sequence and secondary code constraints for GNSS signal acquisition. IEEE Trans Aerosp Electron Syst 47(2):928–945
17. Shivaramaiah NC, Dempster AG Rizos C (2008) "Exploiting the secondary codes to improve signal acquisition performance in Galileo receivers. In: Proceedings of ION GNSS, pp 1497–1506
18. IS-GPS-200D (2006) Navstar GPS space segment/navigation user interfaces
19. Ward PW, Betz JW, Hegarty CJ (2006) Satellite signal acquisition, tracking, and data demodulation. In: Understanding GPS principles and applications, pp 174–175
20. Julien O (2005) Carrier-phase tracking of future data/pilot signals. In: Proceedings of ION GNSS, Long Beach, CA, pp 113–124
21. Hegarty C et al (1999) Evaluation of the proposed signal structure for the new civil GPS signal at 1176.45 MHz. In: WN99W0000034, The MITRE Corporation
22. Proakis J (2001) Digital communications, 4th edn. McGraw-Hill, Boston
23. Betz JW, Blanco MA, Cahn CR et al (2006) Description of the L1C Signal. ION GNSS 19th international technical meeting of the satellite division. TX, US, Fort Worth, pp 2080–2091

24. Betz J, Cahn CR, Dafesh PA et al (2006) L1C Signal design options. National technical meeting of the Institute of Navigation. CA, Monterey, pp 685–697
25. Yao Z, Lu M (2012) Dual-frequency constant envelope multiplex with non-equal power allocation for GNSS. Electron Lett 48(25):1624–1625
26. Yao Z, Zhang J, Lu M (2016) ACE-BOC: dual-frequency constant envelope multiplexing for satellite navigation. IEEE Trans Aerosp Electron Syst 52(1)
27. Russian Space Agency (2002) GLONASS interface control document. Ver 5:
28. Paris R, Fernandez V, Cueto M et al (2007) Towards a Galileo navigation message. In: European navigation conference, Switzerland, pp 456–467. Accessed 29 May–1 June 2007
29. Dafesh PA, Valles EL, Hsu J, Skylar DJ, Zapata LF, Cahn CR (2007) Data message performance for the future L1C GPS signal. ION GNSS 20th international technical meeting of the satellite division. TX, US, Fort Worth, pp 2519–2528

Chapter 3
Basic Properties of Direct Sequence Spread Spectrum Signals

Once the headrope of a fishing net is pulled out, all its meshes open.

— Lüshi Chunqiu

3.1 Introduction

Both the legacy and next-generation satellite navigation systems use the triangulation principle, which is based on the accurate TOA measurement, to achieve the functions of positioning, navigation, and timing. Therefore, the measurement accuracy of the TOA largely determines the performance of the system. As discussed in Sect. 2.8, in order to utilize the frequent sign altering in the spreading codes to measure the distance accurately, and to achieve better performance of multiple access, multipath resistance, and anti-interference, DSSS technology—a relatively mature technology in wireless communication—is used in satellite navigation signals.

The study of spread-spectrum signal processing began during the Second World War. Compared with general signals, the spread-spectrum signal has a distinct advantage of being highly resistant to a variety of intentional or unintentional interference. This advantage was first applied in military communications, where the main concern is the security, reliability, and robustness of the communication link. Subsequently, with the advent of multi-user communication and the rapid development of communication technologies, spread-spectrum communication technology has matured, and applications have gradually expanded to civilian and commercial fields. In addition to the DSSS technology, other spread-spectrum technologies include the frequency hopping spread sprectrum (FHSS) and time hopping spread spectrum (THSS) technologies. Many hybrid spread-spectrum technologies consist of two or more spread-spectrum technologies. Because satellite navigation signals mainly use the DSSS

Z. Yao and M. Lu, *Next-Generation GNSS Signal Design*, Navigation: Science and Technology 6, https://doi.org/10.1007/978-981-15-5799-6_3

system, unless otherwise specified, the term spread-spectrum refers specifically to DSSS in this book.

In the development of spread-spectrum communication, the Rake receiving technique [1] proposed in 1958 was a milestone, marking the beginning of research into multipath diversity. In addition, in the 1960s, Golomb [2], Zierler [3], Gold [4], Kasami [5] and others conducted in-depth research into the design of the spread-spectrum sequences and their correlation properties, making important contributions to the further improvement of spread spectrum technology. For those who are interested in the basic concepts, theories and general implementation of spread-spectrum communication, please refer to [6–8].

However, unlike spread-spectrum communication systems, which focus on data transmission, satellite navigation systems use the spread-spectrum signal for distance measurement. Compared to the consideration of bit error rate in communications systems, satellite navigation systems pose much higher requirements for the ranging performance, particularly under the interference of thermal noise and multipath. Therefore, on the one hand, the design and processing technology of satellite navigation signal absorbs the technology of spread-spectrum communication, and on the other hand, it also shows its uniqueness. For example, raised cosine spread-spectrum waveforms are widely used in communication to achieve high energy concentration and efficient utilization of frequency bands. However, they are not suitable for satellite navigation because of their limited ranging capability [9]. Another example is the multipath signal. In spread-spectrum communication systems, the Rake receiver treats multipath signals as useful, and combines them to improve the received SNR. However, in a spread-spectrum navigation system, multipath signals can cause measuring bias in the code phase and carrier phase and are therefore, always treated as harmful interferences to be eliminated.

As mentioned in Sect. 2.8, in the field of satellite navigation, DSSS signals with different spreading chip waveforms are usually referred to as different spreading modulated signals. The legacy navigation ranging signals mostly use chip waveforms with a rectangular NRZ format, and the corresponding spreading modulation is called BPSK with rectangular chips (BPSK-R). However, in theory, the spreading chip of a DSSS signal can take any waveform. In addition to the rectangular NRZ format, many new types of spreading chip waveforms have emerged in recent years in satellite navigation, such as square waves with sine and cosine phases, and Walsh code symbols. With different spreading chip formats, the time domain waveform and power spectrum shape of the signal will exhibit significant differences. These differences will affect the acquisition, tracking, phase measurement, and demodulation performances in environments of thermal noise, interference, and multipath. It will also affect the implementation of receivers, as well as the compatibility and interoperability with other satellite navigation signals on the same frequency band. Therefore, the spreading chip waveform becomes the key factor in determining the intrinsic performance of the satellite navigation signal, and the spreading modulation design has become the core technology of the GNSS signal design.

From this chapter, we will start our journey on the spreading modulation design of satellite navigation systems. In this chapter, we will discuss some basic concepts and properties of direct spread-spectrum sequences, illustrating the mathematical models of spreading modulated signals and their important characteristics in time and frequency domains. These characteristics will be used frequently in the following chapters. In this chapter, several specific spreading modulation methods will be mentioned as examples in discussions of spreading modulation characteristics. A detailed elaboration of these modulation methods can be found in Chap. 4. Those who are not familiar with these modulation methods can skip these parts and review them after finishing Chap. 4 or refer directly to the corresponding section in Chap. 4.

3.2 Spreading Modulated Signal Model

In Sects. 2.8 and 2.9, the basic concepts of DSSS technology were briefly discussed. In order to conduct a more in-depth analysis of the spread-spectrum signal, here we define a more restricted model of the spreading modulated signal.

DSSS technology can be seen as an extension of binary phase shift keying (BPSK) modulation technology. The DSSS technology modulates the data signal with a high-speed pseudo-random signal. As shown in Fig. 3.1, the transmitted signal generated by this technique can be regarded as the product of the unmodulated RF carrier, data signal $D(t)$, and baseband spread-spectrum signal $g(t)$, that is

$$s(t) = A_s D(t)\left[\mathrm{Re}\{g(t)\}\cos(2\pi f_0 t + \theta) - \mathrm{Im}\{g(t)\}\sin(2\pi f_0 t + \theta)\right], \quad (3.1)$$

where A_s is the amplitude of the signal, f_0 is the carrier frequency, θ is the carrier phase, and $\mathrm{Re}\{x\}$ and $\mathrm{Im}\{x\}$ are the real and imaginary parts of the complex number x respectively. The code rate f_c of the baseband spread-spectrum signal $g(t)$ is much higher than the rate of the data signal $D(t)$, such that the transmitted signal $s(t)$ occupies a much wider frequency band than the data signal $D(t)$.

Spread-spectrum signals are often periodic, and a finite length binary sequence that is used to generate the spread-spectrum signal is termed a PRN sequence. The elements of the sequence can be written as c_n, giving $c_n = c_{N+n}$, where N is the period of the sequence. Compared with the characteristic of the PRN sequence itself, we are more interested in the properties related to the spreading chip waveform in this chapter. A more detailed analysis of the generation and characteristics of PRN sequences can be found in Sect. 2.9 and [10].

The baseband spread-spectrum signal $g(t)$ generated by the PRN sequence can be further expressed as

$$g(t) = \sum_{n=-\infty}^{+\infty} (-1)^{c_n} p(t - nT_c), \quad (3.2)$$

where $p(t)$ is a spreading chip waveform, which is only non-zero within the interval of $[0, T_c)$ and takes values from a finite set of waveforms with a given probability

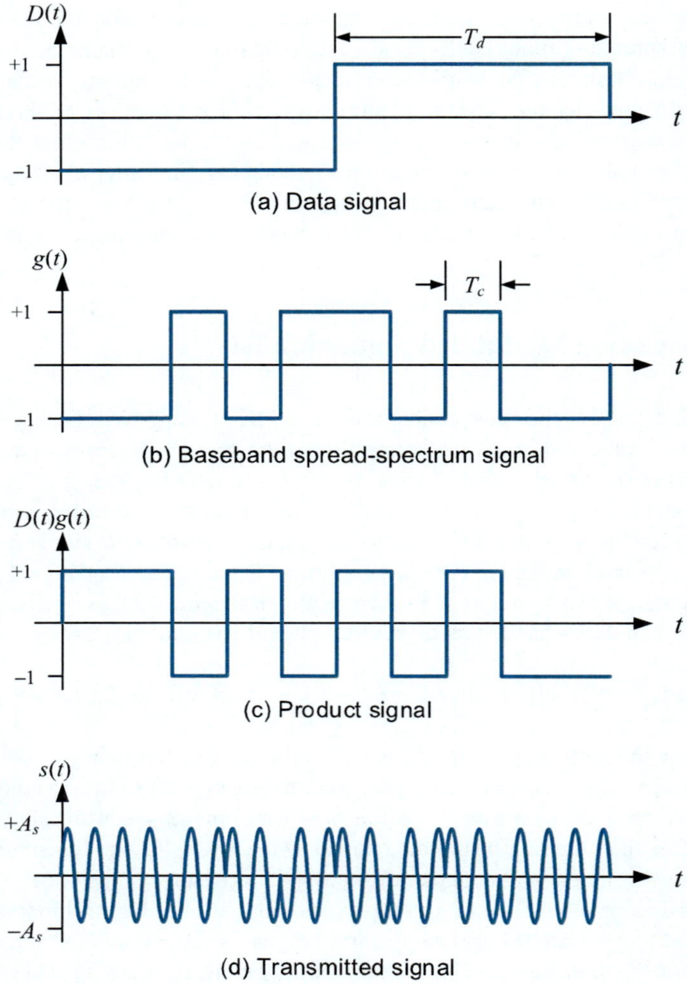

Fig. 3.1 Signals in the DSSS system

$\mathcal{P}_{p_i}$. In theory, the value of $p(t)$ can be continuous, and the amplitude can be real or complex. We have

$$p(t) \in \left\{p_1(t), p_2(t), \ldots, p_{N_p}(t)\right\}, \tag{3.3}$$

and

$$\Pr\{p(t) = p_i(t)\} = \mathcal{P}_{p_i}. \tag{3.4}$$

In most cases, the spread-spectrum signals have the same waveforms at all times, i.e.

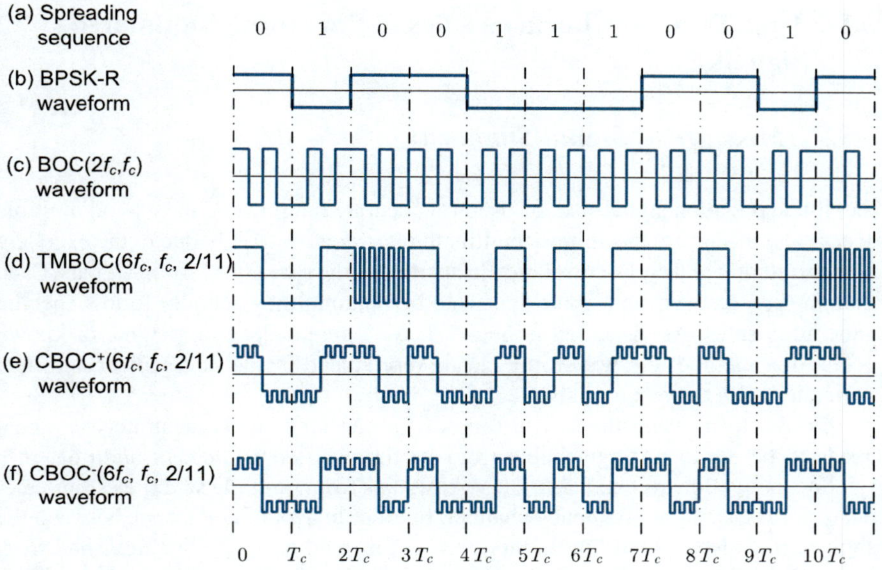

Fig. 3.2 Baseband waveforms of several spread-spectrum modulated signals

$$\Pr\{p(t) = p_1(t)\} \equiv 1. \tag{3.5}$$

However, with (3.4) which is more general, it is convenient to represent and analyze signals with different chip waveforms that are broadcast in different time slots, such as a TMBOC modulated [11] signal (see Sect. 4.7.1).

In the design of the legacy satellite navigation signal, in order to ensure that the signal $g(t)$ has a constant envelope, $p(t)$ is always binary [12] with only two values $+1/\sqrt{T_c}$ and $-1/\sqrt{T_c}$. However, the next-generation GNSS has more complicated chip waveforms whose amplitudes are not necessarily binary, and may not even be mathematically real, such as CBCS modulation, CBOC modulation, and QMBOC modulation (see Sects. 4.5 and 4.7). Figure 3.2 illustrates the baseband waveforms of several typical spreading modulated signals. In Chap. 4, we will analyze the characteristics of these spreading modulation in detail.

It should be mentioned that, considering the level of constant envelope multiplexing technology of more than a decade ago, if the signal used non-binary spreading chip waveforms, it could pose difficulties for constant envelope multiplexing. Therefore, often specific constraints [13] must be addressed in terms of phase relationship and power ratio to apply to various multiplexing techniques. For example, two different polar have to be placed in phase on the carrier with equal power to perform constant envelope multiplexing using the quadrature product subcarrier modulation (QPSM) technique. A detailed discussion of this issue can be found in Sect. 7.5.3.

3.3 Time Domain Characteristics of Spreading Modulated Signals

3.3.1 *Measure of Signal Similarity*

Satellite navigation signals can be used for accurate ranging and have good multiple access and the ability to suppress multipath interference, largely due to the excellent autocorrelation and cross-correlation properties of the spread-spectrum signals. Even though spread-spectrum signals appear to be random, they actually follow specific generation rules. As discussed in Sect. 2.2, for a spread-spectrum signal of known generation method, the waveform can be reproduced inside the receiver and then correlated with the received signal.

In order to measure the arrival time of the received signal accurately, a signal needs to be easily distinguishable from its time-shifted replica. In addition, in a satellite navigation system using CDMA, signals from multiple satellites simultaneously access the same frequency channel. In order to avoid interference between the signals, it is also desired that in the receiver they can be easily distinguished from each other. That is to say, each of the signals in the signal set used by the satellite navigation system on the same frequency band should have as little similarity as possible to the time shifted version of itself and other signals in that signal set.

3.3.2 *Cross-Correlation and Autocorrelation Functions*

The simplest measurement representing the similarity between two signals $x(t)$ and $y(t)$ in a given set is the distance metric. We can calculate the average power of the difference between the two signals, which is

$$P_{\Delta}(\tau) = \lim_{T\to\infty} \frac{1}{T} \int_{-T/2}^{T/2} |x(t) - y(t+\tau)|^2 \,\mathrm{d}t = P_x + P_y - 2R_{x,y}(\tau), \quad (3.6)$$

where τ denotes the time shift variable and

$$P_x \triangleq \lim_{T\to\infty} \frac{1}{T} \int_{-T/2}^{T/2} |x(t)|^2 \,\mathrm{d}t \quad (3.7)$$

$$P_y \triangleq \lim_{T\to\infty} \frac{1}{T} \int_{-T/2}^{T/2} |y(t)|^2 \,\mathrm{d}t \quad (3.8)$$

are the power of $x(t)$ and $y(t)$ respectively. For a given signal, both P_x and P_y are fixed, so the degree of similarity mainly depends on the last term in (3.6), which is called the *cross-correlation function (CCF)* of these two signals and is given by

$$R_{x,y}(\tau) \triangleq \lim_{T\to\infty} \frac{1}{T} \int_{-T/2}^{T/2} x(t)\, y^*(t-\tau)\,\mathrm{d}t, \tag{3.9}$$

in which $*$ represents a complex conjugate. From (3.6) it can also be seen that the average power of the difference between $x(t)$ and $y(t)$ is zero if and only if $x(t) = y(t)$, when the two signals are considered to have the largest similarity.

Similar to the definition of the CCF, we can also define the correlation function between $x(t)$ and its own shifted signal $x(t-\tau)$

$$R_x(\tau) \triangleq \lim_{T\to\infty} \frac{1}{T} \int_{-T/2}^{T/2} x(t)\, x^*(t-\tau)\,\mathrm{d}t, \tag{3.10}$$

which is called the *autocorrelation function (ACF)* of $x(t)$.

3.3.3 Periodic Signal Correlation Function

If the $x(t)$ and $y(t)$ are both periodic signals with periods of T_x and T_y respectively, then we have

$$x(t) = x(t + iT_x), \quad i = 1, 2, \ldots \tag{3.11}$$

$$y(t) = y\left(t + iT_y\right), \quad i = 1, 2, \ldots \tag{3.12}$$

It can be proved that the ACF of the periodic signal $x(t)$ also has the following properties:

$$R_x(\tau) = R_x(\tau \pm iT_x), \quad i = 1, 2, \ldots \tag{3.13}$$

Therefore, for the periodic signal $x(t)$, when calculating its ACF using (3.10), the integration interval no longer needs to be from $-\infty$ to $+\infty$. Rather, the calculation can be conducted on one period of $x(t)$, i.e.

$$R_x(\tau) = \frac{1}{T_x} \int_{t_0}^{t_0+T_x} x(t)\, x^*(t-\tau)\,\mathrm{d}t, \tag{3.14}$$

where the initial time t_0 therein may be any constant value.

Similarly, the CCF of $x(t)$ and $y(t)$ can also be obtained in this way, which is

$$R_{x,y}(\tau) = \frac{1}{T} \int_{t_0}^{t_0+T} x(t)\, y^*(t-\tau)\,\mathrm{d}t, \tag{3.15}$$

where T is the least common multiple of T_x and T_y, and $T \le T_x T_y$. If $x(t)$ and $y(t)$ have the same period, then $T = T_x = T_y$.

The ACF can be regarded as a special case of the CCF when $x(t) = y(t)$. In the following sections, we will focus on the CCF of the spread-spectrum signal.

3.3.4 *Cross-Correlation Function of Spread-Spectrum Signals*

In this section, we first examine the CCF of two baseband spread-spectrum signals of the form (3.2). For simplicity, we first assume that the spreading chip waveform of each signal is a certain waveform with a probability of 1, and assume that both signals have the same chip length T_c and spreading sequence period. These two signals can be expressed as

$$x(t) = \sum_{n=-\infty}^{+\infty} (-1)^{c_n} p(t - nT_c), \tag{3.16}$$

$$y(t) = \sum_{n=-\infty}^{+\infty} (-1)^{b_n} q(t - nT_c), \tag{3.17}$$

where c_n and b_n are the values of the two signals' PRN sequences $\{c_n\}$ and $\{b_n\}$ at the discrete moment n. The chip waveforms $p(t)$ and $q(t)$ are both non-zero only within the interval $[0, T_c)$, and there are no restrictions posed on their shape in this interval. Substitute (3.16) and (3.17) into (3.15), and we have

$$R_{x,y}(\tau) = \frac{1}{T} \sum_{n=-\infty}^{+\infty} \sum_{m=-\infty}^{+\infty} (-1)^{c_n + b_m} \left[\int_0^T p(t - nT_c) q^*(t - mT_c - \tau) \, \mathrm{d}t \right]. \tag{3.18}$$

Note that $p(t)$ and $q(t)$ can only be non-zero within $[0, T_c)$, and the integral in (3.18) is only non-zero in the overlap of $p(t - nT_c)$ and $q(t - mT_c - \tau)$. More clearly, we can write the shift variable τ as $\tau = kT_c + \epsilon$, with $0 \leq \epsilon < T_c$. As shown in Fig. 3.3, the two pulses have overlapping parts only when $n = k + m$ or $n = k + m + 1$, and the integration interval does not exceed the length of one spreading chip T_c. Therefore, (3.18) can be further simplified as

$$\begin{aligned} R_{x,y}(kT_c + \epsilon) = {} & r_{cb}[k] \frac{1}{T_c} \int_\epsilon^{T_c} p(t) q^*(t - \epsilon) \, \mathrm{d}t \\ & + r_{cb}[k+1] \frac{1}{T_c} \int_0^\epsilon p(t) q^*(t + T_c - \epsilon) \, \mathrm{d}t, \end{aligned} \tag{3.19}$$

where

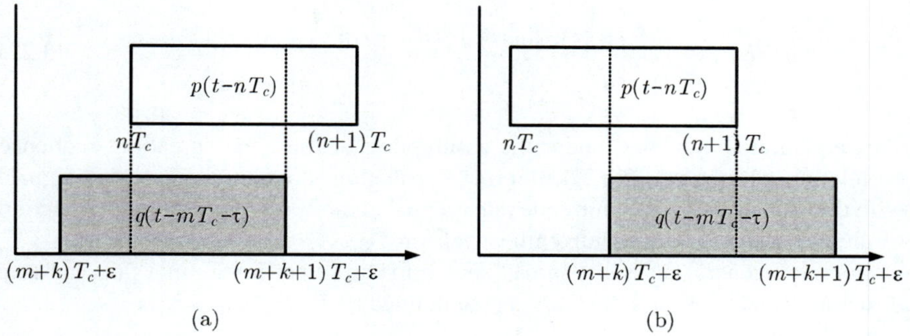

Fig. 3.3 Integral interval of the correlation function

$$r_{cb}[k] \triangleq \frac{1}{N}\sum_{n=0}^{N-1}(-1)^{c_n+b_{n+k}} \tag{3.20}$$

is a discrete periodic cross-correlation function of two PRN sequences. In Sect. 2.9, when discussing the design of PRN codes, the definition of discrete periodic cross-correlation functions is also given. Comparing (3.20) with (2.35), we can see that the two definitions are essentially the same, except that the autocorrelation main peak is normalized in (3.20).

It is easy to calculate the CCF of two spread-spectrum signals using (3.19) where the influence of the PRN sequence and the spreading chip waveform on the value of the final correlation has been decoupled as much as possible. This allows us to analyze the effect of the PRN sequence and chip waveform respectively.

When analyzing the correlation properties of PRN sequences in Sect. 2.9, we used the largest cross-correlation sidelobe as the primary measure of performance. This value is relatively easy to obtain when a sequence family is given. However, for every two PRN sequences and any given k, the precise correlation characteristics of $r_{cb}[k]$ are not easily described in many cases. When we focus on the effects of chip waveforms, it is generally assumed that the correlation properties of PRN sequences are approximately ideal, i.e., the maximum cross-correlation sidelobe of the sequence are low enough to satisfy

$$r_{cb}[k] \approx \begin{cases} 1, & c = b \text{ and } k = 0 \\ 0, & \text{otherwise} \end{cases}. \tag{3.21}$$

Under the assumption that PRN sequences have ideal correlation characteristics, the analysis of the cross-correlation properties of the two spread-spectrum signals becomes clearer. From (3.19), when two signals use different PRN sequences, $R_{x,y}(\tau)$ is always zero. When two signals use the same PRN sequence but their spreading chip waveforms may not be the same, the CCF of $x(t)$ and $y(t)$ is only non-zero within $\tau \in [-T_c, T_c]$, which is

$$R_{x,y}(\tau) = \begin{cases} \dfrac{1}{T_c}\displaystyle\int_0^{T_c} p(t)\, q^*(t-\tau)\,\mathrm{d}t, & \tau \in [-T_c, T_c] \\ 0, & \text{otherwise} \end{cases} . \tag{3.22}$$

This equation shows that, under the assumption of the ideal spreading sequence correlation, when $\tau \in [-T_c, T_c]$, the cross-correlation of two spread-spectrum signals with the same period and same code rate is equal to the aperiodic correlation function of these two signals' spreading chip waveforms.

For more general cases, the spread-spectrum chips in different time slots can take different waveforms, and the CCF can be defined by its expectation, i.e.

$$R_{x,y}(\tau) = \mathbb{E}\left\{\frac{1}{T}\int_{t_0}^{t_0+T} x(t)\, y^*(t-\tau)\,\mathrm{d}t\right\} = \sum_{i,j} \mathcal{P}_{p_i}\mathcal{P}_{q_i} R_{p_i q_i}(\tau) , \tag{3.23}$$

where $R_{p_i q_i}(\tau)$ is the CCF when the chip waveform of $x(t)$ is $p_i(t)$ and the chip waveform of $y(t)$ is $q_i(t)$.

Finally, it should be noted that the ideal correlation characteristics of (3.21) are assumed to be approximately satisfactory in many applications of satellite navigation. Compared with legacy GNSS signals, the modern satellite navigation signals have longer PRN sequence periods and use code families with better correlation characteristics. Therefore, the autocorrelation and cross-correlation performances of modern signals are closer to the ideal. However, we also mentioned in Sect. 2.9 that in some special applications, such as the use of pseudolites [14], or in severely occluded environments, the non-ideal characteristics of auto/cross-correlation of PRN sequences cannot be ignored due to near-far effects.

3.3.5 *Cross-Correlation Function of SCS Modulated Signals*

Although it is very straightforward to use (3.19) and (3.22) to calculate the CCF of the spread-spectrum signal, they involve the slightly complex integral operation. In fact, the modulation techniques currently used in satellite navigation signals, including BPSK-R, BOC, BCS, CBCS and CBOC, can be reduced to step-shape code symbol (SCS) modulation waveforms.

The SCS waveform is divided into a total of M equally spaced sub-chips within a length of the spreading chip. Each sub-chip has a duration of $T_s = T_c/M$, and the waveform value is constant in each, which makes the waveform look like steps, hence the term step-shape. The concept of this general waveform was originally proposed in [15] for the design of unambiguous processing methods of split-spectrum signals (see Sect. 4.8). In-depth analysis of the CCF of spread-spectrum signals using such waveforms is conducted in [16]. For this type of spreading modulated signal, the calculation of the CCF can be much simpler than using (3.22).

3.3.5.1 SCS Spreading Chip Waveform

For given T_c and T_s, we can define a total of M functions $\{\psi_k(t)\}$, $k = 0, 1, \dots M - 1$, where

$$\psi_k(t) = \begin{cases} \frac{1}{\sqrt{T_c}}, & kT_s \leqslant t < (k+1)\,T_s \\ 0, & \text{otherwise} \end{cases} . \tag{3.24}$$

It can be verified that these M functions constitute a set of orthogonal basis functions. We have

$$\int_0^{T_c} \psi_k(t)\,\psi_q(t)\,\mathrm{d}t = \begin{cases} \frac{1}{M}, & k = q \\ 0, & k \neq q \end{cases} \tag{3.25}$$

and any SCS waveform $p(t)$ with the same T_c and M can be represented as a linear combination of this set of basis functions, that is

$$p(t) = \sum_{k=0}^{M-1} \psi_k(t) \cdot \kappa_k , \tag{3.26}$$

where κ_k is the projection of $p(t)$ on $\psi_k(t)$, i.e.

$$\kappa_k = M \int_0^{T_c} p(t)\,\psi_k(t)\,\mathrm{d}t . \tag{3.27}$$

Thus, each SCS spreading chip waveform corresponds to a vector

$$\boldsymbol{\kappa} = \left[\kappa_0, \kappa_1, \dots, \kappa_{M-1}\right]^{\mathrm{T}} \tag{3.28}$$

in the signal space spanned by $\{\psi_k(t)\}$. In this way, when we have the spreading chip length T_c and the vector $\boldsymbol{\kappa}$, the baseband spreading chip waveform used by the signal is uniquely determined. Therefore, $\boldsymbol{\kappa}$ is called the *chip waveform shape vector*, and any SCS spreading chip waveform can be represented by $p(t; \boldsymbol{\kappa})$.

3.3.5.2 Cross-Correlation Function of the SCS Signal

An SCS spread-spectrum signal can be expressed as

$$x(t; \boldsymbol{\kappa}) = \sum_{n=-\infty}^{\infty} \sum_{k=0}^{M-1} (-1)^{c_n} \kappa_k \psi_k(t - nMT_s) . \tag{3.29}$$

Assuming that another signal $y(t)$ has the same T_c and M with $x(t)$, but their PRN sequence $\left\{c'_n\right\}$ and the chip waveform shape vector $\boldsymbol{\kappa}'$ may be different, i.e.

$$y\left(t;\boldsymbol{\kappa}'\right)=\sum_{m=-\infty}^{\infty}\sum_{q=0}^{M-1}(-1)^{c'_n}\kappa'_q\psi_q\left(t-mMT_s\right). \tag{3.30}$$

Then by (3.15), the CCF of these two signals can be written as

$$\begin{aligned}R_{xy}\left(\tau;\boldsymbol{\kappa},\boldsymbol{\kappa}'\right)&=\frac{1}{T}\int_0^T x\left(t\right)y\left(t+\tau\right)\mathrm{d}t\\&=\frac{1}{T}\sum_n\sum_m\sum_{k=0}^{M-1}\sum_{q=0}^{M-1}(-1)^{c_n+c'_m}\kappa_k\kappa'_q\\&\times\int_0^T\psi_k\left(t-nMT_s\right)\psi_q\left(t-mMT_s+\tau\right)\mathrm{d}t.\end{aligned} \tag{3.31}$$

Similar to the analysis in Sect. 3.3.4, if the shift τ is decomposed to $\tau=aT_c+bT_s+\epsilon$, where a is an integer indicating the integer shift on the spreading chip scale, $b=0,1,\ldots,M-1$ is the sequence number of the sub-chip where the shift is located and $\epsilon\in[0,T_s)$ is the fractional part of the shift's location in this sub-chip, then, (3.31) can be simplified as [16]

$$\begin{aligned}R_{xy}\left(\tau\right)&=R_{xy}\left(aT_c+bT_s+\epsilon\right)\\&=r_{cc'}\left[a\right]\left[\check{r}_{\kappa\kappa'}\left[b\right]\left(1-\frac{\epsilon}{T_s}\right)+\check{r}_{\kappa\kappa'}\left[b+1\right]\left(\frac{\epsilon}{T_s}\right)\right]\\&+r_{cc'}\left[a+1\right]\left[\check{r}_{\kappa\kappa'}\left[b-M\right]\left(1-\frac{\epsilon}{T_s}\right)+\check{r}_{\kappa\kappa'}\left[b-M+1\right]\left(\frac{\epsilon}{T_s}\right)\right],\end{aligned} \tag{3.32}$$

where $r_{cc'}\left[a\right]$ is defined by (3.20), and

$$\check{r}_{\kappa\kappa'}\left[b\right]\triangleq\begin{cases}\dfrac{1}{M}\displaystyle\sum_{i=0}^{M-1-b}\kappa_i\kappa'_{b+i}, & 0\leqslant b\leqslant M-1\\ \dfrac{1}{M}\displaystyle\sum_{i=0}^{M-1+b}\kappa_{i-b}\kappa'_i, & 1-M\leqslant b<0\\ 0, & |b|\geqslant M\end{cases} \tag{3.33}$$

is the aperiodic cross-correlation function of $\boldsymbol{\kappa}$ and $\boldsymbol{\kappa}'$.

For a spread-spectrum signal whose PRN sequence has ideal correlation characteristics, when $c'_k=c_k$, $r_{cc'}\left[a\right]$ is non-zero only when $a=0$ in a PRN sequence period. At this time, when $x\left(t\right)$ and $y\left(t\right)$ have the same PRN sequence, their cross-correlation expression can be simplified to

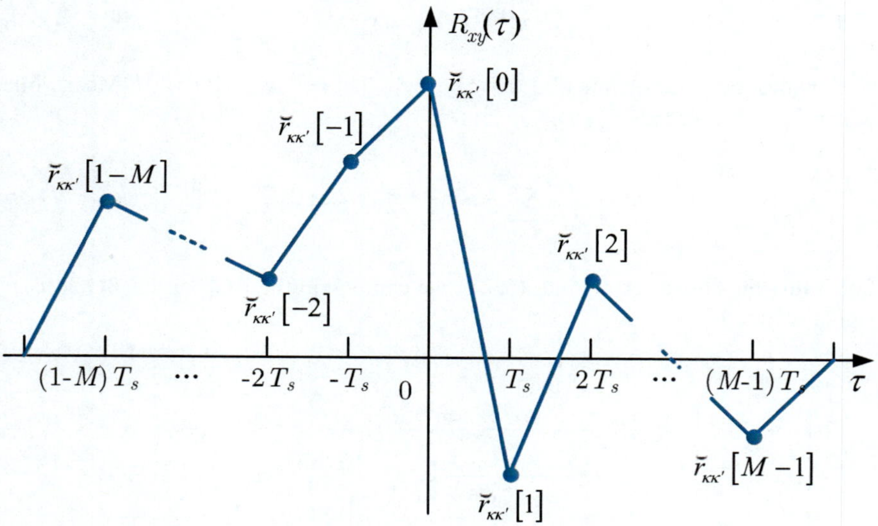

Fig. 3.4 Schematic plot of $R_{xy}(\tau)$ of SCS spread-spectrum signal

$$R_{xy}(\tau) = \begin{cases} \breve{r}_{\kappa\kappa'}[b]\left(1-\frac{\epsilon}{T_s}\right) + \breve{r}_{\kappa\kappa'}[b+1]\left(\frac{\epsilon}{T_s}\right), & \tau \in [0, T_c] \\ \breve{r}_{\kappa\kappa'}[b-M]\left(1-\frac{\epsilon}{T_s}\right) + \breve{r}_{\kappa\kappa'}[b-M+1]\left(\frac{\epsilon}{T_s}\right), & \tau \in [-T_c, 0) \\ 0, & \text{otherwise}. \end{cases} \tag{3.34}$$

Notice that when τ is an integer multiple of T_c/M, i.e. $\epsilon = 0$, from the above equation we have

$$R_{xy}(kT_c/M) = \breve{r}_{\kappa\kappa'}[k] . \tag{3.35}$$

Figure 3.4 is a schematic diagram of the CCF of the SCS spread-spectrum signal. The value of $R_{xy}(\tau)$ is $\breve{r}_{\kappa\kappa'}[k]$ at $\tau = kT_c/M$, and is connected by line segments between each point of kT_c/M [16]. This characteristic of the SCS spread spectrum signal cross-correlation function makes it very convenient for us to analyze the time domain characteristics of the signal. When two SCS spread-spectrum signals use the same spreading sequence, the CCF is completely determined by the two signals' chip waveforms shape vectors $\boldsymbol{\kappa}$ and $\boldsymbol{\kappa}'$. As long as the chip waveform shape vectors are known, we can calculate $\breve{r}_{\kappa\kappa'}[k]$ by (3.33) to obtain the CCF.

Example 3.1 (*Autocorrelation function of sine-phased BOC signal*) A definition and detailed description of the BOC signal can be found in Sect. 4.3. Figure 3.2c shows the time domain waveform of a BOC$(2n, n)$ signal. Here, as an example of applying the cross-correlation calculation method of the SCS spread-spectrum signal, we calculate the ACF of the BOC$(\ell n, n)$ signal, where $M = 2\ell$ is defined as the modulation order of BOC$(\ell n, n)$ signal, and its shape vector is

$$\kappa_k^{(\mathrm{BOC})} = (-1)^k . \tag{3.36}$$

Substituting this term into (3.33), we have $\check{r}_{\kappa\kappa'}[b] = 0$ when $|b| \geqslant M$. Meanwhile, when $|b| < M$, we have

$$\check{r}_{\kappa\kappa'}[b] = \frac{1}{M} \sum_{k=0}^{M-|b|} (-1)^{2k+|b|} = (-1)^b \frac{M-|b|}{M} . \tag{3.37}$$

Substitute the above results into (3.32), we can obtain the ACF of the BOC($\ell n, n$) signal, which is

$$R_{\mathrm{BOC}}(\tau) = \begin{cases} (-1)^{k+1} \left[\frac{\tau(2M-2k-1)}{T_c} - \frac{2(M-1)k-2k^2+M}{M} \right], & \frac{kT_c}{M} \leqslant \tau < \frac{(k+1)T_c}{M} \\ (-1)^{k+1} \left[\frac{\tau(2k-1)}{T_c} + \frac{(M-k)(2k-1)-k}{M} \right], & \frac{(k-M)T_c}{M} \leqslant \tau < \frac{(k-M+1)T_c}{M} \\ 0, & |\tau| > T_c \end{cases} \tag{3.38}$$

where $k = 0, 1, \ldots, M-1$.

Figure 3.5 shows the shape of the ACF of the BOC($4n, n$) signal. The ACF of this signal has multiple peaks. As we will see in Sect. 4.8, such multi-peak characteristic of BOC signals' autocorrelation function poses a certain challenge to the stable and reliable processing of this type of signal. ■

Example 3.2 (*Autocorrelation function of the in-phase CBOC signal*) The baseband signal plotted in Fig. 3.2e is referred to as the in-phase CBOC signal CBOC$^+$ (6, 1, γ) (see Sect. 4.7.2 for a more detailed definition). For an in-phase CBOC signal, defining $\Pi \triangleq \sqrt{1-\gamma}$ and $\Xi \triangleq \sqrt{\gamma}$ are the relative amplitude of the components of BOC(1, 1) and BOC(6, 1), respectively, we have

$$\kappa_{k;\gamma}^{(\mathrm{CBOC})} = (-1)^{\lfloor k/6 \rfloor} \Pi + (-1)^k \Xi . \tag{3.39}$$

Due to the influence of high-frequency BOC(6, 1) component, the ACF of the CBOC$^+$ (6, 1, γ) signal has a complicated shape, which is zero when $|\tau| > T_c$ but a polyline consisting of 24 segments in $[-T_c, T_c]$, as shown in Fig. 3.6. Using (3.32) we can obtain that, for $\tau \geq 0$, when $kT_c/12 \leqslant \tau < (k+1)T_c/12$, the ACF of the CBOC$^+$ (6, 1, γ) signal is

$$\begin{aligned} R_{\mathrm{CBOC}}(\tau) = & (-1)^{k+1} \Pi^2 \left[\frac{\tau(24-2k-1)}{T_c} - \frac{11k-k^2+6}{6} \right] \\ & + (-1)^{\lfloor k/6 \rfloor+1} \Xi^2 \left[\frac{\tau(4-2\lfloor k/6 \rfloor-1)}{T_c} - \lfloor k/6 \rfloor + \lfloor k/6 \rfloor^2 - 1 \right] \end{aligned} \tag{3.40}$$

where $\lfloor \cdot \rfloor$ is the floor function.

For $\tau < 0$, when $(k-12)T_c/12 \leqslant \tau < (k-11)T_c/12$, the ACF of the CBOC$^+$ (6, 1, γ) signal is

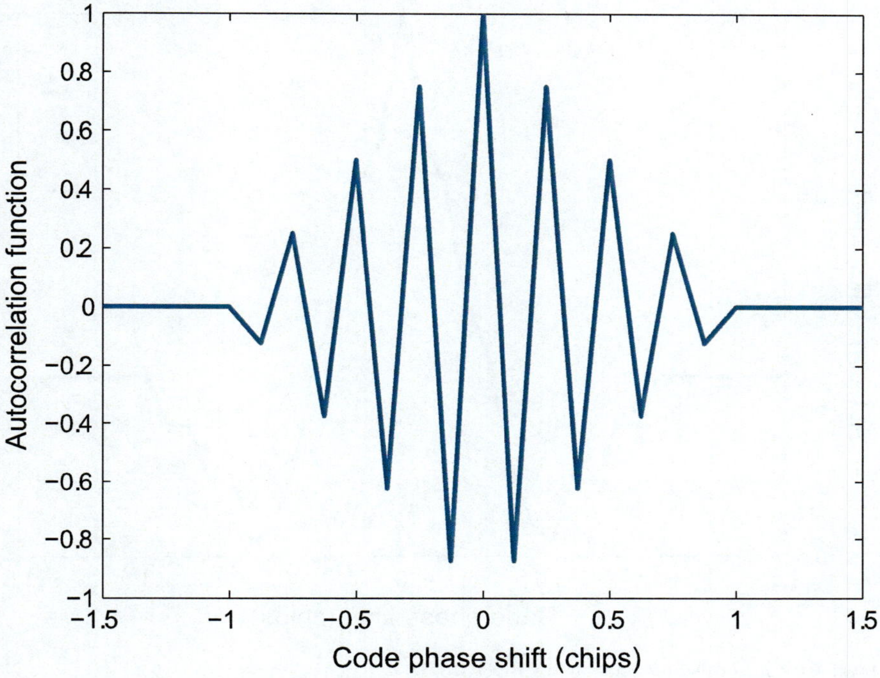

Fig. 3.5 Autocorrelation function of the BOC$(4n, n)$ signal

$$R_{\text{CBOC}}(\tau) = (-1)^{k+1}\,\Pi^2\left[\frac{\tau(2k-1)}{T_c} + \frac{12k-k^2-6}{6}\right] + (-1)^{\lfloor k/6\rfloor+1}\,\Xi^2\left[\frac{\tau(2\lfloor k/6\rfloor-1)}{T_c} + 2\lfloor k/6\rfloor - \lfloor k/6\rfloor^2 - 1\right], \quad (3.41)$$

where $k = 0, 1, \ldots, 11$. ■

3.4 Frequency Domain Characteristics of Spreading Modulated Signals

3.4.1 Power Spectral Density

The content of various frequency components of a signal can be characterized by its PSD. As the discussion in the following chapters will show, the PSD of the spread-spectrum signal determines the bandwidth required for transmitting and receiving the signal. The PSD can also be used to analyze the ranging and demodulating performances of the signal in thermal noise and interference environments. Therefore,

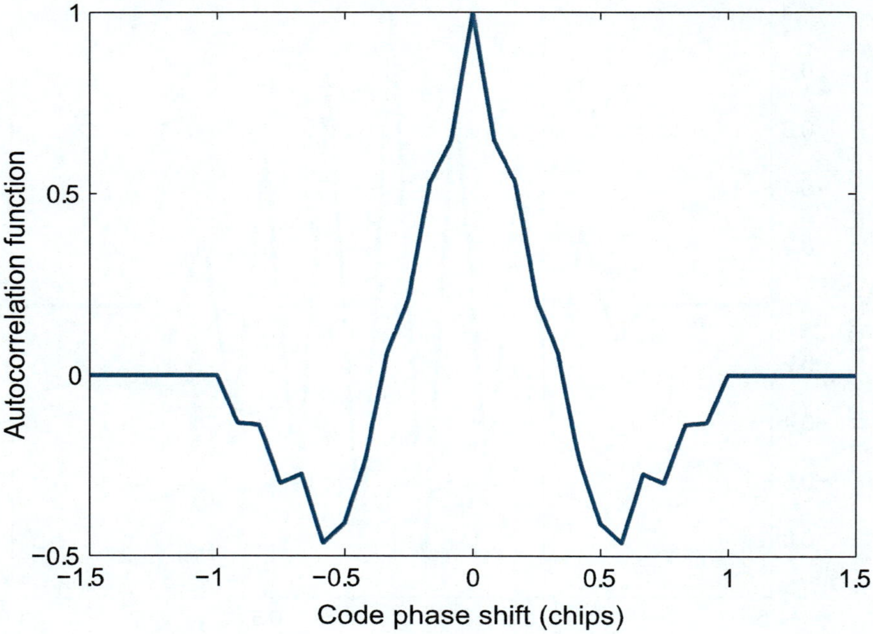

Fig. 3.6 Autocorrelation function of CBOC$^+$ (6, 1, γ) signal

PSD plays an important role in the spreading modulation design and performance analysis.

The PSD of the signal can be obtained either by the ACF of the signal or directly by the spectrum of the signal itself. Suppose a general complex-valued signal $x(t)$ has an autocorrelation function $R_x(\tau)$ defined by (3.10). The PSD of the signal can be obtained from the Fourier transform of the autocorrelation function, i.e.

$$S_x(f) = \mathcal{F}\{R_x(\tau)\} = \int_{-\infty}^{+\infty} R_x(\tau)\, \mathrm{e}^{-\mathrm{j}2\pi f\tau} \mathrm{d}\tau\,. \tag{3.42}$$

3.4.2 *PSD of Spread-Spectrum Signals Without Message Modulation*

First, we will temporarily ignore the influence of the message on the signal modulation. The spread-spectrum signal $g(t)$ defined by (3.2) can be written as

$$g(t) = p(t) * \left[\sum_{n=0}^{N-1} (-1)^{c_n}\, \delta(t - nT_c)\right] * \left[\sum_{k=-\infty}^{+\infty} \delta(t - kNT_c)\right], \tag{3.43}$$

where $*$ represents the convolution operation. According to the convolution theorem, the Fourier transform of $g(t)$ is equal to the product of the respective Fourier transforms of the three terms in the above equation, so we have

$$G(f) = \frac{1}{NT_c} P(f) \sum_{k=-\infty}^{+\infty} \sum_{i=0}^{N-1} (-1)^{c_i} \mathrm{e}^{-\mathrm{j}2\pi i f T_c} \delta\left(f - \frac{k}{NT_c}\right), \tag{3.44}$$

where $P(f) = \mathcal{F}\{p(t)\}$ is the Fourier transform of the chip waveform. Therefore, the PSD of $g(t)$ is

$$S_g(f) = G(f)\, G^*(f) = \frac{1}{NT_c^2} |P(f)|^2\, \Gamma(fT_c) \sum_{k=-\infty}^{+\infty} \delta\left(f - \frac{k}{NT_c}\right), \tag{3.45}$$

where

$$\Gamma(f) = \sum_{s=0}^{N-1} r_c[s]\, \mathrm{e}^{-\mathrm{j}2\pi s f} \tag{3.46}$$

is the discrete time Fourier transform (DTFT) of the truncated sequence in the first period of the ACF $r_c[s]$ of the PRN sequence which is defined by (3.20). It can be seen from (3.45) that if a spread-spectrum signal is periodic, its PSD is a series of line spectra with a spacing of $1/(NT_c)$, whose reciprocal corresponds to the repetition period of the spreading sequence. The envelope shape of the line spectrum is determined by the DTFT of the periodic ACF of the PRN sequence and $|P(f)|^2$, which is the energy spectral density of the spreading chip waveform.

If it is assumed that the correlation of the spreading sequence is ideal, then in the sequence period, $r_c[s]$ can only take a non-zero value of 1 at $s = 0$, and (3.45) can be simplified as

$$S_g(f) = \frac{1}{NT_c^2} |P(f)|^2 \sum_{k=-\infty}^{+\infty} \delta\left(f - \frac{k}{NT_c}\right). \tag{3.47}$$

At this point, the envelope shape of the line spectrum is completely dependent on the spectral characteristics of the spreading chip waveform.

Example 3.3 (*PSD of BPSK-R signals using binary m-sequences*) Suppose there is a PSD of a BPSK-R signal using a binary m-sequence as a spreading code. The m-sequence is short for the "maximum length sequence", and its periodic ACF is

$$r_c[s] = \begin{cases} 1, & s \equiv 0 \bmod N \\ -\dfrac{1}{N}, & \text{otherwise} \end{cases}$$

while $\Gamma(fT_c)$ can be expressed as

$$\Gamma(fT_c) = \frac{N+1}{N} - \frac{1}{N}\frac{\sin(\pi f T_c N)}{\sin(\pi f T_c)} e^{-j\pi f T_c (N-1)} ,$$

which is a continuous function with a period of $1/T_c$. Figures 3.7 and 3.8 show the shapes of $r_c[s]$ and $\Gamma(fT_c)$ when $N = 15$. The spread-spectrum chip waveform of the BPSK-R signal is

$$p_{\text{BPSK-R}}(t) = \frac{1}{\sqrt{T_c}} \text{rect}_{T_c}(t) ,$$

where

$$\text{rect}_T(t) = \begin{cases} 1, & t \in [0, T) \\ 0, & \text{otherwise} \end{cases}$$

is a rectangular pulse function with a duration of T_c. Then, it is easy to obtain the energy spectral density of $p_{\text{BPSK-R}}(t)$, which is

$$|P(f)|^2 = T_c \, \text{sinc}^2(\pi f T_c) .$$

Thus, the PSD of the BPSK-R signal using the binary m-sequence as the spreading code can be written as

$$S_{\text{BPSK-R}}(f) = \frac{1}{N^2 T_c}\delta(f) + \frac{N+1}{N^2 T_c} \text{sinc}^2\left(\frac{\pi k}{N}\right) \sum_{k \neq 0} \delta\left(f - \frac{k}{N T_c}\right) .$$

Figure 3.9 shows the shape of $S_{\text{BPSK-R}}(f)$ and its envelope $\left(N T_c^2\right)^{-1} |P(f)|^2 \Gamma(fT_c)$. It can be seen that the envelope shape of $S_{\text{BPSK-R}}(f)$ is primarily determined by the energy spectral density of the spreading chip waveform $|P(f)|^2$. However, due to the non-ideal correlation characteristics of the m-sequence, its envelope has a certain jitter and the value is smaller in the vicinity of $f = 0$. As N increases, the spectral lines will become denser, and the shape of the envelope will become increasingly similar to $\left(N T_c^2\right)^{-1} \text{sinc}^2(\pi f T_c)$. ■

3.4.3 PSD of Spread-Spectrum Signals with Message Modulation

If the spread-spectrum signal is modulated by the data message, its periodicity will be disrupted. Assuming that one bit length of the data message $\{d_m\} \in \{-1, 1\}$ is T_B, the spread-spectrum signal modulated by the data message can be regarded as the product of $g(t)$ and an infinitely long random pulse train, i.e.

$$D(t) = \sum_{m=-\infty}^{+\infty} d_m \, \text{rect}_{T_B}(t - m T_B) . \tag{3.48}$$

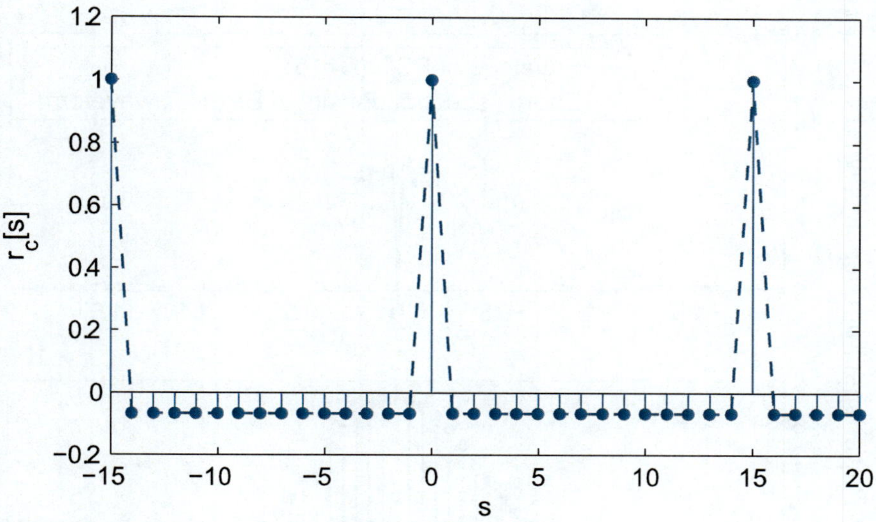

Fig. 3.7 Cross-correlation function of the m-sequence with $N = 15$

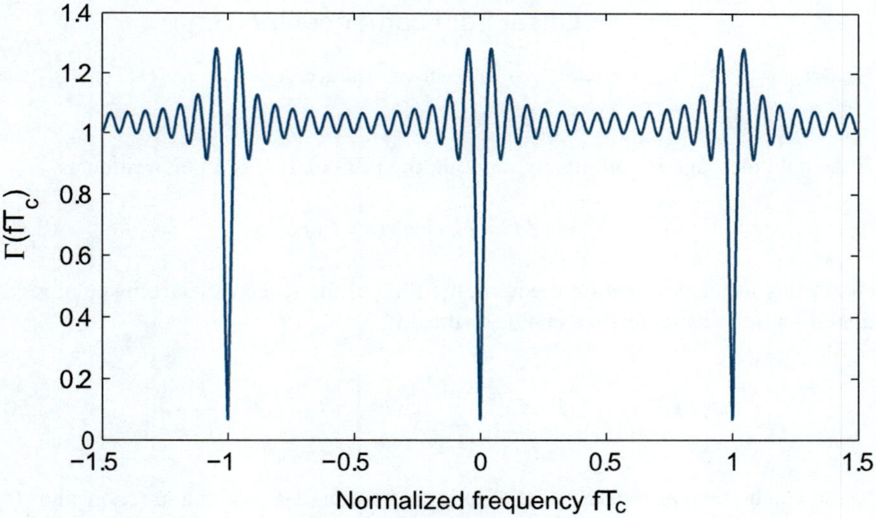

Fig. 3.8 $\Gamma\,(f T_c)$ corresponding to the m-sequence with $N = 15$

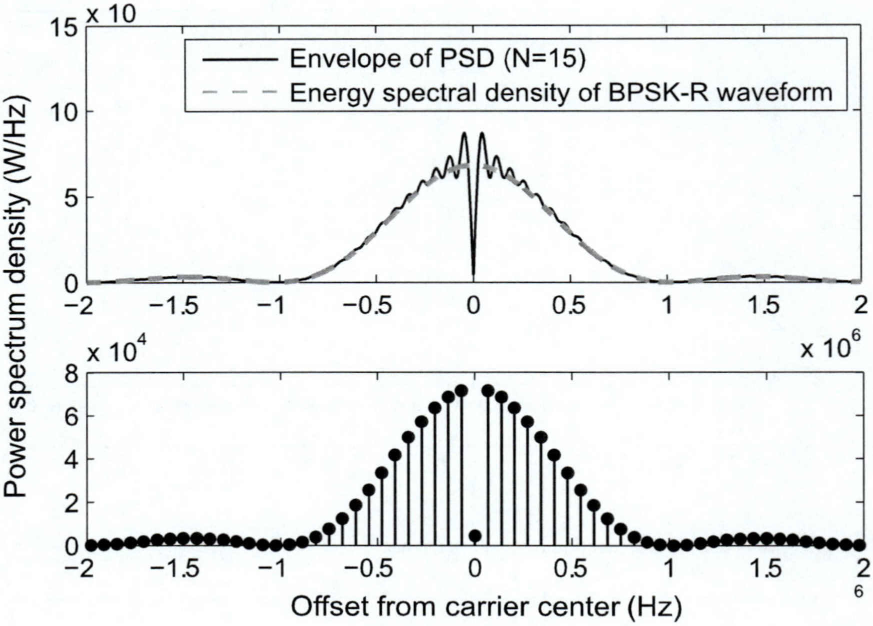

Fig. 3.9 $S_{\text{BPSK-R}}(f)$ and its envelope shape with m-sequence, $N = 15$

If the data message is completely random, the PSD of $D(t)$ can be written as

$$S_d(f) = T_B \operatorname{sinc}^2(\pi f T_B) . \tag{3.49}$$

According to the modulation theorem, the PSD of the spread-spectrum signal modulated with the data message can be written as

$$S_g(f) = \frac{T_B}{NT_c^2} |P(f)|^2 \sum_{k=-\infty}^{+\infty} \operatorname{sinc}^2\left[\pi T_B\left(f - \frac{k}{NT_c}\right)\right] . \tag{3.50}$$

It also can be seen as replacing each line spectrum in (3.47) with a function $\operatorname{sinc}^2(t)$ with the main lobe bandwidth of $1/T_B$. If the bit length of the data message is greater than the period length of the spread-spectrum signal, i.e. $T_B^{-1} < (NT_c)^{-1}$, the PSD of the spread-spectrum signal modulated by the data message is composed of a series of comb-like functions. However, if the bit length of the data message is less than or equal to the period length of the spread-spectrum signal, the signal has a continuous spectrum.

The case where the spread-spectrum signal is modulated by the secondary code is similar to the case where it is modulated by a data message. The secondary coding can also make the spectrum of spread spectrum signal more continuous (see also Sect. 2.9.4).

3.4.4 PSD of Aperiodic Spread-Spectrum Signals

When the spreading sequence is aperiodic, or when its period is close to infinity, let N tend to infinity in (3.47), then the PSD of the spread-spectrum signal can be simplified as

$$\begin{aligned} S_g(f) &= \left\{ \lim_{N \to \infty} \frac{1}{NT_c} \sum_{k=-\infty}^{+\infty} \delta\left(f - \frac{k}{NT_c}\right) \right\} \frac{1}{T_c} |P(f)|^2 \\ &= \frac{1}{T_c} |P(f)|^2 . \end{aligned} \tag{3.51}$$

Under this assumption, the PSD of the signal is entirely determined by its spreading modulation chip waveform.

Equation (3.51) is a very important conclusion, as it establishes a simple correspondence between the PSD of the spread-spectrum signal and its spreading chip waveform under a series of reasonable assumptions.

By combining the analysis of the ACF in Sects. 3.3.4 and 3.3.5, the signal's time domain characteristics and frequency domain characteristics can be completely determined by the spreading modulation chip waveform under the assumption that the spreading sequence is ideal. In this way, we can focus on the spreading chip waveform when studying different modulated signals. It should be noted that although the actual spreading sequence is not completely random and does not have an infinite period, the above approximation is considered reasonable for most analyses of signal design. Moreover, from (3.47) we can see that (3.51) reflects the envelope characteristics of the line spectrum, even considering the line spectrum characteristics caused by the periodicity of the code sequence.

3.4.5 PSD of SCS Modulated Signals

From literature on the spectrum occupancy of BCS signals [12], the PSD of the SCS modulated signal can be obtained as

$$S(f) = T_s \operatorname{sinc}^2(\pi f T_s) \left\{ 1 + \frac{1}{M} \sum_{k=0}^{M-2} \sum_{\ell=k+1}^{M-1} \kappa_k \kappa_\ell \cos\left[(\ell - k) 2\pi f T_s\right] \right\} . \tag{3.52}$$

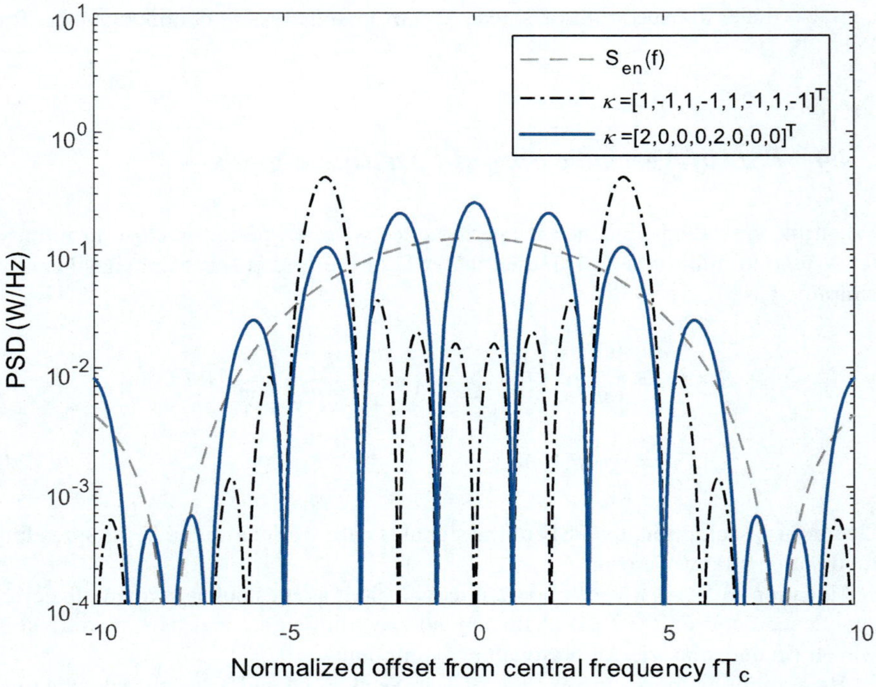

Fig. 3.10 PSD of the SCS signal, where the horizontal axis is normalized by the spreading code rate

It can be seen that the PSD of the SCS modulated signal is the product of two terms. The first term is

$$S_{\text{en}}(f) \triangleq T_s \operatorname{sinc}^2(\pi f T_s) , \tag{3.53}$$

which is the energy spectral density of a BPSK-R spreading chip waveform with a spreading rate $f_s = 1/T_s$. The second term is related to the shape vector $\boldsymbol{\kappa}$, which is the sum of the direct current and a total of $M(M-1)/2$ cosine functions.

Figure 3.10 shows the PSD of the SCS modulated signal when $M = 8$, where the shape vectors of the SCS modulated signal are respectively $(1, -1, 1, -1, 1, -1, 1, -1)^{\mathrm{T}}$ and $(2, 0, 0, 0, 2, 0, 0, 0)^{\mathrm{T}}$. The shape of $S_{\text{en}}(f)$ in this case is also shown in the same figure. From Fig. 3.10 one can see that different spreading chip waveforms make the spectrum distribution different, but due to the rapid attenuation of $S_{\text{en}}(f)$ outside $[-M/T_c, M/T_c]$, the main power of the SCS modulated signals with the same modulation order of M is concentrated in $[-M/T_c, M/T_c]$.

3.4.6 Cross-Power Spectral Density

In this subsection, we discuss the cross-power spectral density between two signals with the same PRN sequence but different chip waveforms. The Fourier transform of the cross-correlation of the two signals is

$$\mathcal{H}_{xy}(f) = \int_{-\infty}^{+\infty} R_{x,y}(\tau)\, \mathrm{e}^{-\mathrm{j}2\pi f\tau} \mathrm{d}\tau \,, \tag{3.54}$$

which is called cross-power spectrum density (CSD), or cross spectrum for short. By using derivations as in the previous section, when the signal $x(t)$ and $y(t)$ have the same PRN sequence and have an ideal correlation characteristic, the shape of the cross spectrum is determined only by their spreading chip waveforms, which can be written as

$$\mathcal{H}_{xy}(f) = \frac{1}{T_c} P_x(f)\, P_y^*(f) \,, \tag{3.55}$$

where $P_x(f)$ and $P_y(f)$ are the Fourier transform of the spreading chip waveform $x(t)$ and $y(t)$ respectively.

3.4.7 Normalization of PSD

When comparing the PSDs of different spread-spectrum signals, the difference in signal amplitude will cause a scale difference. In order to analyze and compare the signals of different modulation modes under a unified condition, the PSD of the signal $S(f)$ can be further decomposed into the product of the signal carrier power and the normalized signal PSD, i.e.

$$S(f) = C_s G_s(f) \,, \tag{3.56}$$

where $C_s = \int_{-\infty}^{+\infty} S(f)\mathrm{d}f$ is referred to as signal carrier power, and $G_s(f) = S(f)/C_s$ is the normalized signal PSD at infinite bandwidth which satisfies

$$\int_{-\infty}^{+\infty} G_s(f)\, \mathrm{d}f = 1 \,. \tag{3.57}$$

In this way, under the premise that the carrier power is the same, we can compare the performance of signals using different modulation methods.

In the following discussion, unless otherwise specified, we always agree that the carrier power of the DSSS signal defined by (3.1) is $C_s = A_s^2/2$, and, the baseband spread-spectrum signal $g(t)$ defined by (3.2) has a normalized signal PSD. This is equivalent to requiring the energy E_g of the spreading chip waveform of $g(t)$ to satisfy

$$E_g = \int_0^{T_c} |p(t)|^2 \, \mathrm{d}t = \int_{-\infty}^{+\infty} |P(f)|^2 \, \mathrm{d}f = T_c \,. \tag{3.58}$$

Additionally, this means that when $\tau = 0$, the ACF of $g(t)$ under the infinite bandwidth is

$$R_g(0) = \int_{-\infty}^{+\infty} G_s(f) \, \mathrm{d}f = 1 \,. \tag{3.59}$$

Furthermore, if $g(t)$ is a SCS spread-spectrum signal as defined in Sect. 3.3.5, the chip waveform energy constraint of (3.58) is also equivalent to the constraint of the shape vector norm of the spreading chip waveform, which is

$$\|\kappa\| = \sqrt{M} \,. \tag{3.60}$$

References

1. Price R, Green PE (1958) A communication technique for multipath channel. In: Proceedings of the IRE, pp 555–570
2. Golomb SW (1967) Shift register sequences. Holden-Day, San Francisco
3. Zierler N (1959) Linear recurring sequences. J. Soc. Ind. Appl. Math. 7:31–48
4. Gold R (1966) Characteristic linear sequences and their coset functions. SIAM J. Appl. Math. 14(5):980–985
5. Kasami T (1966) Weight distribution formula for some classes of cyclic codes, coordinated science laboratory. Technical Report R-285(AD632574), University of Illinois, Urbana
6. Simon M et al (1994) Spread-spectrum communications handbook. Mc Graw-Hill, New York
7. Holmes JK (1990) Coherent spread-spectrum systems. Krieger Publishing, Malabar, FL
8. Holmes JK (2007) Spread-spectrum systems for GNSS and wireless communications. Artech House, Norwood, MA
9. Gaudenzi R, Hoult N, et al (2000) Galileo signal validation development. Wiley, Hoboken
10. Zepernick H, Finger A (2005) Pseudo random signal processing. Theory and application. Wiley, Hoboken
11. Hein GW, Avila-Rodriguez J-A, Wallner S, Pratt AR, Owen JIR, Issler J-L, Betz JW, Hegarty CJ, Lenahan S, Rushanan JJ, Kraay AL, Stansell TA (2006) MBOC: the new optimized spreading modulation recommended for Galileo L1 OS and GPS L1C. In: Proceedings of the international technical meeting of the institute of navigation, IEEE/ION PLANS (2006) April. Loews Coronado Bay Resort, San Diego, California, USA, pp 24–27
12. Hegarty CJ, Betz JW, Saidi A (2005) Binary coded symbol modulations for GNSS. In: Proceedings of the national technical meeting of the institute of navigation, ION-NTM (2005) January 2005. San Diego, California, USA
13. Ries L, Issler J-L, Lestarquit L, Avila-Rodriguez J-A, Hein GW (2006) Spread-spectrum signals, patent number WO/2006/075018, international application no.: PCT/EP2006/050179, Publication date: 20 July 2006
14. Cobb HS (1997) GPS Pseudolites: theory, design and applications. PhD thesis, Stanford University
15. Yao Z (2008) A new unambiguous tracking technique for Sine-BOC (2n, n) signals. In: Proceedings of ION GNSS, Savannah, GA, pp 2432–2438
16. Yao Z, Cui X, Lu M, Feng Z (2010) Pseudo-correlation-function-based unambiguous tracking technique for Sine-BOC signals. IEEE Trans Aerosp Electron Syst 46:1782–1796

Chapter 4
Spreading Modulation Techniques in Satellite Navigation

In my opinion, no single design is apt to be optimal for everyone.

— Donald Norman

4.1 Introduction

In the early days of GPS and GLONASS, the L-band was not as crowded as it is today, so the signal design for traditional satellite-navigation systems did not need to be particularly concerned about spectrum sharing. Moreover, given the design requirements of the time, the signal needed to achieve only a relatively modest accuracy by today's standards. However, when the signal design of the next-generation GNSS was carried out, the situation was very different. On the one hand, as discussed in Sect. 2.4, the L-band has become increasingly crowded, causing interference between signals to become a major problem that must be addressed and solved in system construction. At the same time, for security reasons, it is hoped that sufficient spectrum separation can be achieved between military and civilian signals. These requirements introduce many new constraints to the spectrum occupancy of the signal. However, demand for navigation performance is constantly increasing. It is expected that the new signals will be significantly improved in accuracy, robustness, and diversity of receiving methods compared to traditional signals. These new demands and constraints have posed challenges to the signal design of the next-generation GNSS.

The optimized design of the spreading modulation techniques is considered to be the major way to achieve both spectrum compatibility and performance improvement. In Chap. 3, we presented a detailed analysis of the relationship between the spreading modulation and the time-domain and frequency-domain characteristics of a signal. The most direct effect of spreading modulation design is to change the shape of the

Z. Yao and M. Lu, *Next-Generation GNSS Signal Design*, Navigation: Science and Technology 6, https://doi.org/10.1007/978-981-15-5799-6_4

signal spectrum so that the energy of the signal can be allocated to specific frequency positions.

For the traditional satellite navigation systems, there is often a lack of incentive to pursue innovation in the design of the spreading modulation. In order to simplify the signal transmission and reception, early satellite navigation signals, such as P(Y) and C/A signals of GPS, as well as all signals of GLONASS and BDS-2 system, all used BPSK-R modulation with relatively short spreading code periods and simple data message modulations. In the modernization of GPS, although the spreading code length and the channel structure were adjusted in the GPS L5 and L2C signals, BPSK-R modulation is still used.

With the development of next-generation satellite navigation systems, the deficiency of BPSK-R modulation became increasingly apparent in its spectral compatibility, ranging accuracy, and anti-interference performance. To allow a variety of signals to share the limited frequency band of GNSS and further improve the signals' ranging accuracy and anti-interference performance, several new spreading modulation techniques have been proposed in recent years.

According to the major milestones in the development of the spreading modulation method of navigation signals, research in this field can be roughly divided into the following three stages.

1. The BPSK period (before 2000)

Since the GPS was built in the 1970s, navigation signals have been using BPSK-R technology [1]. Although the square-root raised cosine (SRRC) waveform, which is widely used in communications, was proposed for navigation signals [2], subsequent studies have shown that the ranging performance of such signals is severely limited. Hence it is not used for satellite navigation. This also reflects one difference between the design of signal modulations used in communications versus navigation systems.

2. The BOC period (2000–2004)

The concept of BOC modulation was proposed around the year 2000 [3, 4], marking the beginning of new-generation navigation signal spreading modulation technology. In theory, compared to the BPSK-R signal, the BOC signal has higher pseudorange measurement accuracy, better multipath suppression and anti-interference ability, and more flexible spectrum sharing capability with other signals at the same frequency band. Cosine phase BOC modulation was first introduced into engineering applications in 2002. In 2003, the constant envelope AltBOC modulation technique was proposed [5], which can combine four different spread spectrum signal components into a constant envelope form and transmit them by a single carrier. In addition, 8-PSK BOC modulation [6] has also been considered for satellite navigation signals.

3. The MBOC period (after 2005)

In 2005, the concept of BCS modulation was proposed [7], which further improved the design flexibility of navigation spreading modulation. On the basis of BCS modulation, the designers of the Galileo system proposed a series of new spreading modulations such as "Crazy"-BPSK [8], CBCS [9, 10], and alternative BCS (ABCS) [8]. In 2006, the CBOC modulation [11, 12] was proposed for the modulation of the Galileo E1 open service signal. Correspondingly, in the same year, the GPS III L1C signal introduced the TMBOC modulation, which has the same spectral shape as CBOC modulation but a different time domain realization method [13]. In 2010, another implementation of the MBOC modulation, QMBOC modulation [19], was proposed for BDS-3 B1C signal. These MBOC modulation techniques mark the entry of navigation signal spreading modulation technology into the realm of multi-subcarriers.

With so many new modulation methods emerging in just over a decade, some readers might wonder: Which signal is the best? Do the latecomers always surpass the formers? Is the oldest and simplest BPSK-R signal outdated, and is it going to be replaced by those split-spectrum signals such as BOC and MBOC signals?

It is important to recognize that we cannot discuss which modulation technique is more advanced without considering specific application requirements and engineering constraints. Compared with the BPSK-R signal, split-spectrum signals have higher potential ranging accuracy, but at the cost of increased processing complexity and reduced tracking reliability. In other words, split-spectrum signals are a result of designs considering increasing constraints on spectrum sharing and relaxing constraints on processing complexity, under circumstances where the frequency resources are severely limited while the level of electronic manufacturing is rapidly developing.

What is clear, however, is that the new modulation techniques that have emerged in recent years offer greater design flexibility, providing more adjustable parameters in the waveform. The emergence of these modulation techniques offers signal designers more choices and makes the adjustment of the signal more flexible in terms of ranging performance, spectrum occupancy, and processing complexity. But at the same time, such flexibility poses higher requirements for the signal designer's understanding of the technology and user requirements. Only with a detailed understanding of the characteristics of these general modulation techniques can signal designers choose the most appropriate modulation techniques and parameter configurations for signals in the face of specific requirements and constraints.

In this chapter, we will explain a variety of typical spreading modulation techniques. Some have already been used in existing systems, and have received widespread attention in recent years, while others are less well-known. However, considering that in the next-generation GNSS implementations, new requirements and new constraints will constantly emerge and new modulation methods will be needed, this chapter is not meant to be regarded as a review of the development of navigation signal modulation technology, or an interpretation of the signal formats being used by several major satellite navigation systems. Instead, we regard the spreading modulations used in satellite navigation signals as general techniques and focus our discussion on the principles of these modulation techniques and the design ideas behind their generation processes.

4.2 BPSK-R Modulation

BPSK-R modulation is considered to be the most basic spreading modulation for satellite navigation. Since the performance of BPSK-R signals had met the needs of early navigation applications, system designers lacked incentive to develop more complex spreading modulation techniques. That is why, for quite a long time, almost all satellite navigation signals used BPSK-R modulation.

The time-domain and frequency-domain characteristics of BPSK-R modulation are elaborated in detail in the literature of spread-spectrum communication and satellite navigation signal processing. Moreover, after decades of research in academia and industry, GNSS receiver signal processing technology of BPSK-R signals has grown quite mature.

As indicated by the letter R in its name, BPSK-R modulation uses a rectangular-pulse spreading chip waveform, which can be mathematically expressed as

$$p_{\text{BPSK-R}}(t) = \frac{1}{\sqrt{T_c}} \operatorname{rect}_{T_c}(t) , \tag{4.1}$$

where

$$\operatorname{rect}_{T_c}(t) = \begin{cases} 1, & t \in [0, T_c) \\ 0, & \text{otherwise} \end{cases} . \tag{4.2}$$

In (4.1), the amplitude $1/\sqrt{T_c}$ is limited to comply with the convention of the normalization of the chip waveform energy, as stated in (3.58).

The BPSK-R-modulated spreading chip waveform has a constant value throughout the chip period T_c. Therefore, only T_c or its reciprocal, i.e. the rate of the spreading code, can be adjusted in this modulation technique. According to convention in the satellite navigation community, we can use BPSK-R(n) to represent the BPSK-R signal with a spreading code rate of $f_c = 1/T_c = n \times 1.023$ MHz. For example, the modulation used by the GPS L1 C/A code signal with a code rate of 1.023 MHz can be denoted as BPSK-R(1), and the modulation used by BDS-2 B1I signal with a code rate of 2.046 MHz can be denoted as BPSK-R(2).

The time domain waveform of the BPSK-R signal was shown in Fig. 3.2b. According to the definition of the SCS modulation waveform (see Sect. 3.3.5), the BPSK-R signal can be viewed as an SCS-modulated signal with the chip waveform vector $\boldsymbol{\kappa} = [1, 1, \ldots, 1]^{\mathrm{T}}$. In this chip waveform vector, the number of 1 can be arbitrary. Thus, with (3.34), when the PRN code has an infinitely long period and an ideal auto/cross-correlation property, the ACF of the BPSK-R(n) signal can be expressed as

$$R_{\text{BPSK-R}}(\tau) = \begin{cases} 1 - \dfrac{|\tau|}{T_c}, & |\tau| \le T_c \\ 0, & \text{otherwise} \end{cases} , \tag{4.3}$$

the shape of which is an isosceles triangle, as shown in Fig. 4.1.

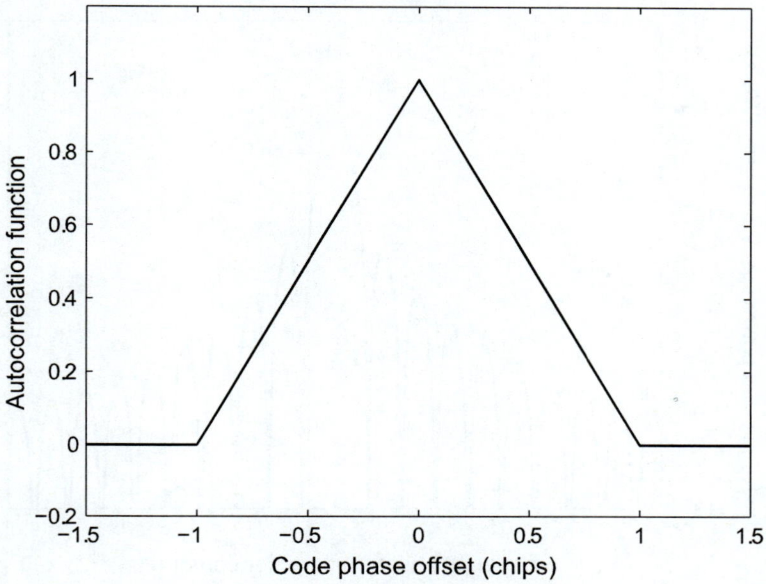

Fig. 4.1 Autocorrelation function of the BPSK-R signal under infinite bandwidth

Recalling from Example 3.3 in Chap. 3, there we have derived the PSD of the BPSK-R(n) signal. When the PRN code has an infinitely long period and an ideal auto/cross-correlation property, the normalized PSD of the BPSK-R(n) signal is

$$G_{\text{BPSK-R}}(f) = T_c \operatorname{sinc}^2(\pi f T_c) \ , \tag{4.4}$$

the shape of which is shown in Fig. 4.2.

The energy of the baseband BPSK-R(n) signal is mainly concentrated near the zero frequency. The null-to-null bandwidth of the main lobe is $2f_c$, and the null-to-null bandwidth of each sidelobe is f_c. The height of the sidelobes gradually decreases as their distance from the main lobe increases.

The spectrum of an ideal BPSK-R signal is infinitely wide, so it needs to be bandlimited and filtered during transmission and reception. Such filtering not only causes signal power loss, but also changes the shape of the signal correlation function actually obtained by the receiver to no longer be the triangle given by (4.3). Consider an ideal equivalent low-pass filter whose frequency response function is

$$H_L(f) = \begin{cases} 1, & |f| \le \beta/2 \\ 0, & |f| > \beta/2 \end{cases} \ . \tag{4.5}$$

Then, the proportion of signal power that can pass through the filter is

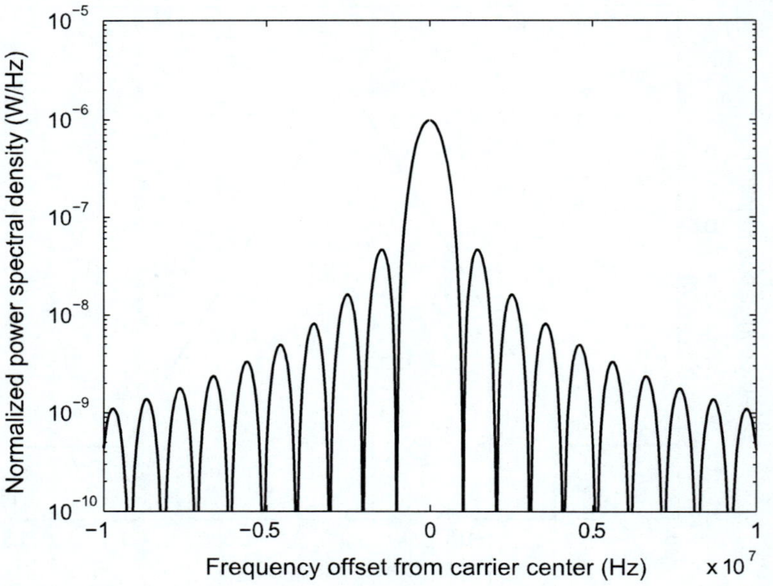

Fig. 4.2 Power spectral density of the BPSK-R(1) signal

$$\eta = \int_{-\beta/2}^{+\beta/2} G_{\text{BPSK-R}}(f)\,\mathrm{d}f\,. \tag{4.6}$$

Figure 4.3 plots the signal power proportion passing through the filter in the total BPSK-R signals at different bandwidths, where the horizontal axis is normalized with f_c. The main lobe of the BPSK-R signal contains about 90% of the signal power. Then, when receiving the first pair of sidelobes, one can obtain about 5% more signal power.

In the receiver, the acquisition and tracking of the signal is based on a CCF between the received signal and the local signal replica. We assume that the received signal and the local signal replica use exactly the same spreading modulation and spreading sequence. The only difference is that the input signal passes through the filter with the frequency response function defined by (4.5), while the local signal replica is not filtered, but is a BPSK-R signal with indefinite bandwidth. Then, the CCF of the two signals can be written as

$$R(\tau) = \int_{-\beta/2}^{+\beta/2} G_{\text{BPSK-R}}(f)\,\mathrm{e}^{\mathrm{j}2\pi f\tau}\mathrm{d}f\,. \tag{4.7}$$

We plot $R(\tau)$ at several different filter bandwidths in Fig. 4.4. It can be seen that when the filter bandwidth is narrow, the correlation function becomes smooth and

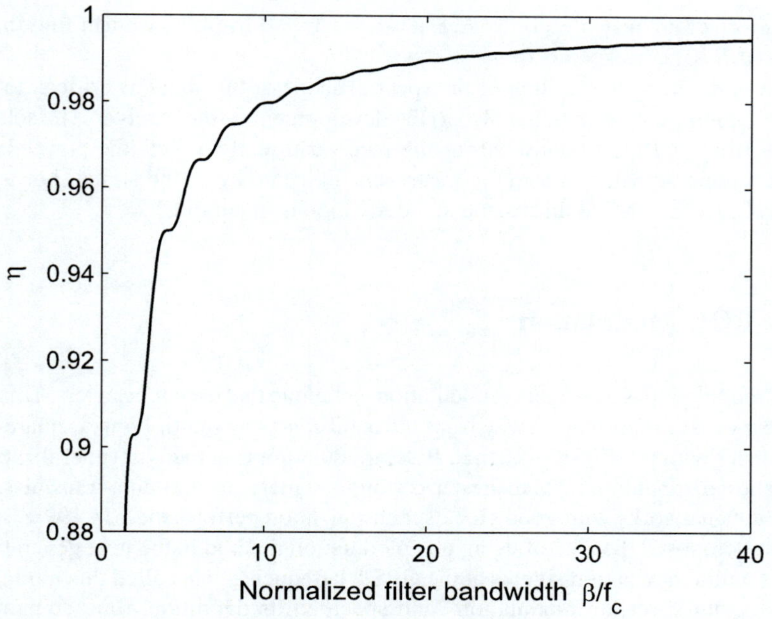

Fig. 4.3 Power proportion of the BPSK-R signal passing through the filter under different receiving bandwidths

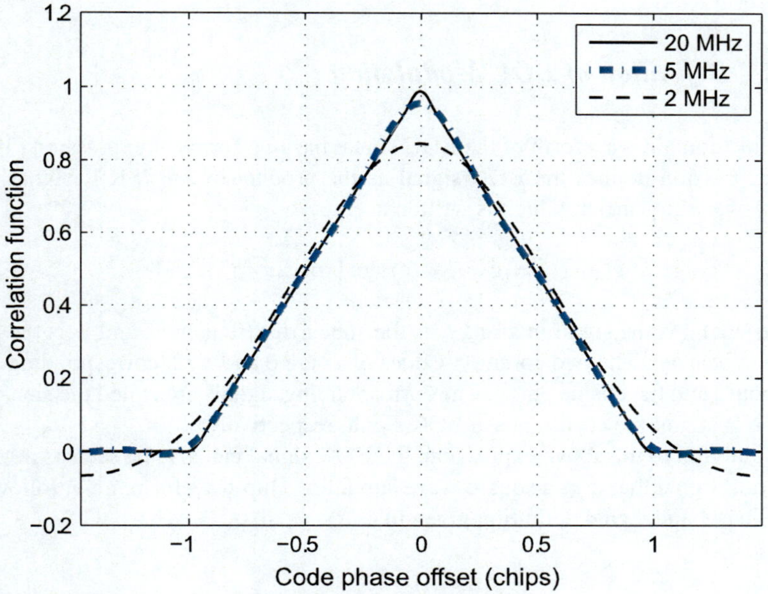

Fig. 4.4 Autocorrelation function of the BPSK-R signal under different bandwidths

the height drops near $\tau = 0$. In fact, if we set $\tau = 0$ in (4.7), we can find that the value of $R(0)$ corresponds to η.

An in-depth understanding of the spectral shape of the signal is critical not only to the design of the signal, but also to the development of the receiver. The selection of the front-end filter bandwidth of the receiver directly affects the power loss of the receiving signal, and also the measurement accuracy of the signal. For a more general case, a detailed discussion and derivation is given in Chap. 5.

4.3 BOC Modulation

BOC modulation is a spreading modulation technique that uses subcarriers. This technique moves signal energy away from the band center by multiplying a square-wave subcarrier with the BPSK-R signal. BOC modulation is actually a generalization of Manchester modulation. Manchester coding is widely used in data transmission in local area networks with good clock synchronization performance. In 1998, Spilker [3] first proposed the idea of using this modulation method in the next generation of civil satellite navigation signals at the GPS L1 frequency. He called this modulation signal "split-spectrum modulation" and specified its definition, time domain, and frequency domain characteristics. In 2001, Betz [4] elaborated on the characteristic analysis of this signal more systematically and named it BOC modulation.

4.3.1 Definition of BOC Modulation

The time domain waveform of the BOC signal has two forms of expression [3]. The first expression defines the BOC signal as the product of a BPSK-R signal and a square-wave subcarrier, which is written as

$$g_{\mathrm{BOC}}(t) = g_{\mathrm{BPSK\text{-}R}}(t)\,\mathrm{sgn}\left[\sin\left(2\pi f_s t + \psi\right)\right] , \tag{4.8}$$

where $\mathrm{sgn}(\cdot)$ is the sign function, f_s is the subcarrier frequency, and ψ is the phase of the subcarrier. The two common values of ψ are 0 and $\pi/2$, corresponding to the sine phase and the cosine phase. The corresponding signals are called the sine-phase BOC signal and the cosine-phase BOC signal respectively.

In addition to the above expression, the BOC signal can also be seen as a spread-spectrum signal that uses a square-wave spreading chip waveform, i.e. it follows the DSSS baseband signal definition given in (3.2):

$$g_{\mathrm{BOC}}(t) = \sum_{n=-\infty}^{+\infty} (-1)^{c_n}\, p_{\mathrm{BOC}}\left(t - nT_c\right) , \tag{4.9}$$

where

$$p_{\mathrm{BOC}}(t) = \frac{1}{\sqrt{T_c}} \operatorname{sgn}\left[\sin\left(2\pi f_s t + \psi\right)\right] \operatorname{rect}_{T_c}(t) \tag{4.10}$$

is a square wave of duration T_c.

As mentioned in Sect. 4.2, the specific form of a BPSK-R modulation can be described using only one parameter, i.e. the spreading code rate. For BOC modulation, due to the presence of the subcarrier, the specific form of a BOC signal needs to be described by three key parameters: the spreading code rate f_c, the subcarrier frequency f_s, and the subcarrier phase ψ.

In the satellite navigation community, when we talk about a specific BOC modulation, we can use a simplified notation: $\mathrm{BOC_j}(m, n)$, where the subscript j can take the value of s or c respectively indicating whether the subcarrier is with sine phase or cosine phase, m represents the subcarrier frequency normalized by 1.023 MHz and n represents the spreading code rate normalized by 1.023 MHz, i.e. $f_s = m \times 1.023$ MHz, $f_c = n \times 1.023$ MHz, where $m \geq n$. For example, if the spreading code rate and subcarrier frequency of a sine-phase BOC signal are both 1.023 MHz, the modulation used can be abbreviated as $\mathrm{BOC_s}(1, 1)$. If the spreading code rate of a cosine-phase BOC signal is 2.5575 MHz and the subcarrier frequency is 15.345 MHz, the modulation used can be abbreviated as $\mathrm{BOC_c}(15, 2.5)$.

In addition to these three key parameters, the BOC signal has another parameter that is critical to the performance of analysis. The ratio $M = 2m/n$ is called the order of BOC modulation, which is generally restricted to integers. It identifies the number of subcarrier semi-cycles in a spreading chip, i.e. $M = T_c/T_s$, where $T_s = 1/(2f_s)$ is the chip width of the subcarrier. Different BOC signals with the same modulation order have the same number of subcarrier semi-cycles in a spreading chip. They differ from each other only in the code rate.

The BOC modulation order M can be either odd or even. Classified by the parity of M, the BOC signals can also be divided into odd-order BOC signals and even-order BOC signals. For example, the modulation order of the above mentioned $\mathrm{BOC_s}(1, 1)$ is 2, while the $\mathrm{BOC_c}(15, 2.5)$ modulation has an order of 12. Both of these signals are even-order BOC signals.

When M is even, as shown in Fig. 4.5, an integer number of subcarriers are included in one spreading chip, so the chip waveform of each spreading code is the same. In this case, the definitions of the two BOC signals given by (4.8) and (4.9) are equivalent. However, when M is odd, the two definitions are no longer equivalent, as demonstrated in the following example.

Example 4.1 *(Time domain waveform of odd-order BOC signals)* Consider a third-order sine-phase BOC signal where $f_s = 1.5 f_c$, i.e. 1.5 sub-chips are included in one PRN chip. The time domain waveform shown in Fig. 4.6 is for a BOC signal generated under the first definition (4.8) that has a PRN sequence of {0, 1, 1, 0}. However, with the same PRN sequence {0, 1, 1, 0}, when the BOC signal is generated under the second definition (4.9), its time domain waveform is shown in Fig. 4.7.

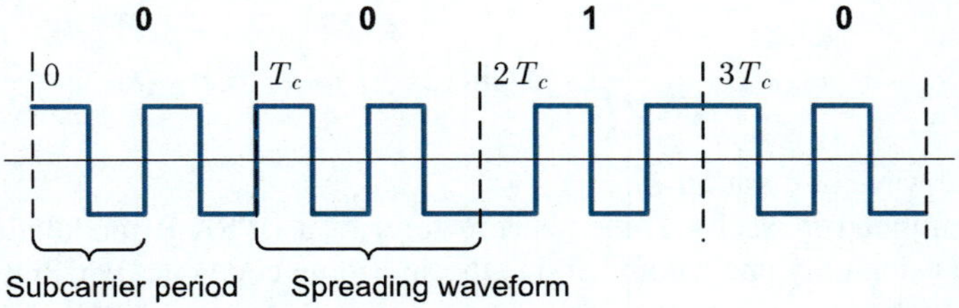

Fig. 4.5 Time domain waveform of even-order BOC signals

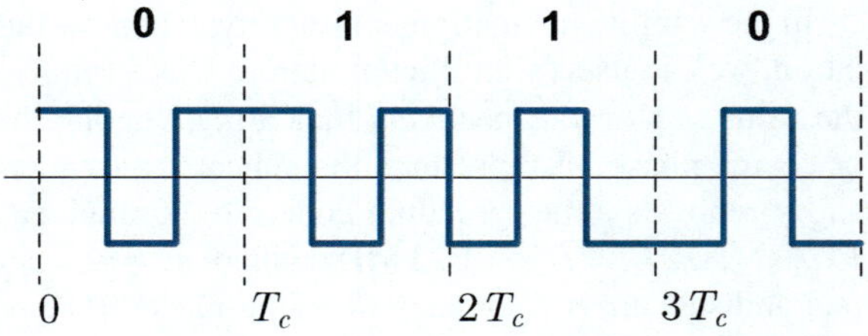

Fig. 4.6 Time domain waveform of odd-order BOC signals generated under the first definition

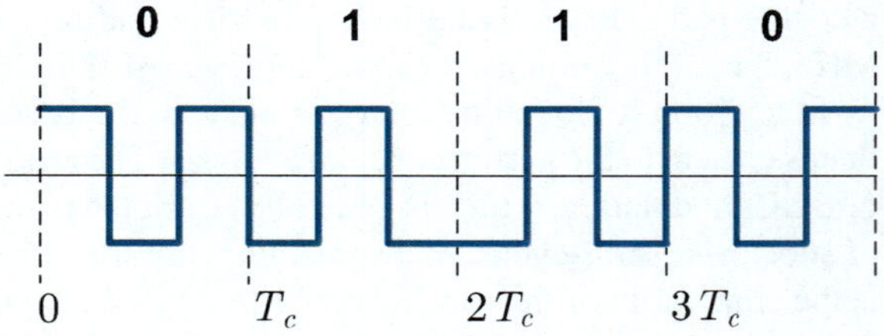

Fig. 4.7 Time domain waveform of odd-order BOC signals generated under the second definition

Comparing Figs. 4.6 and 4.7, it is clear that for odd-order BOC signals, the time domain waveforms generated under the two definitions differ at the transition position of the PRN code. ■

In order to unify the two expressions of the odd-order BOC signal, when M is odd, the second definition should be corrected as

$$g_{\mathrm{BOC}}(t) = \sum_{n=-\infty}^{+\infty} (-1)^{c_n} (-1)^n \, p_{\mathrm{BOC}}(t - nT_c) \ . \tag{4.11}$$

4.3.2 PSD of BOC Signals

As in the case of BPSK-R modulation, PSD and ACF are the most important characteristics of the BOC signal. Here, we only discuss the BOC signal of the sine and the cosine phases. The case of the square wave with other phases is not included in discussion due to its rarity in real situations.

We can obtain the expression of the PSD of the BOC signal from both of its two definitions. Given the first definition (4.8), the BOC signal in the time domain is the product of the BPSK-R signal multiplied by a square wave subcarrier

$$\gamma(t) \triangleq \operatorname{sgn}\left[\sin\left(2\pi f_s t + \psi\right)\right] . \tag{4.12}$$

Therefore, according to the convolution theorem, the BOC signal spectrum is the convolution of the BPSK-R signal spectrum and the Fourier transform of $\gamma(t)$, i.e.

$$P_{\mathrm{BOC}}(f) = P_{\mathrm{BPSK\text{-}R}}(f) * \Gamma(f) , \tag{4.13}$$

where $\Gamma(f) = \mathcal{F}\{\gamma(t)\}$. Because $\gamma(t)$ is a periodic and odd harmonic signal, it satisfies

$$\gamma(t) = -\gamma\left(t \pm \frac{T_s}{2}\right) . \tag{4.14}$$

Therefore, its Fourier transform is a series of line spectral components, and only the odd harmonic terms appear at $\pm(2k+1) f_s$ with the line spacing of $2f_s$ and the envelope $\operatorname{sinc}(\pi f T_s)$. Then, the PSD of the BOC signal is

$$G_{\mathrm{BOC}}(f) = P_{\mathrm{BOC}}(f) P^{*}_{\mathrm{BOC}}(f) . \tag{4.15}$$

When we use the second definition of BOC signals along with the theories elaborated in Chap. 3, it is much easier to obtain the PSD from the general case of SCS modulation (see Sect. 3.4.5). Either way, we can obtain the same expression of the PSD of the BOC signal. Assuming that the PRN code period is infinite and has completely ideal correlation characteristics, then for a sine-phase BOC signal, when M is even, the normalized PSD under infinite bandwidth can be expressed as

$$G_{\mathrm{BOC_s}(f_s,f_c)}(f) = \frac{\sin^2(\pi f T_c)\sin^2(\pi f T_s)}{T_c\left[\pi f \cos(\pi f T_s)\right]^2} . \tag{4.16}$$

When M is odd, the normalized PSD under infinite bandwidth can be expressed as

$$G_{\mathrm{BOC_s}(f_s,f_c)}(f) = \frac{\cos^2(\pi f T_c)\sin^2(\pi f T_s)}{T_c\left[\pi f \cos(\pi f T_s)\right]^2} . \tag{4.17}$$

For the cosine-phase BOC modulation signal, when M is even, the PSD is

$$G_{\mathrm{BOC_c}(f_s,f_c)}(f) = \frac{\sin^2(\pi f T_c)\left[1 - \cos(\pi f T_s)\right]^2}{T_c\left[\pi f \cos(\pi f T_s)\right]^2} . \tag{4.18}$$

and when M is odd, the PSD is

$$G_{\mathrm{BOC_c}(f_s,f_c)}(f) = \frac{\cos^2(\pi f T_c)\,[1-\cos(\pi f T_s)]^2}{T_c\,[\pi f \cos(\pi f T_s)]^2}. \tag{4.19}$$

Using a series of identity relationships such as $T_c = MT_s = 1/f_c$, the above four equations may have several different forms in various literature [3, 4]. However, the given expressions (4.16)–(4.19) can show more clearly the differences and similarities in the form of the power spectrum of different BOC signals.

We can see that in the given expressions, the common denominators of the BOC signals' PSD expressions are $T_c\,[\pi f \cos(\pi f T_s)]^2$. The two terms in the numerators are the square factors of the trigonometric functions of $\pi f T_c$ and $\pi f T_s$ respectively. Whether the BOC signal is sine-phase or cosine-phase determines the form of the related factor of $\pi f T_s$. For the sine-phase BOC, the factor takes the form of a sine square $\sin^2(\pi f T_s)$ and for the cosine-phase BOC, it takes the form $[1-\cos(\pi f T_s)]^2$. Whether the modulation order of the BOC signal is odd or even determines the form of the related factor of $\pi f T_c$. For the odd-order BOC signal, this factor takes the form $\cos^2(\pi f T_c)$, and for the even-order BOC, this factor takes the form $\sin^2(\pi f T_c)$.

Figure 4.8 shows the PSD of both the baseband $\mathrm{BOC_s}(1,1)$ and $\mathrm{BOC_c}(6,1)$ signals. Compared with Fig. 4.2, it can be seen that the maximum PSD of the BOC signal is no longer at the carrier frequency but is with frequency offsets $\pm f_s$. The simple reason for this phenomenon is that a square wave subcarrier with a frequency of f_s has a strong harmonic component near $\pm f_s$. If we regard the BOC signal as

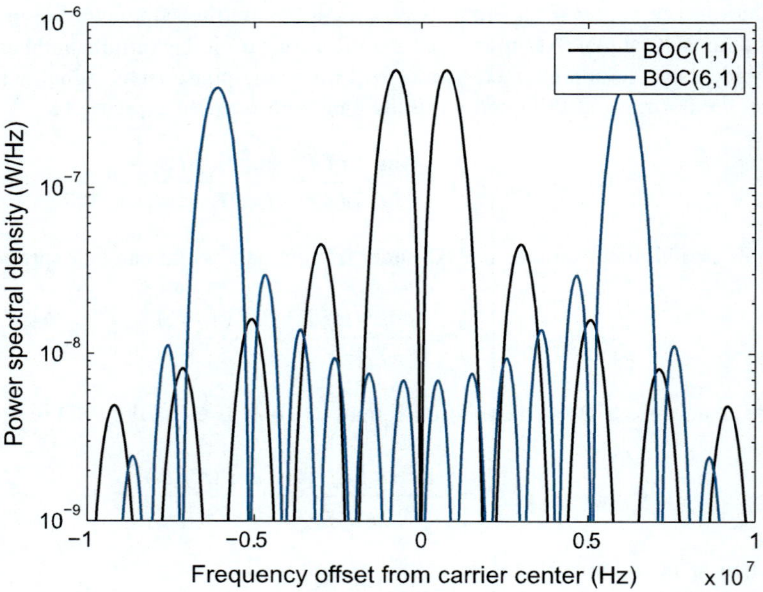

Fig. 4.8 Power spectral density of the BOC signals

the product of the BPSK-R signal and a square wave subcarrier, then it corresponds to the result of a further modulation of the original BPSK-R signal.

For the signal whose spectrum splits on both sides of the carrier frequency center, the definitions of the PSD main lobe and the sidelobe are different from conventional notation. In this case, the two power spectral density lobes with the largest amplitude are called the main lobes, and the ones with relatively small amplitudes are called the sidelobes. The main lobe width of the BOC signal is $2f_c$, and the width of the sidelobe between the main lobes is f_c. For a BOC signal with a modulation order of M, the number of sidelobes between the two main lobes is $M-2$. These characteristics can be easily verified from Fig. 4.8. For the second-order BOC signal, there are no sidelobes between the two main lobes.

4.3.3 The Autocorrelation Function of BOC Signals

Due to the importance of the ACF in the spread spectrum signal receiving, many studies such as [8, 14, 15] have been conducted on the calculation of the BOC signal's ACF. These studies give different forms of expression for the ACF via various derivations, but they are essentially the same. In fact, when M is even, the BOC signal can be seen as a special case of the SCS modulated signal that we introduced in Chap. 3.

When M is even, for the sine-phase BOC signal, the corresponding SCS chip shape vector is

$$\boldsymbol{\kappa} = \left[\underbrace{1,\ldots,1}_{N},\underbrace{-1,\ldots,-1}_{N},\underbrace{1,\cdots,1}_{N},\cdots\right]^{\mathrm{T}}, \tag{4.20}$$

where N can be any positive integer, and the total length of the vector is MN. For the cosine-phase BOC signal, the corresponding chip shape vector is

$$\boldsymbol{\kappa} = \left[\underbrace{1,\ldots,1}_{N/2},\underbrace{-1,\ldots,-1}_{N},\underbrace{1,\cdots,1}_{N},\cdots\underbrace{1,\ldots,1}_{N/2}\right]^{\mathrm{T}}. \tag{4.21}$$

Therefore, the ACF of these two BOC signals can be conveniently calculated using (3.34).

In Example 3.1, as an example of calculating the correlation functions of the SCS signal, we have given the calculation method and the expression of the ACF of the even-order sine-phase BOC signal. Readers are welcome to practice solving the ACF of the even-order cosine-phase BOC signal.

When M is odd, the matter becomes a little more complicated. There is a difference between (4.11) and the definition equation of the SCS modulated signal. Therefore, at this time, the ACF of the BOC signal cannot be directly obtained by (3.34). Instead, we can rewrite (4.11) as

$$g_{\mathrm{BOC}}(t) = \sum_{n=-\infty}^{+\infty} (-1)^{c_n+n}\, p_{\mathrm{BOC}}(t - nT_c)\ . \tag{4.22}$$

In this rewriting, the value reversal of the chip waveform in the original definition is equivalent to the change of the PRN sequence value, the effect of which is equivalent to creating a new set of PRN sequences $c'_n = (c_n + n) \bmod 2$. That is to say, in the new PRN sequence, the even-numbered PRN sequence elements remain unchanged, but the odd-numbered sequence elements get negative values of the original sequence elements. Strictly speaking, the statistical properties of the new sequence actually differ from the original. However, theoretically speaking, if we assume that the period of the PRN code is infinite and has completely ideal correlation characteristics, the modified PRN sequence also satisfies the above hypothesis. In this case, the ACF of the odd-order BOC signal can still be calculated using (3.34).

Figure 4.9 shows the ACFs of $\mathrm{BOC_s}(2n, n)$ and $\mathrm{BOC_c}(2n, n)$ signals. Just as for a general SCS modulation signal, the ACF of the BOC signal is a polyline that has a plurality of zero crossings and peaks.

Whether the BOC signal is either the sine phase or cosine phase, the total number of positive and negative peaks is always $2M - 1$ within the interval $\tau \in [-T_c, T_c]$, and the distance between each peak is T_s. When $\tau = iT_s$, where $i =$

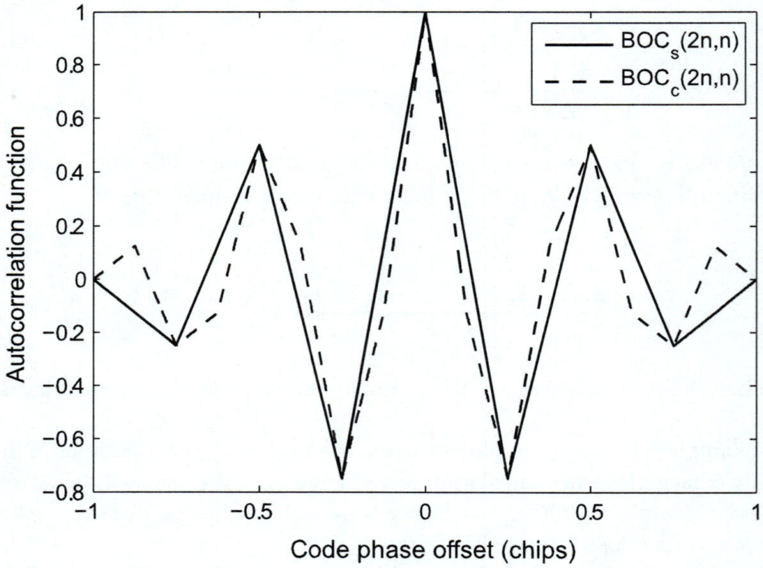

Fig. 4.9 Autocorrelation function for the sinusoidal-phase and cosine-phase BOC$(2n, n)$ signal

$-(M-1), -(M-2), \ldots, -1, 0, 1, \ldots, M-1$, the peak values for both types of BOC signals are

$$R_{\mathrm{BOC}}(iT_s) = (-1)^i \frac{M-1-|i|}{M}. \tag{4.23}$$

However, the ACF of the cosine-phase BOC signal has more line segments between the peaks.

In addition to using the general calculation formula of SCS signal's ACF to express the ACF of the BOC signal, we can also write it as a combination of triangle function

$$\Lambda_L(t) \triangleq \begin{cases} 1-\frac{|t|}{L}, & |t| \le L \\ 0, & \text{otherwise} \end{cases} \tag{4.24}$$

and its time shift. For the sine-phase BOC signals, we have

$$R_{\mathrm{BOC_s}}(\tau) = \Lambda_{T_s}(\tau) + \sum_{k=1}^{M-1} (-1)^k \left(1-\frac{k}{M}\right) \Lambda_{T_s}(|\tau| - kT_s), \tag{4.25}$$

and for the cosine-phase BOC signal, we have

$$R_{\mathrm{BOC_c}}(\tau) = \Lambda_{T_s/2}(\tau) + \sum_{k=1}^{M-1} (-1)^k \left(1-\frac{k}{M}\right) \Lambda_{T_s/2}(|\tau| - kT_s)$$
$$+ \frac{1}{2M}\sum_{k=1}^{M} (-1)^k \Lambda_{T_s/2}(|\tau| - (2k-1)T_s/2). \tag{4.26}$$

4.3.4 *Characteristic Differences Between Sine-Phase and Cosine-Phase BOC Modulations*

Comparing (4.16) and (4.18), and also (4.17) and (4.19), it can be seen that, although the time domain expression of the sine-phase and cosine-phase BOC signals differs only in the subcarrier phase by $\pi/2$, their PSDs as well as ACFs show a much greater difference. To display this difference more directly, Fig. 4.10 plots the PSDs of the sine-phase $\mathrm{BOC_s}(10,5)$ signal and the cosine-phase $\mathrm{BOC_c}(10,5)$ signal, and Fig. 4.11 plots the PSDs of the sine-phase $\mathrm{BOC_s}(15,2.5)$ and the cosine-phase $\mathrm{BOC_c}(15,2.5)$. It can be seen from these two figures that the power of the sine-phase BOC signal is more concentrated on the inner side of the two main lobes, while the power of the cosine-phase BOC signal is more concentrated on the outer side of the main lobes.

We have seen the difference between the PSDs of these two BOC signals through formula derivation. Now we can also explain the cause of this difference through an intuitive method. Figure 4.12 shows the situation near the transition point of two

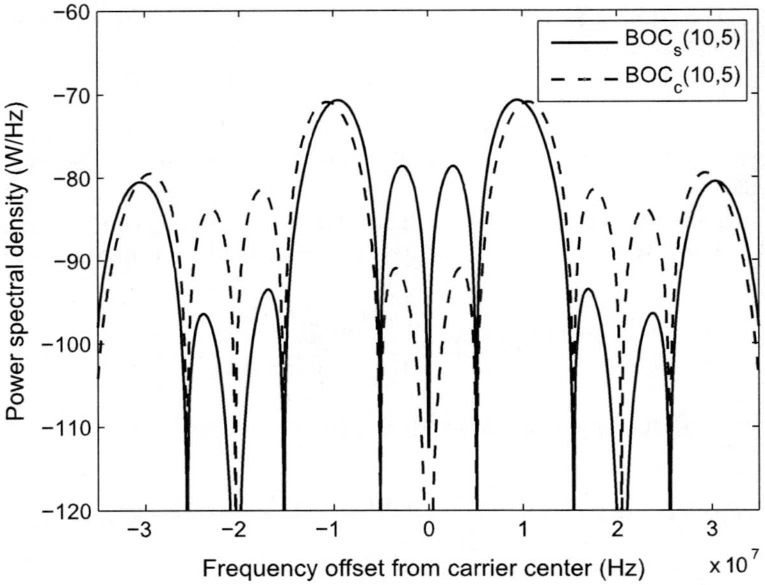

Fig. 4.10 PSDs of the sinusoidal phase $\text{BOC}_\text{s}(10, 5)$ signal and the cosine phase $\text{BOC}_\text{c}(10, 5)$ signal

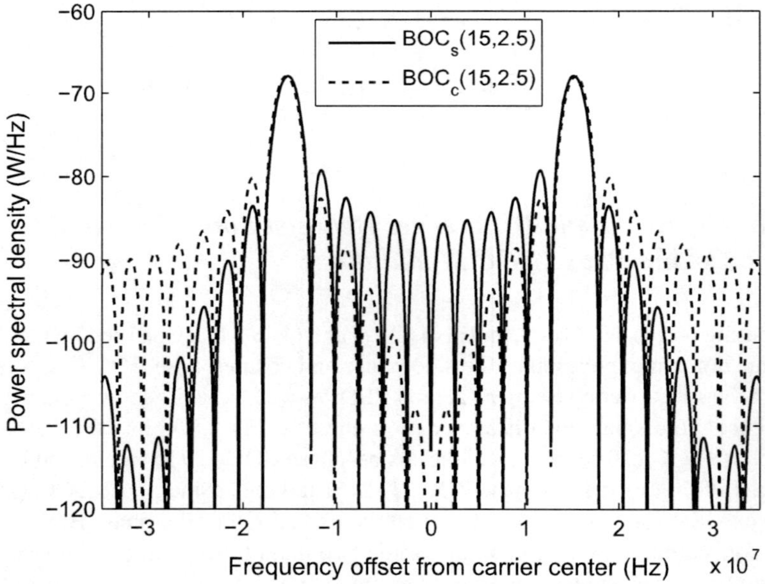

Fig. 4.11 PSDs of the sinusoidal phase $\text{BOC}_\text{s}(15, 2.5)$ signal and the cosine phase $\text{BOC}_\text{s}(15, 2.5)$ signal

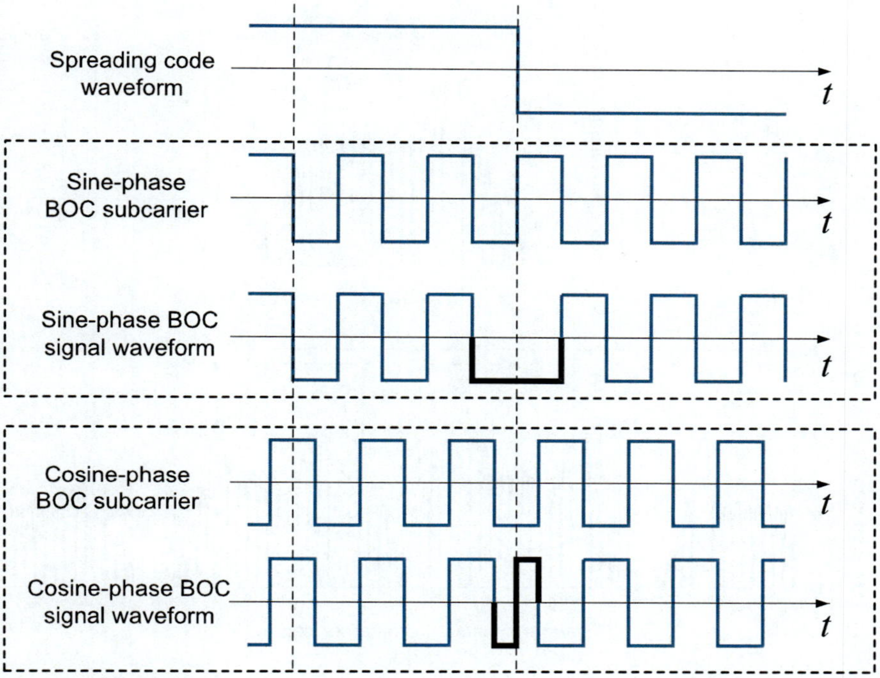

Fig. 4.12 Comparison of waveforms of sine and cosine phase BOC signals near the polarity reversal point of the spreading code

adjacent spreading chips. If the signs of the two spreading chips change, it can be easily verified that a pulse with a frequency twice the subcarrier frequency, i.e. a frequency of $2f_s$, will appear on the waveform of the cosine-phase BOC signal. In the waveform of the entire cosine-phase BOC signal, the probability of such a polarity inversion occurring is 50%. Thus, the probability of the above-mentioned high-frequency pulse occurring is also 50%. For sine-phase BOC signals, the high-frequency pulse does not occur in any case. However, the width of the pulse is twice as large when the polarity of the spreading code is reversed. This explains why the cosine-phase BOC signal of the same order contains more high frequency components than the sine-phase BOC signal.

We can also explain this phenomenon in another way. It is easy to prove that the cosine-phase BOC subcarrier with a frequency of f_s can be obtained by multiplying a sine-phase subcarrier with a frequency of f_s by a sine-phase subcarrier with a frequency of $2f_s$, i.e.

$$\mathrm{sgn}\left[\cos\left(2\pi f_s t\right)\right] = \mathrm{sgn}\left[\sin\left(2\pi f_s t\right)\right] \cdot \mathrm{sgn}\left[\sin\left(4\pi f_s t\right)\right] . \tag{4.27}$$

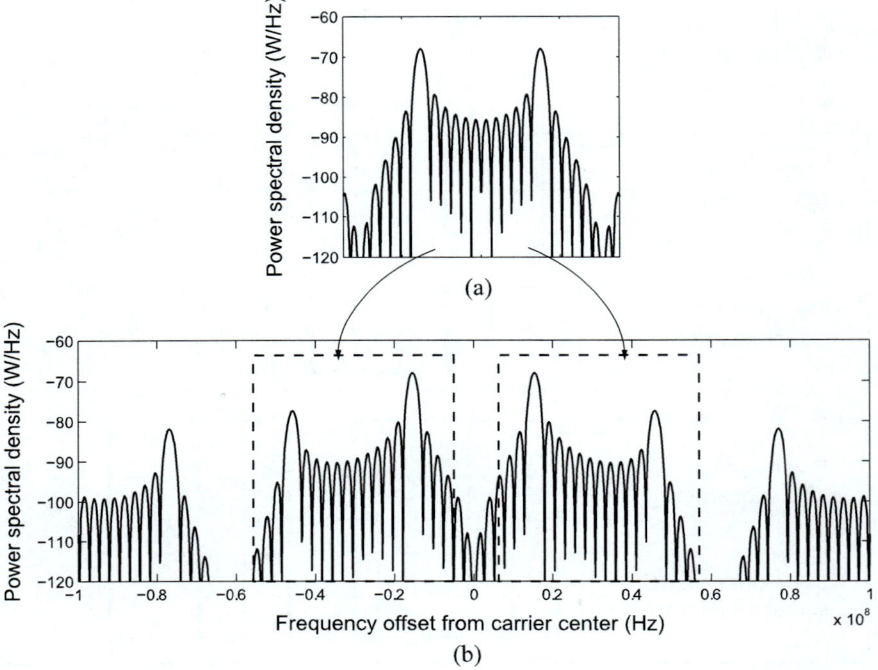

Fig. 4.13 The cosine-phase BOC signal as a further modulation of the sine-phase BOC signal

This means that a cosine-phase BOC signal is equivalent to a sine-phase BOC signal with the same order being further modulated by a sine-phase subcarrier with a frequency of $2f_s$. This modulation process is represented as a shift of spectrum of the sine-phase BOC signal in the frequency domain, as shown in Fig. 4.13a. This shift of the spectrum causes the power of the cosine-phase BOC signal to deviate further from the center frequency. Figure 4.13b shows that the cosine-phase BOC signal still has significant power spectrum components on positions even several times of $2f_s$ far from the center frequency.

In Sect. 5.3, we will derive the relationship between signal PSD and code tracking performance. From this relationship, it can be proved by theory that more high-frequency components result in a cosine-phase BOC signal having a better potential tracking performance. However, this also means that the processing of a cosine BOC signal requires wider transmitting and receiving bandwidths. If the transmitting or receiving bandwidths are too narrow to allow the high-frequency components of the cosine-phase BOC signal to pass through sufficiently, the performance of the cosine-phase BOC signal will deteriorate significantly.

Here we take the BOC(15, 2.5) signal as an example to illustrate this point. When the transmitting and the receiving bandwidths are 40.92 MHz, no matter whether it is a sine-phase or cosine-phase BOC signal, the main lobe and the two sidelobes outside

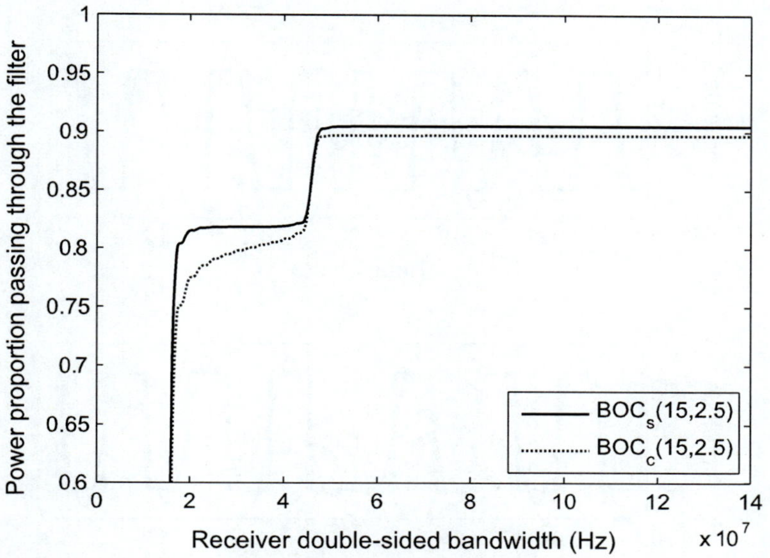

Fig. 4.14 Received power loss of sine and cosine phase BOC(15, 2.5) signals as a function of bandwidth

the main lobe can be received. However, it is clear that the cosine-phase BOC loses more useful power after band limiting, as shown in Fig. 4.14, where the horizontal axis is the single-side receiving bandwidth. Moreover, from the time domain waveform, as shown in Fig. 4.15, after the bandlimited filtering, the double-frequency pulse of the cosine-phase BOC signal at the phase reversal point of the PRN code is severely suppressed. This means that if the cosine-phase BOC signal is transmitted or received only under the main lobe bandwidth, its acquisition and tracking performance will be significantly degraded compared to the sine-phase BOC signal.

The ranging performance of BOC modulated signals and the effect of the multi-peak characteristic of their ACF on the receiving processing will be discussed further in the following sections.

4.4 BCS Modulation

As described in Sect. 4.3, in addition to changing the spreading code rate, BOC modulation also provides the signal designer with two additional adjustable parameters, i.e. subcarrier frequency and subcarrier phase, in order to adjust the bandwidth of the signal. Changing the frequency of the subcarriers can adjust the offset from the spectrum main lobe of the signal to the center frequency of the carrier, thereby allowing radio frequency compatibility between multiple signals sharing the same center fre-

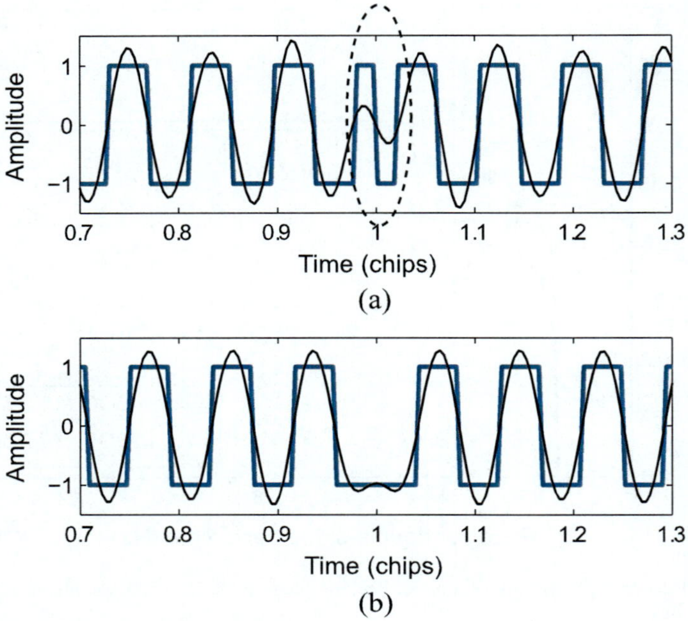

Fig. 4.15 Time-domain waveform of the filtered BOC signal

quency. In 2005, on the basis of BOC modulation, Hegarty et al. [7] generalized the shape of the spreading code chip waveform and proposed the binary coded symbol (BCS) modulation technique, which further improved the design flexibility of the spreading chip.

BCS modulation can be seen as a special case of SCS modulation. The notation of a BCS modulation can be BCS($\boldsymbol{\kappa}$, n), where $\boldsymbol{\kappa}$ is the shape vector, and n represents the spreading code rate normalized by 1.023 MHz. In order to ensure that the envelope of the signal is constant, BCS limits the chip amplitude of the SCS modulated signal to being bipolar, i.e. sets $\kappa_i \in \{1, -1\}$ in (3.28). However, unlike the constant chip waveform of BPSK-R modulation and the positive-and-negative-alternating chip waveform of BOC modulation, the chip waveform of BCS modulation has fewer constraints. In the shape vectors of the BCS signal, the sign of each κ_i can be freely selected, as shown in Fig. 4.16. Therefore, the characteristics of the BCS signal in the time domain and the frequency domain can be more flexibly adjusted by changing its chip waveform, thereby providing the signal designer with more parameters. The ACF and PSD of BCS modulation can be directly obtained by time-domain and frequency-domain analysis of the SCS modulated signals given in Chap. 3.

For a spread-spectrum signal with an ideal PRN sequence correlation characteristic, the autocorrelation expression of the BCS signal with a shape vector $\boldsymbol{\kappa}$ can be written as

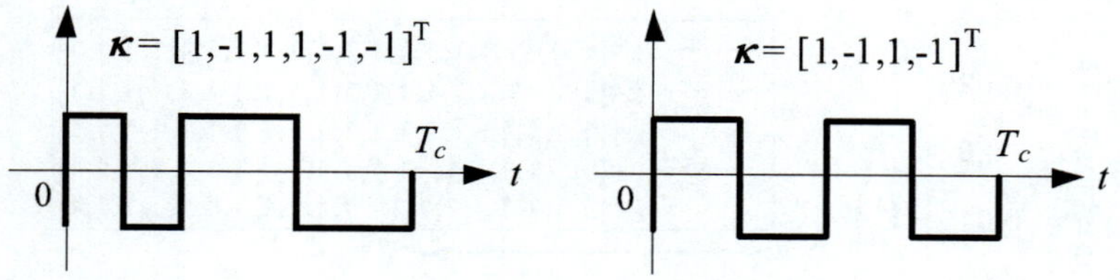

Fig. 4.16 BCS chip waveforms with different shape vector values

$$R_{\mathrm{BCS}}(\tau)=\begin{cases}\breve{r}_{\kappa}[b]\left(1-\dfrac{\epsilon}{T_s}\right)+\breve{r}_{\kappa}[b+1]\left(\dfrac{\epsilon}{T_s}\right), & \tau\in[0,T_c]\\ \breve{r}_{\kappa}[b-M]\left(1-\dfrac{\epsilon}{T_s}\right)+\breve{r}_{\kappa}[b-M+1]\left(\dfrac{\epsilon}{T_s}\right), & \tau\in[-T_c,0)\\ 0, & \text{otherwise}\end{cases}, \tag{4.28}$$

where $\tau = aT_c + bT_s + \epsilon$, a is an integer, $b = 0, 1, \ldots, M-1$, $\epsilon \in [0, T_s)$, and

$$\breve{r}_{\kappa}[b]\triangleq\begin{cases}\dfrac{1}{M}\displaystyle\sum_{i=0}^{M-1-b}\kappa_i\kappa_{b+i}, & 0\leqslant b\leqslant M-1\\ \dfrac{1}{M}\displaystyle\sum_{i=0}^{M-1+b}\kappa_{i-b}\kappa_i, & 1-M\leqslant b<0\\ 0, & |b|\geqslant M\end{cases} \tag{4.29}$$

is the aperiodic autocorrelation function of $\boldsymbol{\kappa}$.

When τ is an integer multiple of T_c/M, i.e. $\epsilon = 0$, we have

$$R_{\mathrm{BCS}}(kT_c/M)=\breve{r}_{\kappa}[k]\,. \tag{4.30}$$

As for general SCS modulated signals, $R_{\mathrm{BCS}}(\tau) = \breve{r}_{\kappa}[k]$ when $\tau = kT_c/M$ and is connected by line segments between each kT_c/M.

The PSD calculation formula for BCS signals is identical to that of the SCS modulated signal in Sect. 3.4.5.

Example 4.2 *(Multipath resistant capability of different BCS signals)* The multipath average range error envelope and the running error envelope of a series of BCS signals with an early-late space of 0.1 chip are shown in Fig. 4.17 and Fig. 4.18, respectively, of which the chip shape vectors are given by simple shifting the same initial chip shape vector

$$\boldsymbol{\kappa}=[-1,1,1,1,1,1,1,1,1,1,1,1,1,1,1,1,1,1,1]^{\mathrm{T}}\,. \tag{4.31}$$

The definition of the multipath error envelope can be found in Sect. 5.7. ■

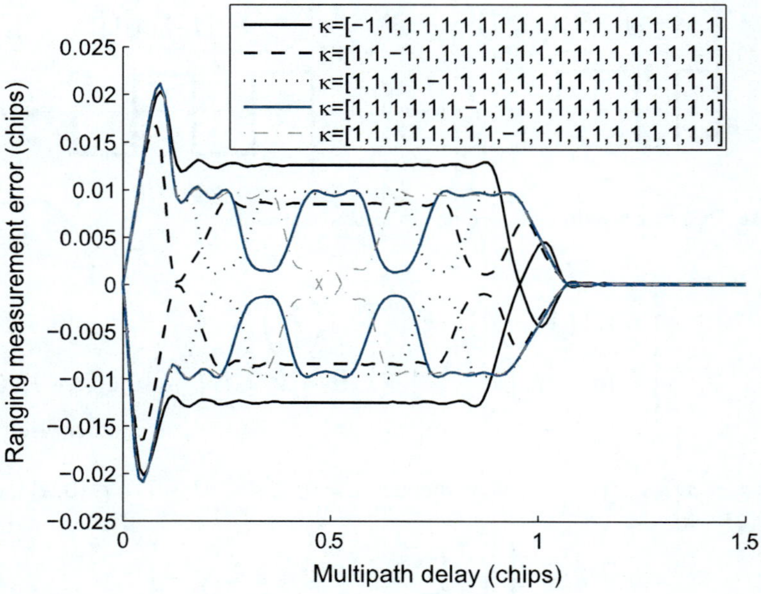

Fig. 4.17 Multipath error envelope of different BCS modulated signals

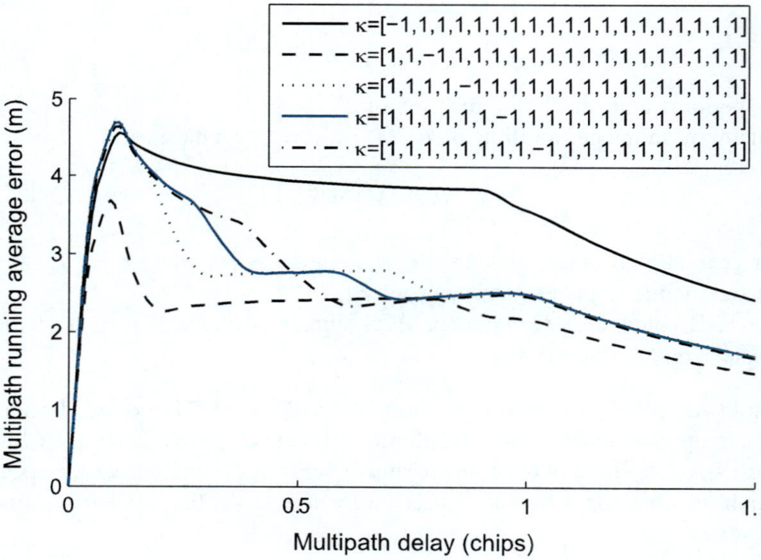

Fig. 4.18 Multipath running average error envelope for different BCS modulated signals

The results of Example 4.2 show that a simple shift in the chip shape vector causes a significant change in the final anti-multipath performance of the signal. This feature of the BCS signal allows for more flexibility in the adjustment of the signal in terms of its potential performance and spectrum planning. However, this also makes it almost impossible to qualitatively analyze the impact of the value of each BCS symbol on performance in general cases. Usually, the design of the BCS chip waveform is performed either by the trial-and-error method or by exhausting the value of each element of κ under certain constraints.

4.5 CBCS Modulation

4.5.1 *Definition of CBCS Signals*

The narrow definition of composite BCS (CBCS) modulation was given by Hein et al. in 2005 [9]. As the modulation selected for the interoperable signal of the Galileo system at the E1 band, the chip waveform $p_{\text{CBCS}}(t)$ of the CBCS signal is defined as the superposition of a BOC(1, 1) chip waveform and a high-order BCS chip waveform with the same spreading code rate, i.e.

$$p_{\text{CBCS}}(t) = \sqrt{1-\gamma}\, p_{\text{BOC}(1,1)}(t) \pm \sqrt{\gamma}\, p_{\text{BCS}(\kappa,1)}(t)\ , \tag{4.32}$$

where the coefficient γ is used to control the power ratio of the BOC(1, 1) component and the BCS component in the composite signal. According to the sign relationship between BOC(1, 1) and BCS(κ,1) components, whether plus or minus, the CBCS modulation can be divided into in-phase CBCS modulation and anti-phase CBCS modulation, which are respectively denoted as CBCS^{+} and CBCS^{-}. The above definition of CBCS stipulates that the CBCS signal must have the BOC(1,1) component, and its spreading code rate must be 1.023 MHz. This is because the joint interoperable signal design agreement of the GPS and Galileo programs stipulates that the interoperable signal must use the BOC(1,1) modulation as a baseline, i.e. the signal must be able to be received using a BOC(1,1) receiver.

In fact, BOC(1,1) is a type of BCS signal. If we leave aside the historical background of when CBCS was proposed and see it as a general spreading modulation technique, the generalized CBCS spreading chip waveform can be defined as the superposition of two arbitrary BCS chip waveforms with the same spreading code rate, i.e.

$$p_{\text{CBCS}}(t) = \sqrt{1-\gamma}\, p_{\text{BCS}(\kappa_1,n)}(t) \pm \sqrt{\gamma}\, p_{\text{BCS}(\kappa_2,n)}(t)\ . \tag{4.33}$$

Figure 4.19 is a schematic diagram of the generation of the CBCS chip waveform.

Readers may ask why we need the more complex CBCS signals since we already have the BCS signal. Even though the BCS modulation offers sufficient freedom for signal designers in terms of spectrum sharing and potential receiving performance

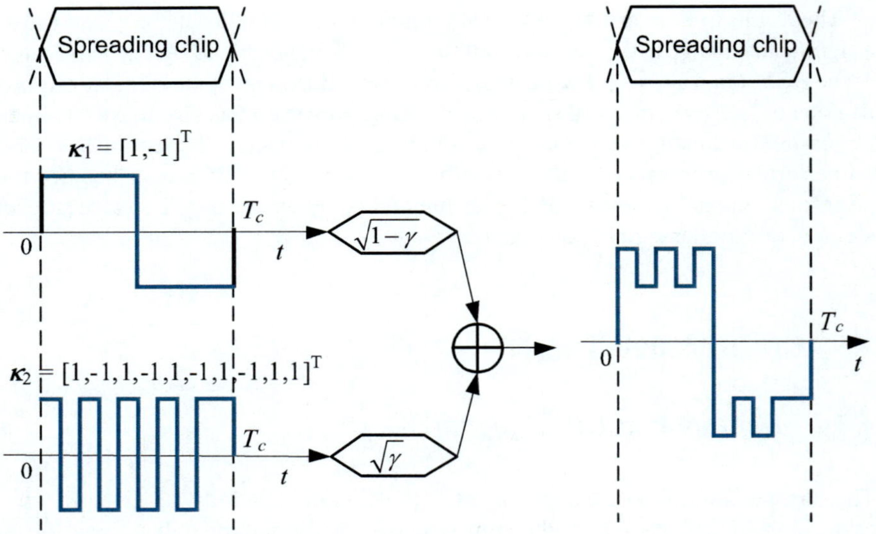

Fig. 4.19 Generation of the CBCS chip waveform

adjustment, it still has certain deficiencies in the diversity of receiving strategies, which is especially important for interoperable signals.

The so-called interoperable signal refers to the open service signal that each GNSS broadcasts in the same satellite navigation frequency band and is optimized for joint receiving. Interoperable signals are designed to allow the user equipment to simultaneously receive satellite signals from multiple GNSSs with the lowest possible cost, and to achieve better performance by integrating the information carried by these signals than any single signal. For example, the interoperability of the modernized civilian signal L1C of GPS III, the E1 OS signal of Galileo, and the B1C signal of BDS-3 in the L1/E1/B1 band was considered and incorporated in their initial design. They have the same center frequency, similar bandwidth, and similar receiving methods, which enables the receiver to share the RF front-end when receiving and processing these three signals simultaneously, and even share the baseband processing channel, thereby effectively reducing the implementation complexity of the receiver, to achieve the effect of the whole is greater than the sum of its parts.

Such good characteristics of interoperable signals make their wide range of applications promising in the future. It can be expected that interoperable signals of each satellite navigation system will be the basis of the major civil navigation and positioning services. Therefore, their design must meet a diversity of application requirements including low-end consumer electronics, middle-end vehicle navigation, and high-end precision measurement. The various application directions have different emphases on signal performance characteristics. For example, for users of high-precision surveying, the measurement accuracy of signals is the most important performance indicator, while processing complexity is of less significance. However, for users of consumer electronics where the navigation functions are embedded into mobile devices such as mobile phones, the complexity of processing is a major concern, but the ranging accuracy is not critical. Therefore, as pointed out in [16], the design of modulation schemes is more constrained for interoperable signals, as they should allow multiple receiving strategies at the beginning of signal design.

It is difficult for the BCS signal to meet the above requirements. Although its spectral shape and the ranging performance in thermal noise and multipath environments can be adjusted by changing the chip shape, the scope of potential receiving strategies is extremely limited. In theory, a wider signal spectrum obtained by chip shaping can provide better resistance for thermal noise and multipath, but it also requires the receiver to use a wide receiving bandwidth. Moreover, the complex chip waveform also significantly increases the implementation complexity of the receiver baseband channel.

CBCS modulation and subsequent spreading modulation methods such as MBOC modulation can be classified as composite subcarrier spreading modulation, i.e. multiple types of subcarriers are combined first, and then the spreading code is secondarily modulated. The composite subcarrier spreading modulations are frequently used for civil interoperable signals. This is not only because BOC(1,1) is set as the common baseline by GPS and Galileo in the joint design of L1/E1 interoperable signals but more importantly, this composite modulation has the inherent advantages of multiple processing modes.

The emergence of composite subcarrier spreading modulation provides more flexible choices for receiver processing strategies. When the signal satisfies a form in which a wideband component is superimposed on a narrowband component, a wideband matching receiving mode can be used to obtain the performance advantage of the entire wideband signal for receivers of high-end applications. For low-end applications, the receiver can use a narrow front-end filter that allows only the narrowband components to pass through, while the local de-spreading signal uses only the chip waveform of the corresponding narrowband component (e.g. the BOC_{11}-like processing mode discussed in Examples 5.1 and 5.3). Although there will inevitably be some performance degradation under such a narrow band mismatch receiving mode, this is still undoubtedly attractive for low-cost receivers that are sensitive to the cost and power consumption.

4.5.2 The Autocorrelation Function of the CBCS Signal

Similar to BCS modulation, CBCS modulation can also be seen as a special case of SCS modulation. The only difference is that the shape vector of the CBCS waveform is no longer limited to being bipolar. In fact, the CBCS chip waveform can take four values, $\kappa_i \in \left\{\pm\sqrt{1-\gamma} \pm \sqrt{\gamma}\right\}$. Therefore, the ACF of the CBCS modulation can still be directly obtained by the general ACF calculation formula of the SCS modulated signals given in Chap. 3.

In addition, the ACF of the CBCS signal can also be obtained by its definition (4.7) and the general calculation formula of the correlation function. According to (4.7), the autocorrelation functions of the CBCS$^+$ modulated signal and the CBCS$^-$ modulated signal are

$$R_+(\tau) = (1-\gamma) R_{\mathrm{BCS}(\kappa_1,n)}(\tau) + \gamma R_{\mathrm{BCS}(\kappa_2,n)}(\tau) + 2\sqrt{\gamma-\gamma^2} R_c(\tau), \quad (4.34)$$

$$R_-(\tau) = (1-\gamma) R_{\mathrm{BCS}(\kappa_1,n)}(\tau) + \gamma R_{\mathrm{BCS}(\kappa_2,n)}(\tau) - 2\sqrt{\gamma-\gamma^2} R_c(\tau), \quad (4.35)$$

respectively, where $R_c(\tau)$ is the CCF of the BCS(κ_1, n) signal and the BCS(κ_2, n) signal. Since the cross-correlation terms of the BCS(κ_1, n) and BCS(κ_2, n) signals are included in these two equations, the ACFs of the in-phase and anti-phase CBCS signals are not the same, which leads to some differences in their performance. In general, the anti-phase CBCS has better thermal noise tracking and multipath resistance, while the in-phase CBCS signal has less received power loss under band-limited conditions, and thus has better demodulation performance. A more detailed discussion of this phenomenon is in the section of CBOC.

4.5.3 PSD of CBCS Signals

Due to the existence of cross-correlation terms, the PSD of the CBCS signal is not simply the weighted superposition of the PSDs of the two BCS signals.

$$(1-\gamma) G_{\mathrm{BCS}(\kappa_1,n)}(f) + \gamma G_{\mathrm{BCS}(\kappa_2,n)}(f). \quad (4.36)$$

In fact, we can obtain the PSD of the in-phase and anti-phase CBCS signals by conducting a Fourier transform of the ACF (4.34) and (4.35), which are

$$G_{\mathrm{CBCS}+}(f) = (1-\gamma) G_{\mathrm{BCS}(\kappa_1,n)}(f) + \gamma G_{\mathrm{BCS}(\kappa_2,n)}(f) + 2\sqrt{\gamma-\gamma^2} G_c(f), \quad (4.37)$$

$$G_{\mathrm{CBCS}-}(f) = (1-\gamma) G_{\mathrm{BCS}(\kappa_1,n)}(f) + \gamma G_{\mathrm{BCS}(\kappa_2,n)}(f) - 2\sqrt{\gamma-\gamma^2} G_c(f), \quad (4.38)$$

respectively, where $G_c(f)$ is the cross-spectrum between the modulated components BCS(κ_1, n) and BCS(κ_2, n). The definition of the cross-spectrum can be found in Sect. 3.4.6. If the total PSD satisfies (4.36), it is apparent that the two CBCS modulated spreading chip waveforms must have the same power and appear in pairs such that the cross-correlation terms cancel each other out. For example, we can synthesize the following composite signals

$$g_{\mathrm{CBCS}}(t) = \sum_{n=-\infty}^{+\infty} \left[(-1)^{c_n} p_{\mathrm{CBCS}+}(t-nT_c) + (-1)^{c'_n} p_{\mathrm{CBCS}-}(t-nT_c)\right] \quad (4.39)$$

or a signal that interleaves in-phase and anti-phase CBCS spreading chips, i.e.

$$\Pr\{p(t) = p_{\mathrm{CBCS}+}(t)\} = \Pr\{p(t) = p_{\mathrm{CBCS}-}(t)\} = \frac{1}{2}. \quad (4.40)$$

Both of these measures can free the PSD of the total signal from cross-spectrum components. The latter implementation is also referred to as alternative BCS (ABCS).

Example 4.3 *(Tracking performance of CBCS signals)* In this example, we consider CBCS([1, −1, 1, −1, 1, −1, 1, −1, 1, 1], 1, 20%) modulation. This signal was once planned as the modulation scheme of the Galileo E1 OS signal. Its waveform can be seen as the superposition of BOC(1,1) and BCS(κ_1, 1), where $\kappa = [1, -1, 1, -1, 1, -1, 1, -1, 1, 1]^{\mathrm{T}}$, and $\gamma = 1/5$.

Figures 4.20 and 4.21 show the multipath error envelope and the running average error envelope of this signal under a receiving bandwidth of 12 MHz, where the multipath direct amplitude ratio is −5 dB, and the early-late correlation space is 0.1 chip. For comparison, the performance curve of the BOC(1, 1) signal is also plotted on same figures.

Although the high-frequency component of BCS accounts for only 1/5 of the total signal power, the anti-multipath performance of the CBCS signal is significantly higher than that of the BOC(1,1) signal. When the running average error envelope is used as a measure indicator, the improvement can reach 40% and above.

The high frequency components in the CBCS signal are also effective at improving the tracking accuracy under thermal noise. Figure 4.22 shows the lower bound of the

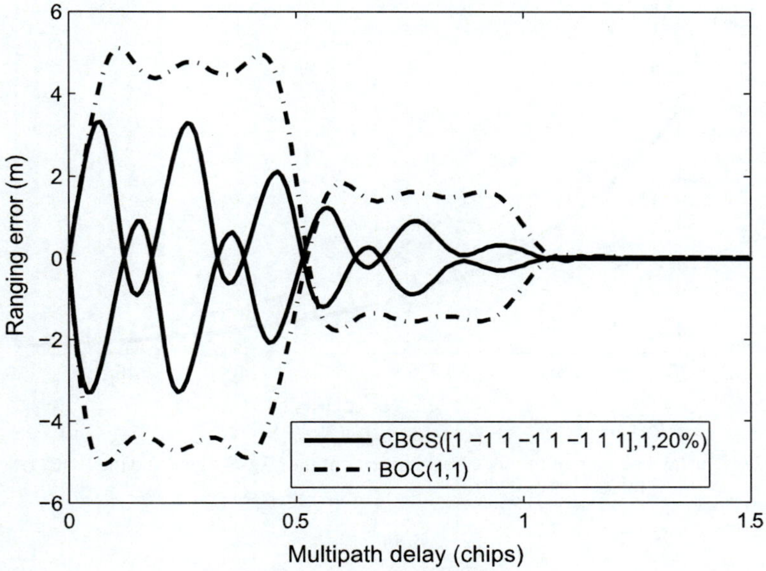

Fig. 4.20 Multipath error envelopes of CBCS signal and BOC(1,1) signal

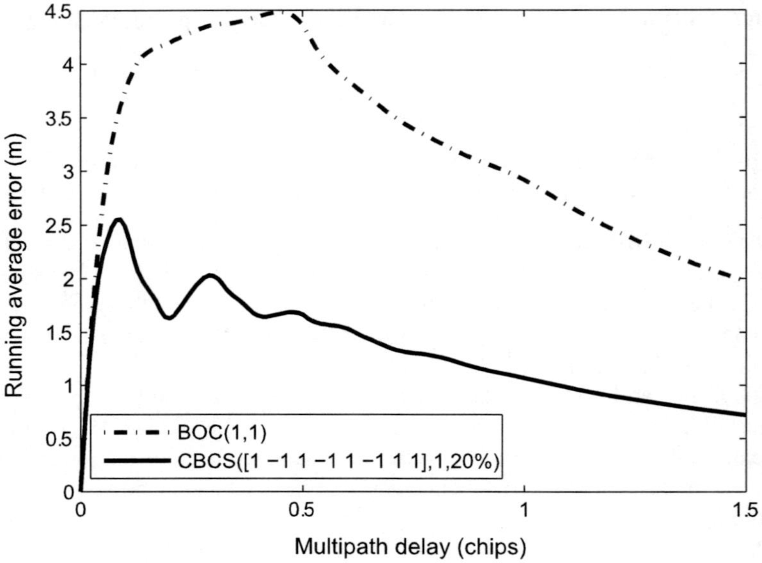

Fig. 4.21 Multipath running average error envelopes of the CBCS signal and the BOC(1,1) signal

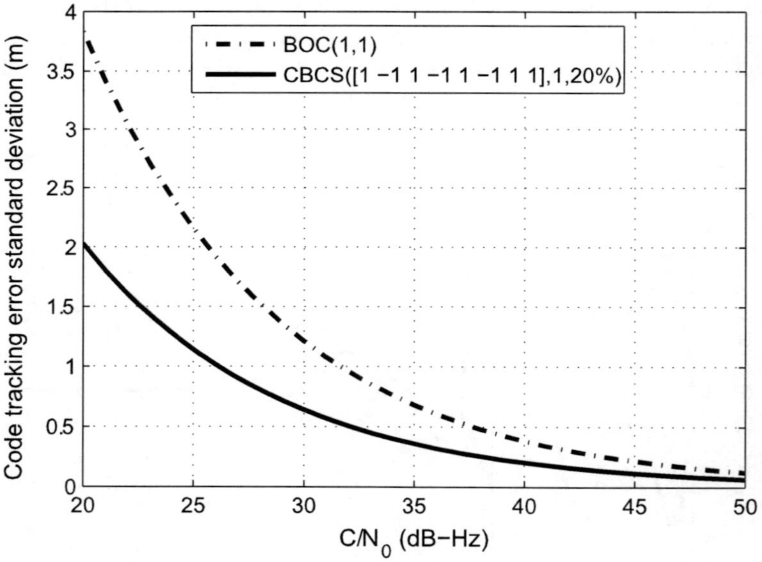

Fig. 4.22 The lower bound of the code tracking error of the CBCS signal and the BOC(1,1) signal

code tracking error for BOC(1,1) and the above CBCS signal in an additive white Gaussian noise (AWGN) environment. The receiver's front-end bandwidth is 12 MHz. The detailed evaluation method used in this analysis is given in Sect. 5.3. It can be seen that a few high-frequency components in the signal waveform can significantly improve the potential ranging accuracy of the signal. The mechanism behind this phenomenon is also elaborated in Sect. 5.3. ■

4.5.4 Cross-Correlation Deviation of CBCS Signals

As mentioned in Sect. 4.5.1, CBCS signals have multiple receiving strategies, some of which achieve optimum performance, while others reduce the processing complexity for the receiver.

The most straightforward processing method is matched receiving. That is to say, the receiver directly generates a CBCS modulated signal replica, and then a correlation operation is performed with the received signal. According to the matched filtering theory [17], when the local signal waveform is the same as the input signal waveform, the correlator can obtain the highest output SNR. However, the matched receiving method requires the receiver to generate a CBCS signal with multiple levels, and the operation in the correlator also uses multi-bit multiplication. In addition, the bandwidth of the receiver must be wide enough to receive the wideband BCS component in the CBCS signal.

If the above receiving method is too complicated for some complexity-sensitive applications, one can use another simpler reception strategy to process only the narrowband BCS component in the received signal. The receiver locally generates a BCS(κ_1,1) signal and performs a correlation operation with the received signal. In this way, the local signal can be quantized by using only 1 bit, which greatly reduces the complexity of the correlation operation, and the sampling rate can be appropriately reduced due to the narrower signal bandwidth. Furthermore, for interoperable signals using a certain narrowband BCS component as the common baseline component, even if the superimposed wideband BCS components are different, the multi-system receiver can still regard these signals as narrowband BCS signals and use the same channel structure for processing. The mismatch between this local waveform and the received waveform greatly simplifies the structure of the receiver, despite causing a certain degree of power loss.

Yet another reception processing strategy is to only receive the wideband BCS components in the signal. Although the existing CBCS signal is not designed with this application in mind, the power of the wideband component accounts for a small proportion of the total signal power, which makes the received power loss larger, but this does not influence it to be a reasonable application. If the proportion of the wideband BCS component in the signal can be appropriately increased, receiving only the wideband component can improve the multipath resistance for a wideband receiver.

Here, we can see that the composite subcarrier spreading modulation represented by CBCS modulation offers great flexibility to designers of the receiver. They can choose the appropriate receiving processing method according to the needs of the application group and have a flexible trade-off between performance and complexity.

When a receiver only processes a certain BCS component in a CBCS signal, without loss of generality, we assume that the local replica is a BCS(κ_1,1) signal. The CCF between the received signal and the local replica signal can be written as

$$R_{\mathrm{mis}}(\tau) = \sqrt{1-\gamma}\, R_{\mathrm{BCS}(\kappa_1,n)}(\tau) \pm \sqrt{\gamma}\, R_c(\tau) \,, \tag{4.41}$$

where $R_{\mathrm{BCS}(\kappa_1,n)}(\tau)$ is the ACF of the BCS(κ_1,1) signal, and $R_c(\tau)$ is the CCF between the signal components BCS(κ_1,1) and BCS(κ_2,1). Since the chip waveform of BCS is a real function, the first term in the above equation is clearly evenly symmetric, but $R_c(\tau)$ is not necessarily evenly symmetric. Thus, for general values of κ_1 and κ_2, when the CBCS signal is received in a single-component mismatch manner, there may be an asymmetric correlation function, i.e. $R_{\mathrm{mis}}(\Delta) \neq R_{\mathrm{mis}}(-\Delta)$.

For a signal with an asymmetric correlation function, when the receiver uses the early-minus-late (E-L) discriminator to track it and when the actual code phase deviation is zero, the output value of the discriminator

$$\delta(\varepsilon)|_{\varepsilon=0} = K_d \left[R_{\mathrm{mis}}\left(\frac{\Delta}{2}\right) - R_{\mathrm{mis}}\left(-\frac{\Delta}{2}\right) \right] \tag{4.42}$$

is non-zero. This phenomenon may cause an inherent deviation in code tracking, which leads to a large deviation in the final positioning result. As the signal component power ratio γ differs, the value of this inherent deviation is also different.

In theory, for a given γ, the receiver design can pre-calculate the magnitude of this inherent deviation and correct it in the pseudorange measurement generation, but this solution lacks operability in practice. For the non-ideal correlation characteristic PRN code used in practice, (3.19) shows that the magnitude of the offset value is also related to the actual correlation characteristics of different PRN sequences. Different correlation E-L spacing, and different filter characteristics will also change this deviation value. Therefore, it is impossible for the manufacturer of the receiver to measure and correct this inherent deviation for each situation and for each satellite's signal. A remedy is that the in-phase and anti-phase CBCS signals are processed simultaneously in the receiver, and their correlation values are coherently combined to cancel out the cross-correlation terms in (4.41) [18]. However, in the design of the signal, this essential flaw should be avoided as much as possible in every case.

In fact, the above defects can be avoided in the design of CBCS modulation. It is only necessary to design the shape vectors of the two BCS chip waveforms so that they have a symmetric CCF. At this time, the value of the element in each BCS chip shape vector must satisfy a certain constraint relationship.

From the previous analysis of the SCS modulated signal's CCF, it can be seen that the requirement for the symmetry of the CBCS CCF is equivalent to the requirement that the cross correlation value of the signal at $\tau = \pm kT_s$ must be equal, i.e. $\check{r}_{\kappa\kappa'}[b] =$

$\check{r}_{\kappa\kappa'}[-b]$ for all $b = 0, 1, \ldots, M-1$. Combined with (3.33), this constraint can be rewritten as

$$\sum_{i=0}^{M-1-b} \kappa_i \kappa'_{i+b} - \sum_{i=b}^{M-1} \kappa_i \kappa'_{i-b} = 0\,, \tag{4.43}$$

where κ_i and κ'_i are the elements of $\boldsymbol{\kappa}_1$ and $\boldsymbol{\kappa}_2$ respectively.

Shape vectors $\boldsymbol{\kappa}_1$ and $\boldsymbol{\kappa}_2$ that satisfy (4.43) can be found by traversal search. In the actual design, in addition to the symmetry of the CCF, there are other constraints on the shape of the correlation function. If we take all of these constraints into consideration at the same time, the actual search space will be greatly reduced. For example, in order to minimize the radio frequency interference of the CBCS signal to other signals at the carrier center frequency, it is desirable to have $G_{\mathrm{CBCS}}(0) = 0$, i.e. the BCS shape vector should be balanced:

$$\sum_{i=0}^{M-1} \kappa_i = 0\,. \tag{4.44}$$

Since κ_i can only be ± 1, only the even-order BCS signal can meet this balance requirement.

Further, if the shape vector $\boldsymbol{\kappa}$ of one of the two BCS components of the synthesized CBCS signal is anti-symmetric, i.e. $\kappa_i = -\kappa_{M-1-i}$ for all $i = 0, 1, \ldots, M-1$, then the symmetry constraint of the CCF (4.43) can be reorganized as

$$\sum_{i=0}^{M-1} \kappa_i \left(\kappa'_{i+b} - \kappa'_{i-b}\right) = 0\,. \tag{4.45}$$

It can be easily proven that a sufficient but not necessary condition to satisfy the above formula is that $\boldsymbol{\kappa}'$ is also anti-symmetric. If the constraint $R_c(0) = 0$ is added, a necessary but not sufficient condition is that M must be a multiple of four.

For CBCS signals with BOC(1,1) component as the baseline, since the shape vector of BOC(1,1) signal is already anti-symmetric, we can restrict the search space of the other BCS shape vector to be anti-symmetric and balance vectors where M is a multiple of four, and verify whether the constraint

$$R_c(0) = \sum_{i=0}^{M/2-1} \left(\kappa'_i - \kappa'_{i+M/2}\right) = 0 \tag{4.46}$$

are met.

When $M = 4$, the only BCS shape vector that meets the requirements is $\boldsymbol{\kappa} = [1, -1, 1, -1]^{\mathrm{T}}$ which corresponds to BOC(2,1). When $M = 8$, there are three shape vectors that meet the requirements, which are $\boldsymbol{\kappa} = [1, 1, -1, -1, 1, 1, -1, -1]^{\mathrm{T}}$, $\boldsymbol{\kappa} = [1, -1, -1, 1, -1, 1, 1, -1]^{\mathrm{T}}$, and $\boldsymbol{\kappa} = [1, -1, 1, -1, 1, -1, 1, -1]^{\mathrm{T}}$, where the last one corresponds to BOC(4,1). When $M = 12$, there are six κ that meet the requirements, which are respectively

$$\begin{aligned}
\kappa_1 &= [1, 1, 1, -1, -1, -1, 1, 1, 1, -1, -1, -1]^{\mathrm{T}} , \\
\kappa_2 &= [1, 1, -1, 1, -1, -1, 1, 1, -1, 1, -1, -1]^{\mathrm{T}} , \\
\kappa_3 &= [1, 1, -1, -1, 1, -1, 1, -1, 1, 1, -1, -1]^{\mathrm{T}} , \\
\kappa_4 &= [1, 1, -1, -1, -1, 1, -1, 1, 1, 1, -1, -1]^{\mathrm{T}} , \\
\kappa_5 &= [1, -1, 1, -1, 1, -1, 1, -1, 1, -1, 1, -1]^{\mathrm{T}} , \\
\kappa_6 &= [1, -1, 1, -1, -1, 1, -1, 1, 1, -1, 1, -1]^{\mathrm{T}} .
\end{aligned}$$

All of the above κ can ensure that the tracking is unbiased when the receiver only processes the BOC(1,1) component in the CBCS signal. By further evaluating the multipath and thermal noise performance of these BCS components and that of the CBCS signal, it emerges that the overall performance of κ_5 is the best, which corresponds to BOC(6,1). The CBCS signal at this time corresponds exactly to the CBOC(6,1) signal (see Sect. 4.7.2).

Finally, since most of the existing CEM techniques are for bipolar signals, the non-bipolar amplitude of the CBCS baseband signal poses some challenges to the realization of CEM of such signals on the transmitter. We will discuss this issue further in Chaps. 6 and 7.

4.6 TMBCS and QMBCS Modulations

The combination of the wideband and narrowband BCS components is not limited to amplitude superposition. In fact, the two BCS components can be combined in any way. In addition to CBCS, typical combining methods of BCS components include time division multiplexing and phase quadrature multiplexing, which are time-multiplexed BCS (TMBCS) modulation and quadrature multiplexed BCS (QMBCS) modulation respectively. They can all be considered as forms of composite subcarrier spreading modulation. The TMBCS signal combines two different BCS components in a time division multiplexing manner, and only one of the components is broadcast at a certain time. QMBCS places the two BCS components on two carrier phases with 90 degrees of difference. There will be no further discussion of these two modulations here. Instead we will be focus on one of their subsets, namely multiplexed BOC (MBOC).

4.7 MBOC Modulations

Strictly speaking, MBOC modulation is not a specific modulation method, but a constraint on the spectral shape of the spread spectrum signal. Alternatively, it can

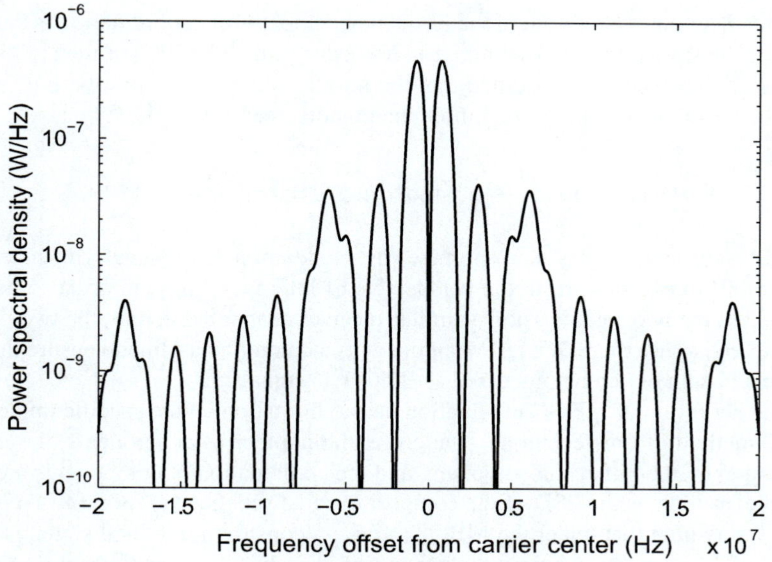

Fig. 4.23 Spectrum of MBOC(6, 1, 1/11) modulated signal

also be considered as a general term for a type of modulation method with a certain PSD. The definition of MBOC modulation only limits the power spectrum shape of the signal. A general MBOC modulated signal can be denoted as MBOC(m, n, γ), and its normalized PSD without bandlimited filtering can be expressed as

$$G_{\mathrm{MBOC}(m,n,\gamma)}(f) = (1-\gamma)\, G_{\mathrm{BOC}(n,n)}(f) + \gamma G_{\mathrm{BOC}(m,n)}(f) \,. \tag{4.47}$$

That is to say, the MBOC modulated signal is obtained by combining a wideband BOC(m, n) signal component and the narrowband BOC(n, n) signal component with a power ratio of $\gamma : (1-\gamma)$. From this point of view, MBOC modulation can be regarded either as a spreading modulation technique or a multiplexing technique of two signal components. Since the definition of MBOC modulation has no limitation on the time domain implementation, the BOC(m, n) component and the BOC(n, n) component can be combined in any multiplexing manner, as long as the PSD of the final composite signal satisfies the constraints of (4.47). Figure 4.23 shows the spectral shape of the MBOC(6,1,1/11) modulated signal in the 20 MHz range around the center frequency.

The concept of MBOC modulation was originally proposed in conjunction with the joint design of the GPS L1C and Galileo E1 OS signals. In that historical context, the USA and the EU coordinated the compatibility and interoperability between the GPS and Galileo systems. One of the most important agreements was to determine the common design constraints of the L1C signal and the E1 OS signal. Since the frequency compatibility coordination precedes the specific modulation mode design, and the main input condition of the frequency compatibility coordination is the total PSD of the signal, the USA and the EU first constrained the PSD of the civil signal on the L1/E1 band. It is specified that the normalized PSD of both the L1C signal and the E1 OS signal under the infinite bandwidth is MBOC(6,1,1/11), i.e.

$$G_{\mathrm{MBOC}(6,1,1/11)} = \frac{10}{11} G_{\mathrm{BOC}(1,1)}(f) + \frac{1}{11} G_{\mathrm{BOC}(6,1)}(f) \,. \tag{4.48}$$

In this way, both parties can optimize the modulation chip waveform under this constraint. In addition, from the perspective of interoperability, in order to be able to process the two signals using a similar receiver channel structure, the two parties also agreed to use the BOC(1,1) component as a common baseline to ensure that the optimized signals can be processed as BOC(1,1) signals.

The definition of MBOC modulation has no limitation on the specific implementation method of this technique. Therefore, the number of components, the implementation method of each component, and the combination of components are very flexible, as long as the PSD of the composite signal satisfies the form of (4.47).

If we assume that under the MBOC(m, n, γ) constraint, the total signal is composed of M channels, wherein the proportion that the ith channel accounts for the total signal power is q_i, and the proportion that BOC(m, n) signal component in the ith channel is η_i. Then, in order to ensure that the total signal satisfies the MBOC(m, n, γ) constraint, q_i, η_i, and γ must satisfy the following relationship

$$\sum_{i=1}^{M} q_i \eta_i = \gamma \,. \tag{4.49}$$

For example, the GPS L1C signal consists of a data channel and a pilot channel with $M = 2$. All BOC(6, 1) components are placed on the pilot channel, while the data channel uses only the BOC(1,1) component. The power ratio of the pilot channel and data channel is $3:1$. According to the above equation, if we want the proportion of the BOC(6, 1) component in the total signal to satisfy the constraint of $\gamma = 1/11$, the power ratio of the BOC(6, 1) component in the pilot channel should be 4/33. If the other conditions are unchanged, the pilot channel and data channel power ratio become $1:1$, then the power of the BOC(6,1) component in the pilot channel should be correspondingly changed to $2/11$.

In addition to the variety of the mixture ratio of BOC(m, n) and BOC(n, n) in different channels, the time domain implementation of the existing MBOC modulation is also not unique. The modulation used by the modernized civil signal GPS L1C at the L1 frequency band is called TMBOC modulation, while the Galileo E1 OS signal uses CBOC modulation. The BDS-3 civil signal B1C uses QMBOC modulation. Although these three modulation implementations have the same total PSD, the modulated signals are distinctive in actual transmission and reception because of their different implementation ways.

4.7.1 TMBOC Modulation

The BOC(n, n) and BOC(m, n) components of the TMBOC modulation are combined in a time-multiplexed manner. Only one of the components is broadcast at one time slot, but the two components use the same spreading sequence. Consider TMBOC(6, 1, 4/33) used for the pilot channel of the GPS L1C signal as an example, where the BOC(6, 1) subcarrier is used for the 1st, 5th, 7th, and 30th positions of every 33 spread spectrum symbols while the remainder use the BOC(1,1) subcarrier. Its chip waveform is shown in Fig. 4.24.

The time multiplexing of TMBOC usually follows fixed rules, but it can be mathematically approximated that in each chip, the signal takes the BOC(n, n) chip with a certain probability of $\mathcal{P}_1$, and takes BOC(m, n) with a probability of $\mathcal{P}_2$. That is, with the model of (3.2), the baseband spread spectrum signal modulated by TMBOC(m, n, γ) can be written as

$$g(t) = \sum_{n=-\infty}^{+\infty} (-1)^{c_n} p(t - nT_c) , \tag{4.50}$$

where

$$p(t) \in \left\{ p_{\mathrm{BOC}(n,n)}(t) , p_{\mathrm{BOC}(m,n)}(t) \right\} \tag{4.51}$$

and

$$\begin{cases} \Pr\left\{ p(t) = p_{\mathrm{BOC}(n,n)}(t) \right\} = 1 - \gamma , \\ \Pr\left\{ p(t) = p_{\mathrm{BOC}(m,n)}(t) \right\} = \gamma . \end{cases} \tag{4.52}$$

Since the time slots of the two components are orthogonal, from (3.23) we can obtain the ACF of the TMBOC signal as

$$R_{\mathrm{TMBOC}}(\tau) = (1 - \gamma) R_{\mathrm{BOC}(n,n)}(\tau) + \gamma R_{\mathrm{BOC}(m,n)}(\tau) . \tag{4.53}$$

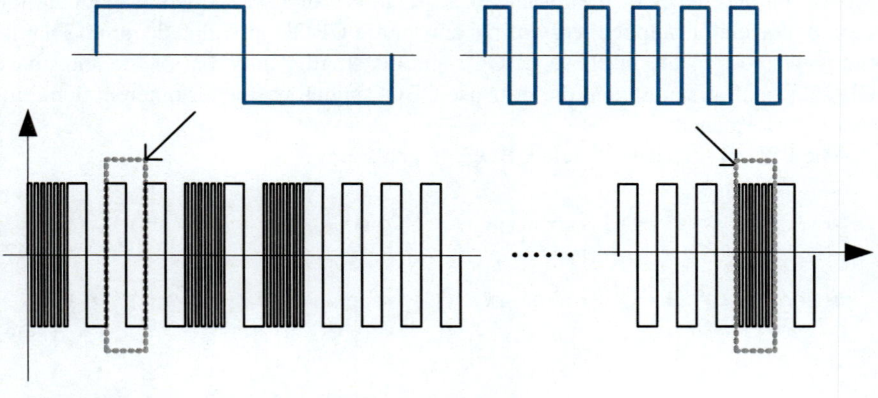

Fig. 4.24 Time domain waveform of TMBOC(6, 1, 4/33)

4.7.2 CBOC Modulation

CBOC modulation is a special case of CBCS modulation. For the CBOC signal, the BOC(n, n) component and the BOC(m, n) component always appear simultaneously, and are directly combined by amplitude addition or subtraction with different weights. The weights of the two components determine their power ratio in the composited spectrum.

A CBOC modulated spreading chip waveform can be expressed as

$$p_{\mathrm{CBOC}}(t) = \sqrt{1-\gamma}\, p_{\mathrm{BOC}(n,n)}(t) \pm \sqrt{\gamma}\, p_{\mathrm{BOC}(m,n)}(t) \ . \tag{4.54}$$

The CBOC modulation can be further divided into in-phase CBOC modulation and anti-phase CBOC modulation, which are denoted as CBOC$^+$ and CBOC$^-$ respectively, depending on whether the BOC(n, n) modulation component and the BOC(m, n) modulation component are added or subtracted during superposition. The ACF of the CBOC$^+$ and CBOC$^-$ modulated signals can be obtained using (4.7), which are respectively

$$R_+(\tau) = (1-\gamma)\, R_{\mathrm{BOC}(n,n)}(\tau) + \gamma R_{\mathrm{BOC}(m,n)}(\tau) + 2\sqrt{\gamma-\gamma^2} R_c(\tau) \ , \tag{4.55}$$

$$R_-(\tau) = (1-\gamma)\, R_{\mathrm{BOC}(n,n)}(\tau) + \gamma R_{\mathrm{BOC}(m,n)}(\tau) - 2\sqrt{\gamma-\gamma^2} R_c(\tau) \ , \tag{4.56}$$

where $R_c(\tau)$ is the cross-correlation term between the BOC(n, n) and the BOC(m, n) component.

In cases of infinite bandwidth, the cross-correlation function $R_c(\tau)$ of the BOC(n, n) component and the BOC(m, n) component can be calculated using (3.34) or (3.35). As shown in Fig. 4.25, the corresponding $R_c(\tau)$ is zero when the chips of these two components are fully aligned, which means that in the receiver processing channel, the energy of the prompt correlator output is lossless. However, when the signal is bandlimited, this feature no longer holds. The value of the CCF near $\tau = 0$ gradually increases as bandwidth decreases, which raises the prompt correlator output energy of the in-phase CBOC modulated signal, while reducing the prompt correlator output energy of the anti-phase CBOC modulated signal. That is, the power loss of the in-phase CBOC signal is smaller than that of the anti-phase CBOC signal, meaning that the in-phase CBOC signal has better inherent demodulation performance.

The PSD of the CBOC$^+$ and CBOC$^-$ signals are

$$G_{\mathrm{CBOC}+}(f) = (1-\gamma)\, G_{\mathrm{BOC}(n,n)}(f) + \gamma G_{\mathrm{BOC}(m,n)}(f) + 2\sqrt{\gamma-\gamma^2} G_c(f) \ , \tag{4.57}$$

$$G_{\mathrm{CBOC}-}(f) = (1-\gamma)\, G_{\mathrm{BOC}(n,n)}(f) + \gamma G_{\mathrm{BOC}(m,n)}(f) - 2\sqrt{\gamma-\gamma^2} G_c(f) \ , \tag{4.58}$$

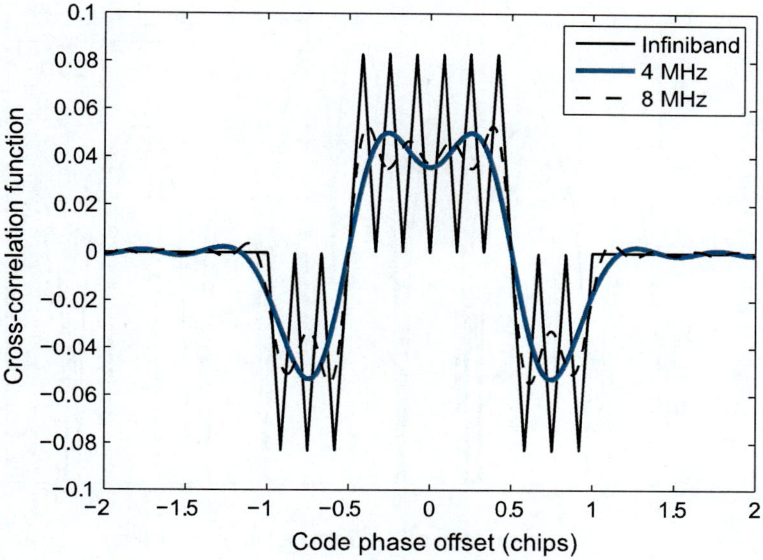

Fig. 4.25 Cross-correlation function between the BOC(1,1) component and BOC(6, 1) component

respectively, where $G_c(f)$ is the cross-spectrum between the BOC(n, n) and the BOC(m, n) modulated components. We can see that the spectrum of each single CBOC component does not satisfy the MBOC definition due to the existence of cross-correlation terms. To ensure that the total PSD meets the definition of (4.47), the modulated signals of the two CBOCs must have the same power and appear in pairs so that the cross-correlation terms cancel each other out.

Figure 4.26 shows the PSDs of CBOC$^+$ and CBOC$^-$ signals. It can be seen that the CBOC$^-$ signal has more high-frequency components, which means that CBOC$^-$ modulation has a higher potential ranging accuracy than the CBOC$^+$ modulation, and so it is more suitable for ranging.

Given such characteristics, in the Galileo system, CBOC$^+$ modulation is allocated on the data channel for better demodulation performance, while CBOC$^-$ modulation is allocated on the pilot channel to ensure better ranging and anti-multipath performance.

In order to ensure that the total PSD satisfies the MBOC definition, it is only when the two CBOC modulated signals have the same power and appear in pairs in the total transmitted signal that the cross-correlation terms can cancel each other out in the total composited signal. In the design of the time domain implementation of the signal, the limitation of the spectrum is just one of many factors to consider. In order to multiplex various signals and use a single carrier to broadcast them, it is also required that each component of MBOC signal can be multiplexed with other signals at the same frequency point in a constant envelope form. At present, the

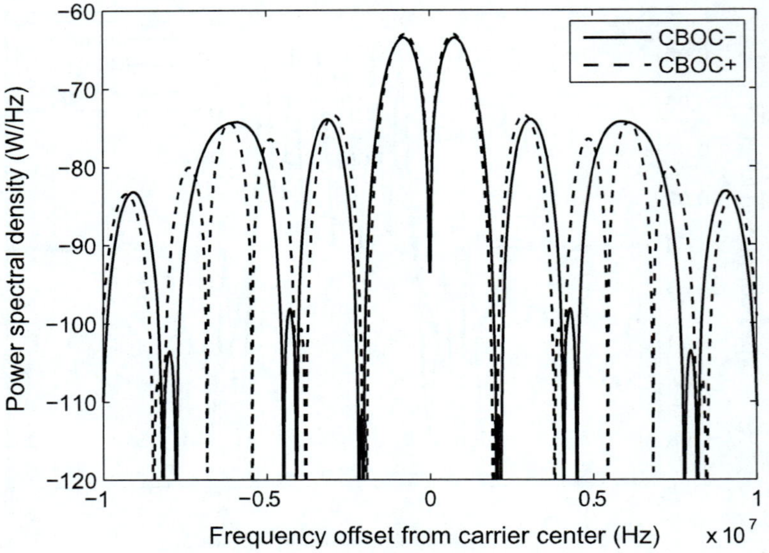

Fig. 4.26 Power spectral density of in-phase CBOC modulation and anti-phase CBOC modulation

multiplexing methods used by major GNSSs are mostly for bipolar signals. Since the CBOC signal has multi-level amplitude, the CBOC^{+} and CBOC^{-} components must be equal in power and must be superimposed in-phase on the carrier so that the BOC(n, n) and BOC(m, n) components alternate in different time slots in the composite signal in order to use a constant-envelope combination technique called quadrature product subcarrier modulation (QPSM) to multiplex it with other bipolar signals (see Sect. 7.5.3 for more detailed discussion). Such restrictions make the use of CBOC modulation less flexible.

As analyzed in Sect. 2.10.2, since ranging accuracy and tracking robustness are key to system performance in satellite navigation systems, more power should be allocated to the pilot channels whenever possible. However, CBOC modulation cannot achieve this due to the spectrum constraints and the limitations of multiplexing. Moreover, unlike BPSK-R, BOC, and TMBOC modulations, the time domain waveform of the CBOC modulated signal is not bipolar but has multiple amplitude levels, which increases the complexity of both the transmitting and receiving devices. The signal amplitude at each moment cannot be represented by a single bit but requires multiple bits. When a logic circuit is used to correlate such signals in a receiver, multi-bit multiplication and accumulation operations greatly increase the implementation complexity of the receiver. Even if the two components are correlated separately and then recombined, the number of correlators is doubled.

4.7.3 QMBOC Modulation

Quadrature multiplexed BOC (QMBOC) modulation is another time domain implementation of MBOC in addition to CBOC and TMBOC modulation [19]. It has been employed by the BDS-3 B1C signal. In QMBOC modulation, the BOC(n, n) and BOC(m, n) components are synthesized differently from the time division multiplexing used by TMBOC and the time domain superposition adopted by CBOC. Instead, the two components are separately modulated on the two quadratured phases of the carrier.

4.7.3.1 Signal Definition

The baseband signal of a QMBOC(m, n, γ) signal can be expressed as

$$p_{\mathrm{QMBOC}}(t) = \sqrt{1-\gamma}\, p_{\mathrm{BOC}(n,n)}(t) \pm \mathrm{j}\sqrt{\gamma}\, p_{\mathrm{BOC}(m,n)}(t) \ . \tag{4.59}$$

In (4.59), according to the positive or negative sign between the two terms, the corresponding QMBOC signals are referred to as in-phase QMBOC and anti-phase QMBOC and are denoted as QMBOC$^+$ and QMBOC$^-$ respectively.

Although the definition of the QMBOC modulated spreading chip (4.59) shares some similarities in form with the definition of CBOC, their characteristics are actually quite different. It is easy to verify that regardless of whether QMBOC is in-phase or anti-phase, the ACF of the QMBOC signal is always

$$R_{\mathrm{QMBOC}}(\tau) = (1-\gamma)\, R_{\mathrm{BOC}(n,n)}(\tau) + \gamma\, R_{\mathrm{BOC}(m,n)}(\tau) \ . \tag{4.60}$$

Since the BOC(n, n) and the BOC(m, n) components are located on two quadratured phases of the carrier, the cross-correlation term does not appear in the ACF of the QMBOC signal. We can see that the PSD of QMBOC not only satisfies the MBOC definition (4.47), but also surpasses the limitation that the in-phase and anti-phase signals must be equal in power and appear at the same time. Thus, for a signal using QMBOC modulation, if it has both data and pilot channels at the same time, the two channels can be allocated different powers as needed. The power ratio of the BOC(n, n) and the BOC(m, n) components of QMBOC modulation in each of the channels may also be different as long as the total power satisfies the MBOC definition (4.47). Therefore, it is concluded that the implementation of QMBOC modulation is more flexible than CBOC modulation.

Considering a data channel as an example, without loss of generality, and assuming that QMBOC$^+$ is used, the baseband signal of QMBOC can be written as

$$\begin{aligned} s(t) &= c_d(t)\, D(t)\, s_{\mathrm{QMBOC}}(t) \\ &= \sqrt{1-\gamma}\, c_d(t)\, D(t)\, s_{\mathrm{BOC}(n,n)}(t) + \mathrm{j}\sqrt{\gamma}\, c_d(t)\, D(t)\, s_{\mathrm{BOC}(m,n)}(t) \ , \end{aligned} \tag{4.61}$$

where $c_d(t)$ is the spreading code of the data channel and $D(t)$ is the modulated data message.

Since the wideband and narrowband signal components are orthogonal, another configuration option for QMBOC modulation is to separately modulate different messages on the BOC(n, n) component and the BOC(m, n) component. At this point, the baseband signal of the data channel can be expressed as

$$s'(t) = \sqrt{1-\gamma}c_d(t)D_1(t)s_{\mathrm{BOC}(n,n)}(t) + \mathrm{j}\sqrt{\gamma}c_d(t)D_2(t)s_{\mathrm{BOC}(m,n)}(t), \quad (4.62)$$

where $D_1(t)$ is the data message modulated on the BOC(n, n) component, and $D_2(t)$ is the data message modulated on the BOC(m, n) component. When $D_1(t) \equiv D_2(t)$, $s'(t)$ is degenerated into an ordinary QMBOC signal. In most cases, given $\gamma \ll 1$, for example in the MBOC modulation used by the GPS L1C signal, Galileo E1 OS signal and BDS-3 B1C signal, the power of the BOC(m, n) component is 10 dB weaker than the BOC(n, n) component. However, the demodulation performance of BOC(m, n) can be improved by lowering the data rate. For example, the same bit error rate can be obtained for $D_2(t)$ by reducing the data rate of $D_2(t)$ to 1/10 of the $D_1(t)$. Therefore, if necessary, it is possible to modulate different messages on the two signal components of the QMBOC to increase the amount of information transmitted by the signal.

4.7.3.2 Matched Receiving of QMBOC Signals

Consider a case when the modulated data messages are the same on the BOC(m, n) component and on the BOC(n, n) component. At this time, without loss of generality, the intermediate frequency (IF) signal of the data channel can be expressed as

$$\begin{aligned} s_{\mathrm{IF}}(t) &= A_s\sqrt{1-\gamma}c_d(t)D(t)s_{\mathrm{BOC}(n,n)}(t)\cos(\omega_{\mathrm{IF}}t) \\ &\quad - A_s\sqrt{\gamma}c_d(t)D(t)s_{\mathrm{BOC}(m,n)}(t)\sin(\omega_{\mathrm{IF}}t), \end{aligned} \quad (4.63)$$

where A_s is the signal amplitude and ω_{IF} is the intermediate frequency carrier's angular frequency. For simplicity, the noise term is ignored here.

Inside the receiver, the IF signal $s_{\mathrm{IF}}(t)$ is first multiplied by the local in-phase (I) and quadrature (Q) carrier replicas, and then multiplied by the local reproduced spreading code modulated by the BOC(m, n) waveform and the BOC(n, n) waveform respectively. The product then goes through an integrate and dump (I&D) filter for coherent integration of length T. Ignoring the influence of the Doppler frequency difference, the output of the I&D filter can be expressed as

$$\begin{cases} I_1 = A_s\sqrt{1-\gamma}DR_{\mathrm{BOC}(n,n)}(\Delta\tau)\cos\phi - A_s\sqrt{\gamma}DR_c(\Delta\tau)\sin\phi \\ I_2 = A_s\sqrt{1-\gamma}DR_c(\Delta\tau)\cos\phi - A_s\sqrt{\gamma}DR_{\mathrm{BOC}(m,n)}(\Delta\tau)\sin\phi \\ Q_1 = A_s\sqrt{1-\gamma}DR_{\mathrm{BOC}(n,n)}(\Delta\tau)\sin\phi + A_s\sqrt{\gamma}DR_c(\Delta\tau)\cos\phi \\ Q_2 = A_s\sqrt{1-\gamma}DR_c(\Delta\tau)\sin\phi + A_s\sqrt{\gamma}DR_{\mathrm{BOC}(m,n)}(\Delta\tau)\cos\phi \end{cases} \quad (4.64)$$

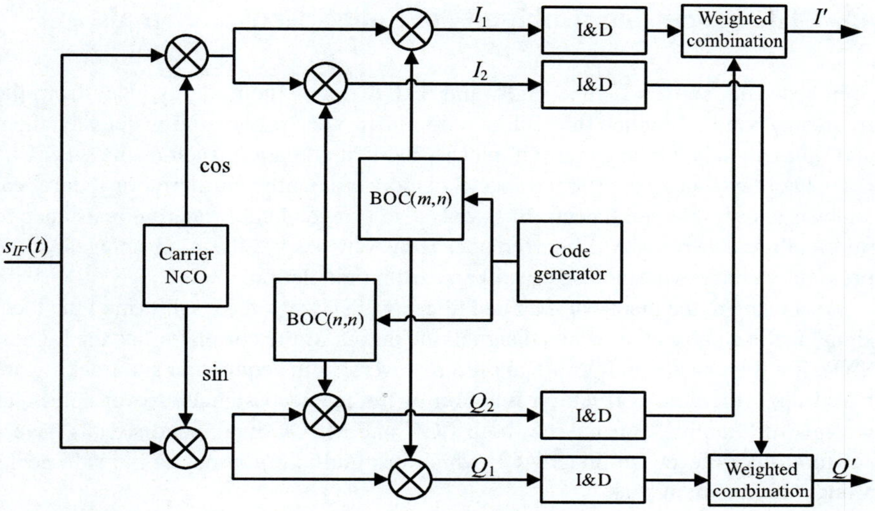

Fig. 4.27 Matched receiving scheme of the QMBOC signal

where ϕ and $\Delta\tau$ are respectively the phase difference between the input carrier and the locally reproduced carrier, and the phase difference between the input spreading code and the local spreading code. Combine the filter output with the following rules

$$\begin{cases} I' = \sqrt{1-\gamma}I_1 + \sqrt{\gamma}Q_2 = A_s D R_{\text{QMBOC}}(\Delta\tau)\cos\phi \\ Q' = \sqrt{\gamma}I_2 - \sqrt{1-\gamma}Q_1 = A_s D R_{\text{QMBOC}}(\Delta\tau)\sin\phi \end{cases}. \tag{4.65}$$

Then the combined I' and Q' can be processed using traditional acquisition methods and tracking loops.

Figure 4.27 shows the matched receiving scheme for the QMBOC signal. For simplicity, only the processing circuit for one spreading code phase is given, and the acquisition and tracking circuits after the combination of correlation values is simplified.

Because the PSD and ACF of the QMBOC signal are the same as those of the TMBOC signal, the QMBOC signal has the same performance as the TMBOC under matched receiving conditions. The detailed analysis of the mismatched receiving performance of the above modulations will be carried out as case studies after the introduction of performance analysis methods in the next chapter. Now it is the time to discuss some of the challenges in the receiving of BOC and MBOC signals.

4.8 The Processing Ambiguity of Split Spectrum Signals

Split spectrum signals such as BOC and MBOC move their energy away from the frequency band center and thus achieve significant spectral separation between them and other signals which use traditional BPSK-R modulation, such as the GPS C/A code, in order to share the L-band spectrum more efficiently. Furthermore, theoretical analysis shows that split spectrum signals also provide better inherent resistance to multipath and narrowband interference. However, despite these advantages, some problems remain with the use of split spectrum modulation.

According to the theory of matched filtering [17], when the waveform of the local signal is the same as the received signal, the output of the correlator has the highest SNR. For this reason, in traditional GPS receivers, both acquisition and tracking are based upon the auto-correlation function of the received signals. Nevertheless, as we saw in Examples 3.1 and 3.2, both BOC and MBOC modulated signals have a sawtooth-like piecewise linear ACF which has multiple non-negligible side peaks along with the main peak.

Since there are significant amounts of signal energy located at the sidepeaks of the BOC/MBOC ACF, during signal acquisition it is quite likely that one of side peak magnitudes may exceed that of the main peak due to noise, resulting in false acquisition. If false acquisition occurs, the code tracking loop will initially lock on the side peak. Similarly, in the code tracking loop, the discriminator characteristic curve of a split spectrum modulated signal has multiple stable false lock points due to the side peaks of ACF. Once the loop locks on to one of the side peaks, it would result in an intolerable bias in pseudorange measurements which is unacceptable for GNSS aiming to provide accurate navigation solutions. This problem is referred to as the *ambiguity problem* for split spectrum modulated signal acquisition and tracking. In order to employ these signals in the next-generation GNSS, solutions have to be found to eliminate or minimize this bias threat.

4.8.1 Description of the Problem

Here, we define code tracking reliability as the ability of the tracking loop to track the main peak of the ACF of the spreading signal. For traditional BPSK-R signals, the most important factor affecting the tracking reliability is the non-ideal cross-correlation between the spreading sequences of different satellites. If the margin between the maximum cross-correlation of different codes and the autocorrelation main peak is not large enough, the tracking loop may mistakenly lock to the cross-correlation peak of a stronger signal in case of signal occlusion and pseudolite applications [20]. The signals of the next-generation GNSS generally use long-period spreading sequences, which significantly improves the cross-correlation properties of the code. However, the special shape of the split spectrum signal's ACF brings new problems for tracking reliability.

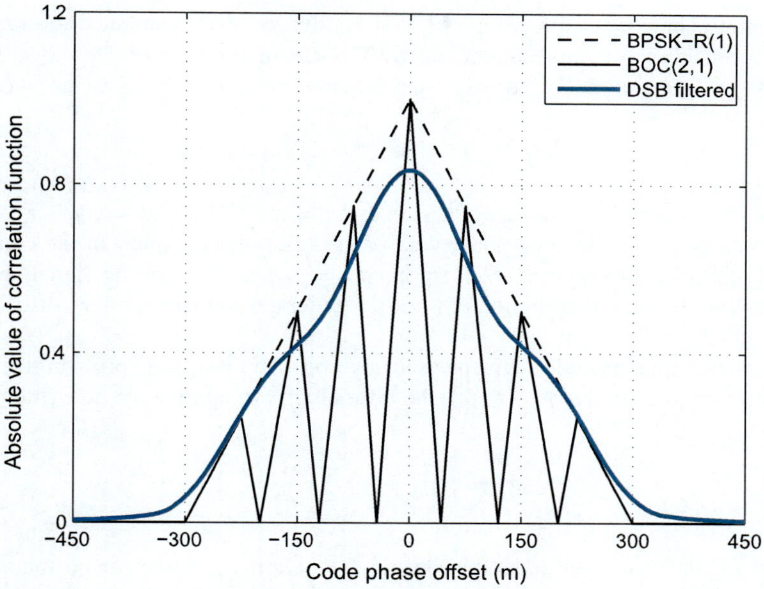

Fig. 4.28 Normalized autocorrelation function envelopes of the BPSK-R(1) signal, the BOC(2, 1) signal, and the sideband filtered correlation function of the BOC(2, 1) signal

Consider the BOC signal as an example. Figure 4.28 compares the ACF of the BPSK-R(1) signal and the BOC(2,1) signal. The spreading code rates of these two signals are the same, i.e. 1.023 MHz, but the shapes of their correlation function are completely different. The ACF of the BOC signal has a sharper main peak, which means better tracking accuracy in thermal noise. However, the ACF of the BOC signal has multiple side peaks within $\tau = \pm 1$ chips. The height of these side peaks is comparable to the main peak. In the acquisition and tracking stages, these side peaks could be mistaken for the main peak.

4.8.2 The False Acquisition Probability Under the Serial Acquisition Strategy

As given in Sect. 5.4.2, the detection statistic used by the acquisition signal model

$$\Psi = \sum_{k=0}^{L-1} \left(I_k^2 + Q_k^2 \right) \tag{4.66}$$

is the non-coherent accumulation of L consecutive pre-detection integration results. If only considering the interference of AWGN, the distribution of Ψ follows a non-central χ^2 distribution of $2L$ degrees of freedom (DOF), with the non-central parameter

$$\lambda = LT_{\mathrm{coh}}C/N_0 \,\mathrm{sinc}^2\left(\pi\Delta f T_{\mathrm{coh}}\right) R_{\mathrm{B}}^2\left(\Delta\tau\right) , \tag{4.67}$$

where $R_{\mathrm{B}}(x)$ is the ACF of the BOC signal, $\Delta\tau = \tau - \hat{\tau}$ and $\Delta f = f_d - \hat{f}_d$ are the estimated errors of the code phase delay and the carrier Doppler. In the complete signal alignment hypothesis (H_1), we have $\lambda = LT_{\mathrm{coh}}C/N_0$. In the signal absence hypothesis (H_0), the distribution of (4.66) degenerates to the central χ^2 distribution of $2L$ DOF.

The detection threshold is given based on a constant false alarm probability. When the decision threshold is γ ($\gamma \geqslant 0$), the false alarm probability can be written as

$$P_{\mathrm{fa}}\left(\gamma\right) = \exp\left(-\frac{\gamma}{2}\right)\sum_{n=0}^{L-1}\frac{1}{n!}\left(\frac{\gamma}{2}\right)^n . \tag{4.68}$$

Therefore, the value of γ under the constant false alarm condition can be obtained by conducting an inverse solution of $P_{\mathrm{fa}}(\gamma)$. For example, when $P_{\mathrm{fa}} = 10^{-6}$, $L = 15$, one can obtain that $\gamma = 82.0$.

Within the serial acquisition strategy, the search algorithm detects the code phase one by one in the early-to-late order. Some side peaks may be detected in advance of the main peak. Therefore, the false acquisition probability under the serial acquisition strategy is the probability that the energy of side peaks exceeds the threshold. For the M-order BOC signal, the energy attenuation coefficient of the ith side peak to the main peak is

$$\xi_i = \left(\frac{M-i}{M}\right)^2 . \tag{4.69}$$

Then, the probability of capturing this side peak is

$$P_{EA} = Q_L\left(\sqrt{\xi_i\lambda}, \sqrt{\gamma}\right) , \tag{4.70}$$

where Q_m is the mth order generalized Marcum Q function [21].

Figure 4.29 shows the probability of falsely capturing the nearest side-peak of the main peak of the BOC(n, n) signal, the BOC$(2n, n)$ signal, and the BOC$(4n, n)$ signal respectively, where $P_{\mathrm{fa}} = 10^{-6}$, $T_{\mathrm{coh}} = 1$ ms, and $L = 10$. It can be seen that under the above acquisition parameters, when the CNR is higher than 40 dB-Hz, the false acquisition probability of the BOC(n, n) signal's side peaks cannot be ignored, especially when the CNR is higher than 42 dB-Hz where the false acquisition probability is close to 1. As the modulation order of the input signal increases, the energy difference between the side peak and the main peak becomes smaller and smaller. For the BOC$(4n, n)$ signal, the difference between the two is less than 1.2 dB. Therefore,

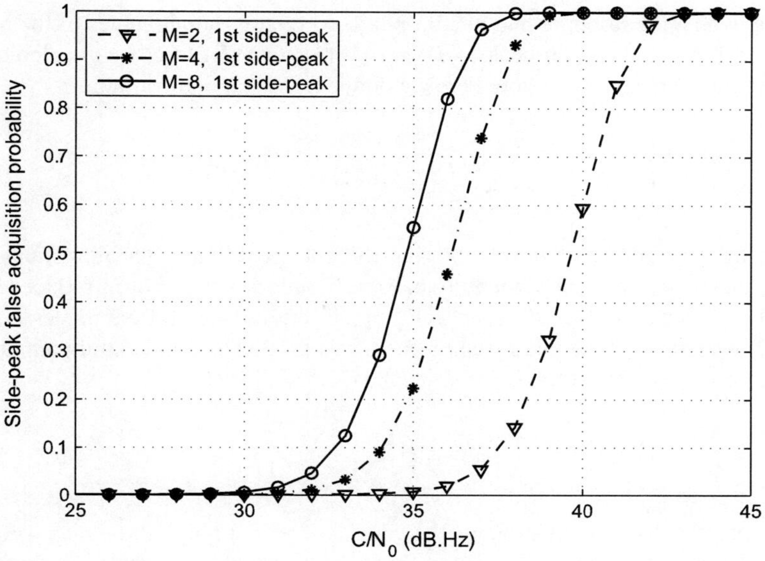

Fig. 4.29 The false acquisition probabilities of the first side peak of the BOC signal when $M = 2, 4, 8$

the traditional acquisition method under the serial strategy has an excessive probability of false acquisition, which cannot meet the application requirements when processing the split spectrum signals.

4.8.3 The False Acquisition Probability Under the Parallel Acquisition Strategy

Compared to the serial acquisition strategy, the parallel acquisition strategy can simultaneously obtain detection statistics at multiple code phase delay locations. Since the signal energy at side peaks is lower than that at the main peak, the maximum value is selected as the peak position among all the output results, which can significantly reduce the probability of false acquisition under high SNR. However, when the influence of noise causes the detected statistic at the side peak to be higher than that at the main peak, false acquisition can still occur.

Within the parallel acquisition strategy, the probability of correct acquisition is the probability that the detection statistic at the main peak is greater than those at the remaining $2M - 2$ side peaks. As the distance from the side peak to the main peak increases, the energy gradually decays. Here we only consider the effects of the two highest peaks on both sides of the main peak, though for $M > 2$, this simplification will result in an overly optimistic result.

We denote the in-phase and quadra-phase correlator output values of the nearest peak position of the main peak as I' and Q' respectively. In order to calculate the probability of false acquisition, here we define a new random variable

$$\Theta = \sum_{k=1}^{L-1} \left[\left(I_k^2 + Q_k^2 \right) - \left(I_k'^2 + Q_k'^2 \right) \right] . \tag{4.71}$$

Then, the probability of false capturing the side peak under the maximum value selection strategy has the probability of $\Theta < 0$. Ignoring the error introduced by the search step size, for the M-order BOC signal, the non-central parameter λ of the detection statistic at the main peak is given by (4.67). The non-central parameter of the detection statistic at the first side peak is

$$\xi_1 \lambda = \left(\frac{M-1}{M} \right)^2 \lambda . \tag{4.72}$$

In the parallel acquisition strategy, the correlator output noise at the main peak and the first side peak position is statistically correlated, with the correlation coefficient

$$\mu_{xy} = -\frac{M-1}{M} . \tag{4.73}$$

Then, the probability of false capturing a specific side peak can be written as

$$P_{\mathrm{ea}} = \begin{cases} Q_1(a,b) - I_0(ab)\,\mathrm{e}^{-\frac{\lambda}{2}} + 2^{1-2L} \cdot I_0(ab)\,\mathrm{e}^{-\frac{\lambda}{2}} \sum\limits_{k=0}^{L-1} \binom{2L-1}{k} \\ +2^{1-2L}\mathrm{e}^{-\frac{\lambda}{2}} \sum\limits_{n=1}^{L-1} \left\{ I_n(ab) \left[\left(\frac{b}{a}\right)^n - \left(\frac{a}{b}\right)^n \right] \sum\limits_{k=0}^{L-1-n} \binom{2L-1}{k} \right\}, & L > 1 \\ Q_1(a,b) - \frac{1}{2} I_0\left(\left(\frac{M-1}{M}\right) \frac{\lambda}{2} \right) \mathrm{e}^{-\frac{\lambda}{2}}, & L = 1 \end{cases} \tag{4.74}$$

where

$$a = \sqrt{\frac{\lambda}{2}\left[1 - \sqrt{1 - \left(\frac{M-1}{M}\right)^2}\right]}, \tag{4.75a}$$

$$b = \sqrt{\frac{\lambda}{2}\left[1 + \sqrt{1 - \left(\frac{M-1}{M}\right)^2}\right]}, \tag{4.75b}$$

and $I_n(x)$ is the n-order modified Bessel function of the first kind.

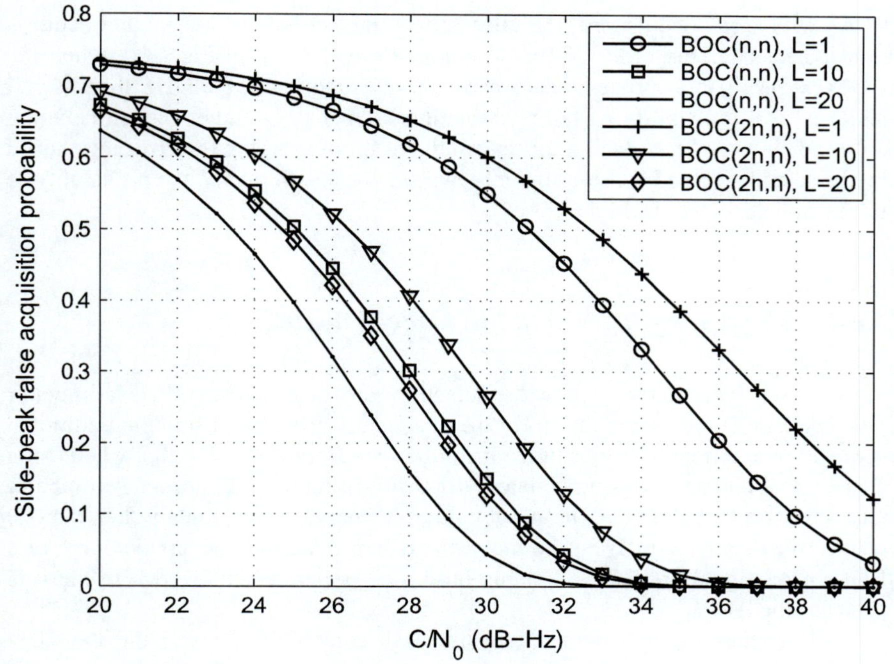

Fig. 4.30 The probability of side-peak false acquisition in parallel acquisition for the BOC(n, n) and BOC($2n, n$) signals

Considering only the influence of the nearest side peaks on both sides of the main peak, the probability of no false acquisition is that the main peak is simultaneously larger than side peaks on both sides, i.e. $(1 - P_{\text{ea}})^2$. Then, the probability of false acquisition is

$$P_{EA} = 1 - (1 - P_{\text{ea}})^2 \ . \tag{4.76}$$

Figure 4.30 shows the false acquisition probabilities for the BOC(n, n) signal and the BOC($2n, n$) signal under the parallel acquisition strategy with $T_{\text{coh}} = 1$ ms, $L = 1$, 10, and 20, respectively. Comparing Figs. 4.29 and 4.30, we see that the parallel search adopts a taking-maximum-value strategy to prevent false acquisition in cases of high SNR. However, the value comparison method is not effective in low SNR environments.

We now consider an extreme case. When the SNR is very low, the contribution of the signal components in the main peak and the side peak is small, and the statistics participating in decision-making are dominated by thermal noise. At this point, the probability of the main peak exceeding a single side peak is 50%, so the probability of false acquisition is close to 75%. The results in Fig. 4.30 also agree with this inference.

As with serial acquisition, the false acquisition probability of parallel acquisition increases as the order of BOC modulation increases. In Fig. 4.30, when the BOC$(2n, n)$ signal is processed at a carrier-to-noise ratio of 30 dB-Hz, there is still about a 15% probability of false acquisition even if the number of non-coherent accumulations is 20. Note that this is still the result of ignoring the error introduced by the search step and considering only two side peaks. In reality, the probability of false acquisition will be higher.

4.8.4 The Ambiguity Thread in Code Tracking

Figure 4.31a, b respectively show the detection statistics and the code discriminator characteristic curves when the BOC$(2n, n)$ signal is processed using the traditional acquisition and tracking algorithms. It can be seen from Fig. 4.31b that when using a traditional narrow early-minus-late (NEML) tracking loop [22], the discriminator characteristic curve of BOC(m, n) signal has a smaller linear domain than the one of the BPSK-R(n) signal. Furthermore, the discriminator characteristic curve of a BOC(m, n) signal has $2M - 2$ stable false lock points which are due to the side peaks of the ACF.

If the detection statistic of a certain side peak exceeds the decision threshold due to noise during the acquisition stage, the corresponding initial phase shift is given to the tracking loop. If the pull-in range of the code discriminator is not wide enough, the tracking loop will lock to the corresponding false lock point of the discriminator characteristic curve. In addition, even if there is no false acquisition, the tracking loop may also be offset from the correct lock point to the false lock point due to high noise, jitter, or short-term loss of locking caused by occlusion.

Figure 4.32 shows an example of a side-peak false lock caused by an excessive initial code phase bias in BOC$(2n, n)$ signal tracking. The CNR in this example is 45 dB-Hz and the predetection integration time is 1 ms. The discriminator of the code tracking loop uses a 0.1 chip early-late correlator space. It can be seen that when the initial code phase bias is 0.14 chips, the tracking loop locks on a false lock point. Although the output of discriminator hovers around zero and the loop maintains lock, the true estimate bias of code phase has reached 0.25 chips and is stable over this deviation. Using the BOC(2, 1) signal as an example, the measurement bias of the corresponding pseudorange can reach about 75 m. For navigation receivers where the precise measurement of the code phase delay is the primary task, false locking will introduce unacceptably large deviations to the final positioning results.

4.8.5 Methods to Eliminate Ambiguity

During the decade from when BOC modulation was initially proposed to the present time, several solutions have been proposed to solve the ambiguity problem. In sum-

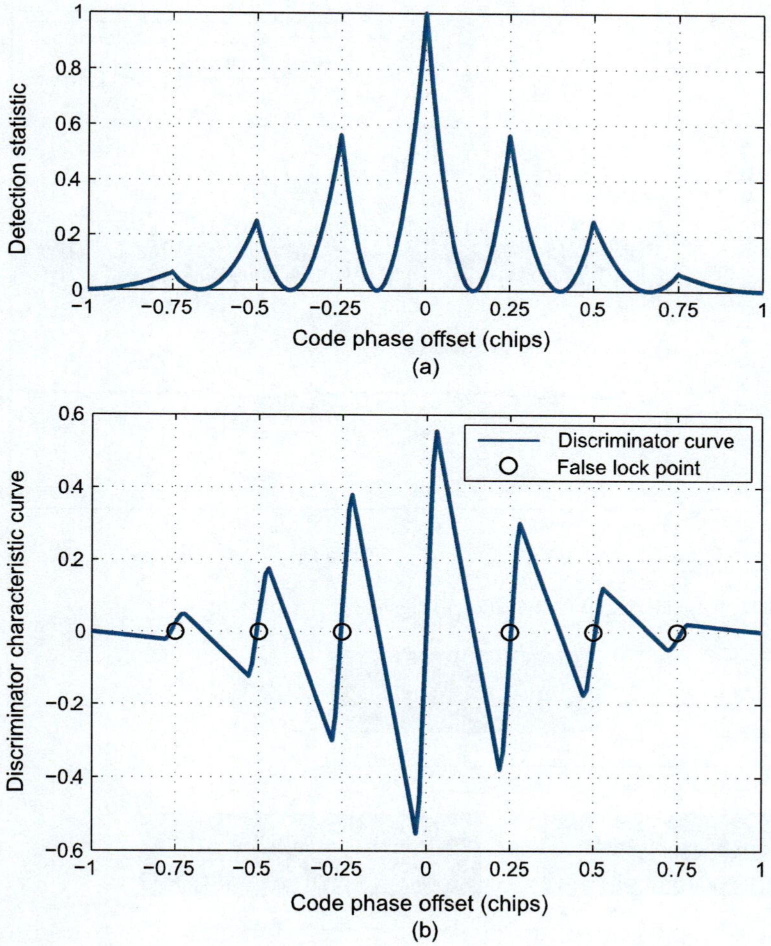

Fig. 4.31 Detection statistics and the DLL discriminator characteristic curve of the BOC$(2n, n)$ signal

mary, the elimination of the ambiguity threat can be achieved in two ways: false lock detection and recovery techniques, as well as unambiguous processing techniques. More specifically, considering the operating domain, unambiguous processing can be further classified into frequency-domain processing and time-domain processing, as shown in Fig. 4.33.

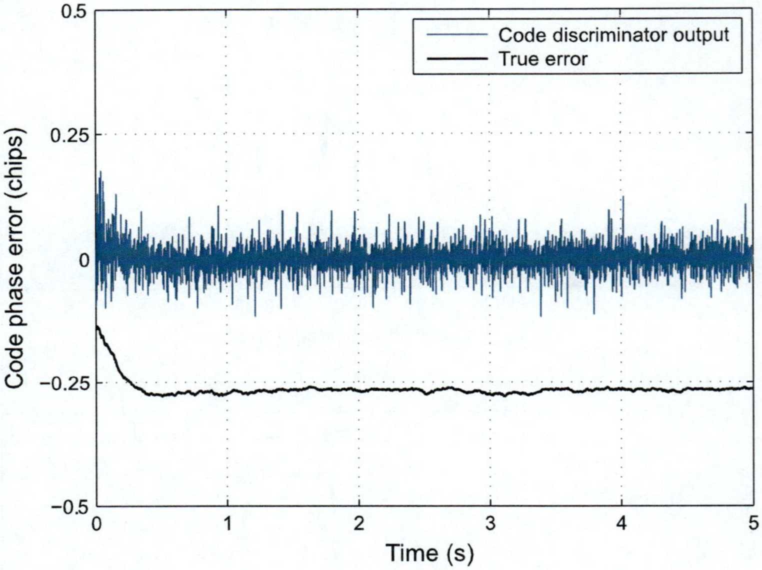

Fig. 4.32 False locking of the BOC($2n, n$) signal

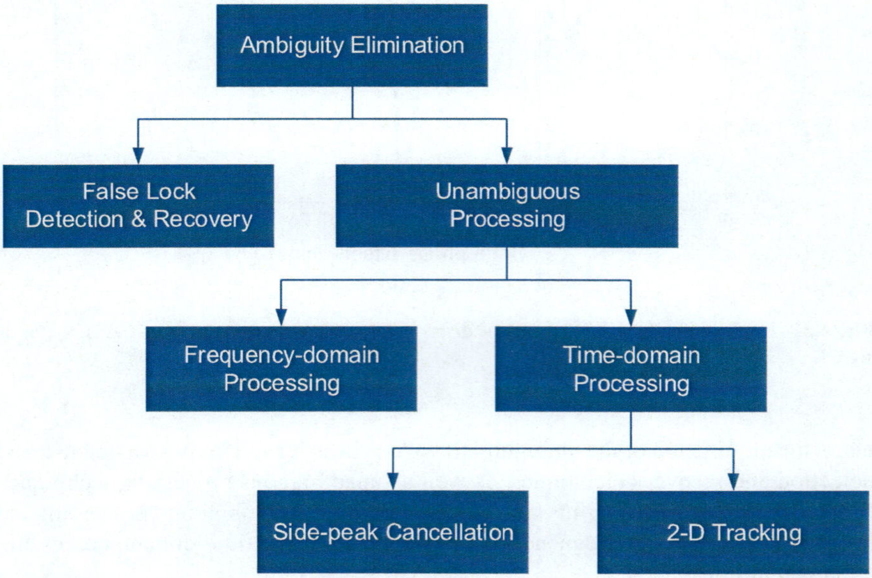

Fig. 4.33 Existing ambiguity elimination solutions

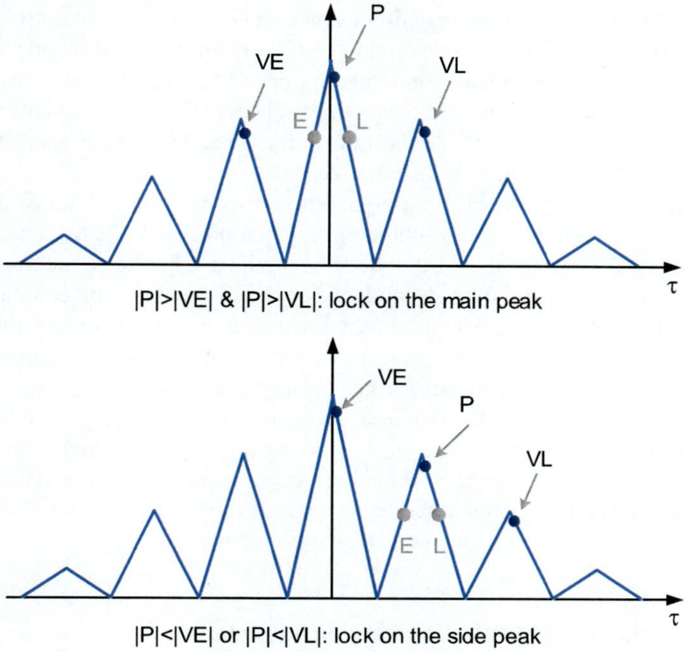

Fig. 4.34 Bump-jumping technique

4.8.5.1 False Lock Detection & Recovery

False lock detection and recovery techniques do not remove ambiguity but rather checks false lock. The most representative detection and recovery technique is referred to as the bump-jumping technique [23]. This technique employs the traditional ambiguous code tracking loop and constantly checks whether this loop is locked on to the main peak of BOC ACF. To do so, the bump-jumping technique uses two additional correlators located at the theoretical location of the two highest side peaks, as shown in Fig. 4.34.

These two correlators are referred to as very early (VE) and very late (VL) correlators. By measuring and comparing the magnitude of the outputs of these two correlators and the prompt correlator, the bump-jumping technique determines whether the false lock occurs. It can be seen from Fig. 4.34 that ignoring the effect of noise, the magnitude of prompt correlator output is the greatest when the code loop locks on the main peak. And if either VE or VL correlator output is the largest, which means that tracking might be biased, the loop will "jump" in the appropriate direction.

When locked on the main peak, this technique has high tracking accuracy. However, since it is based on comparing magnitudes of the main peak and side peaks, the detection may have a high probability of false alarms when the SNR is low. In [23], two up/down counters are employed to reduce this false alarm probability. In each comparison, if one of the magnitudes of VE and VL correlator outputs exceeds that of the prompt one, the corresponding counter is incremented by one, otherwise the corresponding counter is decremented by one. The counter is not decremented below 0 or incremented above the preset threshold N. When the counter reaches the threshold, the loop jumps to the highest peak. By using this counter mechanism, the false alarm probability can be reduced effectively.

Strictly speaking, the bump-jumping method does not truly eliminate the hidden danger of false locking. The advantage of this method is that the tracking accuracy is the same as that of the standard code tracking loop when false locking does not occur. However, due to its heavy dependence on the shape of the correlation function, the shape distortion of the correlation function caused by multipath and filtering will have a major impact on its performance. Moreover, the amplitude comparison used in this method has a high tendency for generating false alarms and missed detections in a noise environment. While a longer sequential detection process can improve the decision-making performance, the response time will be longer [23]. Once a false lock occurs, the loop cannot immediately detect and take action. Therefore, the bump-jumping method is not suitable for applications with high requirements for continuity and security, nor for low SNR environments.

4.8.5.2 Frequency-Domain Unambiguous Processing

Frequency-domain processing techniques are represented by sideband techniques. Betz, one of the proposers of BOC modulation, suggested treating two spectrum main lobes of the BOC signal as two respective BPSK-R signals [24]. Because the BOC signal can be regarded as the secondary modulation of the BPSK-R signal by the square wave subcarrier, the sideband filtering process can be understood as

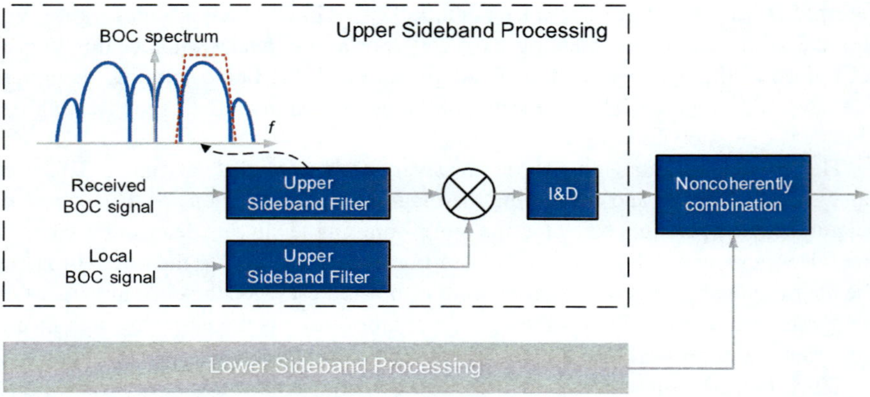

Fig. 4.35 Block diagram of sideband technique

essentially demodulating the original BPSK-R signal from the subcarrier first, and then performing the traditional acquisition and tracking processing.

The earliest sideband technique was described in [24]. As shown in Fig. 4.35, the single sideband technique uses only one of the sidebands (either upper or lower) of the BOC modulated signal. Both the received signal and the local BOC modulated baseband signal are filtered. Only the upper or lower sidebands of the received and local signals are retained.

The solid blue line in Fig. 4.28 shows the correlation function of the BOC(2,1) signal after sideband filtering. It can be seen that the signal loses the multi-peak characteristic of the BOC signal after sideband filtering. Moreover, the main peak width of the correlation function is close to the BPSK-R case. Therefore this correlation function can be used instead of BOC ACF in acquisition and tracking.

The double sideband technique uses both the upper and lower sideband of BOC modulated signal. These two sidebands are processed separately before the output of correlators, and then the correlation values are added non-coherently. Compared with single sideband technique, the double sideband technique suffers lower non-coherent correlation losses. However, it requires twice the sideband-selection filter number of the single sideband technique.

The BPSK-like method [25] is another frequency-domain unambiguous processing technique. This method is also based on the consideration of the BOC spectrum as the sum of two BPSK spectrum shifted by $\pm f_s$. The main difference compared with the method described above is that only one low-pass filter is employed for the received signal. As shown in Fig. 4.36, the filter bandwidth includes the two main lobes of the spectrum. Another difference is that, the local signal is not the filtered BOC-modulated baseband signal but the BPSK-R signal shifted by the sub-carrier frequency f_s. The BPSK-like technique can also be either single or double sideband, according to whether both the sidebands are used and combined non-coherently or only one sideband is used.

The original BPSK-like method can only be used for sine-phased BOC modulations with even BOC order. In [26], a modified version of the BPSK-like method is proposed to extend BPSK-like method to BOC signals with odd order.

Although the correlation functions in sideband techniques do not present any side peak, meaning that they are fully unambiguous, these methods suffer from some drawbacks. The main defect is that they completely remove all of the advantages of BOC signal tracking in terms of Gaussian noise and multipath mitigation, since they cause the RMS bandwidth of the received signal to approach that of the BPSK-R signal. Moreover, two side lobes are combined in non-coherent mode, which introduces correlation losses into the process. It seems that sideband techniques are not appropriate in terms of tracking. However, the correlation functions in this kind of methods have a wide main correlation peak, which allows using a longer code delay step in acquisition to reduce mean acquisition time. Therefore, sideband techniques can be attractive options in BOC modulated signal acquisition.

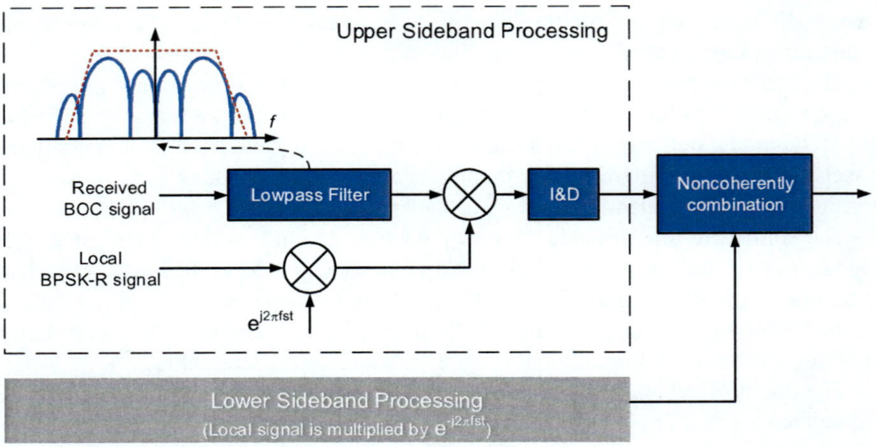

Fig. 4.36 Block diagram of BPSK-like method

4.8.5.3 Time-Domain Unambiguous Processing

The existing time-domain unambiguous tracking methods of BOC signal can be divided into two classes for the treatment of the spreading code and the subcarrier. The first class of methods employs identical delays in the local code and subcarrier, and only one loop is used to track the variation of this delay. This class of methods can be collectively referred to as side-peak cancellation (SC) techniques which solve the ambiguity problem by taking advantage of the geometrical property of correlation functions. This class can also be referred to as the one-dimensional (1-D) tracking technique since the correlation function used is a 1-D function of delay.

Besides 1-D tracking, there is another emerging class of unambiguous tracking methods in which the subcarrier and spreading code component are processed separately. More specifically, local replicas of the subcarrier and code use different delays and two independent loops are employed to track the variations of these two parameters. Representative existing methods in this class mainly include double estimate tracking (DET) [27–29], double phase estimator (DPE) [30, 31], and the robust unambiguous tracking method proposed in [32]. Since the correlation function used in these methods is a function of two variables, code delay and subcarrier delay, these methods can be collectively called the two-dimensional (2-D) tracking techniques.

4.8.5.4 Side-Peak Cancellation Techniques

The basic idea of SC techniques is using a synthesized correlation function (SCF) instead of BOC ACF in acquisition and tracking. Correlation function between the received BOC signal and some local auxiliary signals whose chip waveforms may be different from the received one are combined linearly or non-linearly to form the

SCF with no side-peak. SC methods are flexible. Due to different auxiliary signal chip waveforms and combination modes, SC methods differentiate from each other greatly.

The first side-peaks cancellation technique is proposed in [33]. This approach removes the ambiguities of the correlation function, but one drawback is that this method destroys the sharp peak of the correlation function. For accurate tracking, preserving a sharp peak of the correlation function is a prerequisite. An innovative unambiguous tracking technique for sine phase BOC(n, n) signals, which is referred to as the autocorrelation side-peak cancellation technique (ASPeCT), is proposed in [34]. This technique uses two sets of local signals, one of which is the replica of the input BOC(n, n) signal, and the other is the BPSK-R(n) signal with the same spreading sequence as the input signal. The two sets of signals are respectively correlated with the input signal, and then square subtraction is performed. The two peaks of the CCF of the BOC signal and the BPSK-R signal just eliminate the two side peaks of the BOC(n, n) signal's ACF. Applying the synthesized function to the early-minus-late power (EMLP) discriminator, the discriminator characteristic curve can completely eliminate the two false locking points at 0.5 chips while maintaining good tracking accuracy and anti-multipath performance in cases where the pre-filtering bandwidth is wide enough. However, when the signal bandwidth is limited and the shape of the correlation function becomes smooth, ASPeCT still has a false locking point. And this technique is only applicable to sine-BOC(n, n) signals.

The side peak cancellation method is based on the geometry of the CCF of the signal. In the earlier study of side peak cancellation method design, the waveform of the auxiliary signal was mostly obtained by trial and error. Due to the different modulation orders of the BOC signal, the shape of the correlation function is very different, which makes most of the side peak cancellation methods including ASPeCT be not universal.

In [35], the design method based on side peak elimination is studied in depth, and a general design framework of the side peak cancellation method is given. In this framework, the local auxiliary signal chip waveform can be designed by means of mathematical analysis so that the waveform shape selection can be more flexible. Under the guidance of this design framework, a series of more general unambiguous acquisition and tracking methods have emerged, such as the general removing ambiguity via side-peak suppression (GRASS) method [36] for unambiguous acquisition and pseudo correlation function based unambiguous delay lock loop (PUDLL) [37] for unambiguous tracking. Both of these techniques can be applied to sine phase BOC signals with any modulation order. On this basis, Yao et al. [38] elaborates on the versatility of PUDLL. By replacing the local waveform, it can be directly applied to the cosine phase BOC signal. In recent years, under the guidance of the design framework of the side peak elimination method, new methods of unambiguous acquisition and tracking are still emerging.

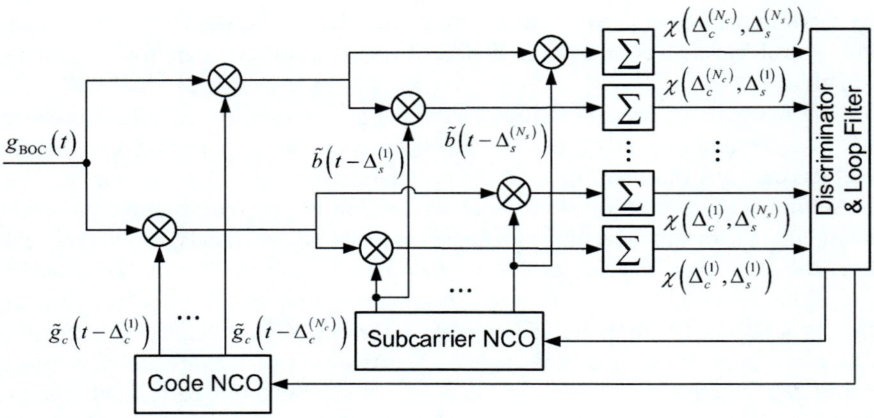

Fig. 4.37 Principal structures of a 2-D tracking loop

4.8.5.5 Two-Dimensional Tracking Techniques

The original concept of the two-dimensional tracking method was proposed by Hodgart et al. [27]. The core idea is to remove the limitation that the delay τ_c of the code phase and the delay τ_s of the subcarrier phase must be exactly the same in the local replica signals. Two different tracking loops are employed to track the spreading code and the subcarrier independently. The final estimation of the signal delay τ is obtained by a combination of the estimates of τ_c and τ_s.

More specifically, in 2-D tracking techniques, as shown in Fig. 4.37, when processing $g_{\text{BOC}}(t) = g_c(t)\,b(t)$ where $g_c(t)$ is the spreading code component and $b(t)$ is the square wave subcarrier, receivers generate local code components $\tilde{g}_c(t - \tau_c - \Delta_c^{(k)})$ with several different extra delays $\Delta_c^{(k)}$ for $k = 1, 2, \ldots, N_c$, and local subcarrier components $\tilde{b}(t - \tau_s - \Delta_s^{(i)})$ with several different extra delays $\Delta_s^{(i)}$ for $i = 1, 2, \ldots, N_s$, respectively. The products of $\tilde{g}_c(t - \tau_c - \Delta_c^{(k)})$ and $\tilde{b}(t - \tau_s - \Delta_s^{(i)})$ are correlated with the received signal $g_{\text{BOC}}(t)$ to form a 2-D CCF which involves the bivariate function

$$\begin{aligned}&\chi\left(\tau - \tau_c - \Delta_c^{(k)}, \tau - \tau_s - \Delta_s^{(i)}\right)\\&= \frac{1}{T}\int_0^T g_{\text{BOC}}(t-\tau)\,\tilde{g}_c\left(t - \tau_c - \Delta_c^{(k)}\right)\tilde{b}\left(t - \tau_s - \Delta_s^{(i)}\right)\mathrm{d}t\,.\end{aligned} \tag{4.77}$$

With appropriate choices of $\Delta_c^{(k)}$ and $\Delta_s^{(i)}$, and combining correlation values observed at multiple points in the plane of the 2-D correlation surface, one can form discriminators which generate error signals, controlling two independent feedback loops to track τ_c and τ_s continuously.

Existing 2-D tracking techniques differ mainly in the selection of local replicas which affect the shape of the CCF, and the combination way of CCF samples in discriminators. The core of designing new 2-D tracking technique also lies in the choice of local waveforms and discriminator forms. Note that in order to despread successfully, the local code is always the same as that of the received signal, i.e., $\tilde{g}_c(t) = g_c(t)$. However, for a local subcarrier $\tilde{b}(t)$, the only constraint is that it should be a periodic signal with the same period as the subcarrier of the received BOC signal. The waveform of $\tilde{b}(t)$ can be the same as that of the received signal, as in [27, 28, 39], or it can be a different waveform. For example, in [30, 31], a single sine wave subcarrier is employed instead of binary subcarrier.

Specific discriminator forms in 2-D tracking techniques can also be varied, but essentially the outputs of all of these techniques are the estimation of tracking errors $\Delta\tau_c = \tau_c - \tau$ and $\Delta\tau_s = \tau_s - \tau$. For example, in DET, the code discriminator and the subcarrier discriminator are respectively

$$\delta_c = \chi(\Delta\tau_c - \Delta_c/2, \Delta\tau_s) - \chi(\Delta\tau_c + \Delta_c/2, \Delta\tau_s) \tag{4.78}$$

and

$$\delta_s = \chi(\Delta\tau_c, \Delta\tau_s - \Delta_s/2) - \chi(\Delta\tau_c, \Delta\tau_s + \Delta_s/2)\ , \tag{4.79}$$

where Δ_c and Δ_s are the spacing between early and late replicas for code and subcarrier discriminators respectively.

For all of the existing 2-D tracking techniques, the 2-D CCF (4.77) has a unique peak within a period of the spreading code in the code dimension, while it also has an infinite number of peaks with a period of $2T_s$ in the subcarrier dimension. Using a sine phase BOC(2, 1) signal as an example results in its 2-D CCF as shown in Fig. 4.38.

For receivers that use a 2-D tracking structure, any one of the peaks of the 2-D CCF can be tracked. When both of these loops are locked, the code loop can provide an unambiguous but noisier delay estimation

$$\hat{\tau}_c \approx \tau + n_c\ , \tag{4.80}$$

where n_c is tracking noise jitter of the code phase delay dimension, and the subcarrier loop can provide a high accuracy but ambiguous delay estimation

$$\hat{\tau}_s \approx \tau + NT_s + n_s\ , \tag{4.81}$$

where N is an arbitrary integer representing the tracking ambiguity of the subcarrier dimension, and n_s is tracking noise jitter of the subcarrier phase delay dimension. Since the chip rate of the subcarrier is greater than the code rate, the variance of n_s is generally less than the variance of n_c. If n_s and n_c satisfy

$$|n_c - n_s| < 0.5T_s\ , \tag{4.82}$$

an unambiguous and high-accuracy signal delay estimation can be obtained through the following non-linear combination of the two delay estimations:

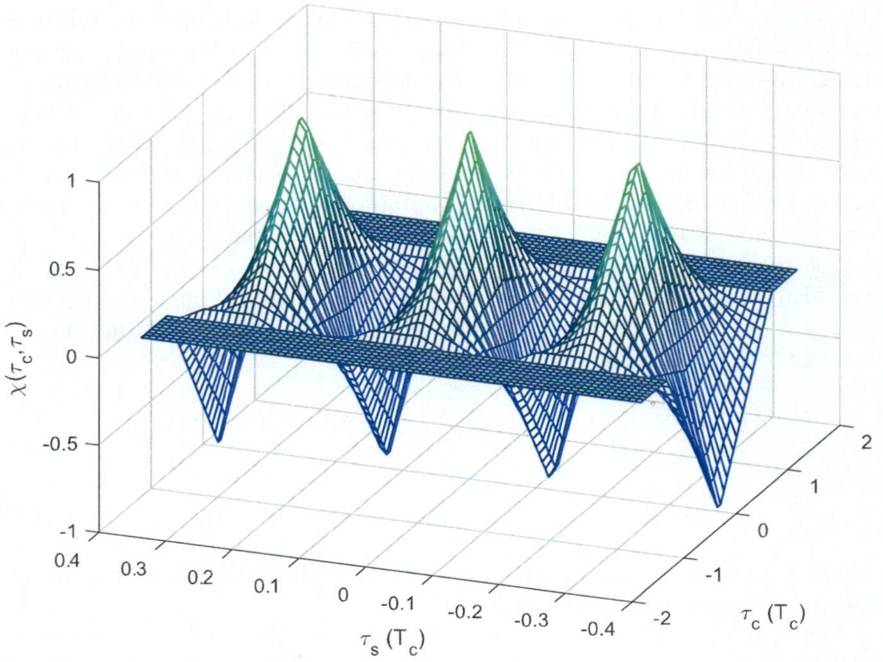

Fig. 4.38 The two-dimensional cross-correlation function of the BOC(2,1) signal

$$\hat{\tau}^* = \hat{\tau}_s - \text{round}\left(\frac{\hat{\tau}_s - \hat{\tau}_c}{T_s}\right) \cdot T_s \,, \tag{4.83}$$

where round(·) represents the rounding operation.

It can be seen that the 2-D tracking methods observe the BOC signal received in two dimensions, which has a higher observation dimension than 1-D methods. It is also easy to prove that the 1-D correlation function can be regarded as a diagonal profile of the 2-D correlation function. Therefore, in theory, the 2-D tracking methods have more flexible and diverse implementations in the processing of the BOC signal.

Although the current 2-D tracking methods will still have a large error when the condition (4.82) cannot be satisfied in the case of dynamic or serious noise interference, the exploration of more robust tracking methods under the 2-D processing framework is a promising research direction. In [40, 41], the key properties of the general 2-D correlation function are quantified. The exact expressions of 2-D CCF of BOC signals and the general theoretical prediction approaches of tracking jitter and multipath mitigation performance in 2-D tracking architecture, considering the effects of subcarrier waveform mismatching and front-end filtering, are also proposed. Employing these tools, 2-D correlation functions are understandable not only on their shape, but also on the coupling relation between dimensions, and the effects of bandlimiting. These results provide the basis for the theoretical analysis and design of future 2-D loops.

4.9 Other Spreading Modulations

4.9.1 AltBOC Modulation

Alternative BOC (AltBOC) modulation is conceptually different from the other modulation methods described above. Although we are accustomed to referring to it as a type of spreading modulation, this technique is more like a multiplexing technique for multiple signal components. The method and process of AltBOC modulation is basically similar to that of the BOC modulation. It also re-adjusts the spectral position and shape of the spread spectrum signal by using subcarriers, but the upper and lower sidebands of the AltBOC modulated signal spectrum correspond to different signal components. The two-component AltBOC signal carries one signal component in the upper sideband and one in the lower sideband respectively, while the four-component AltBOC signal can carry two signal components in each sideband, thereby achieving joint multiplexing of four different baseband signal components located at two frequencies.

Similar to the representation of the BOC signal, an AltBOC modulated signal with subcarrier rate $f_s = m \times 1.023$ MHz and spreading code rate $f_c = n \times 1.023$ can be expressed as AltBOC(m, n). The subcarriers used by AltBOC modulation are complex, allowing the PSD of one of its components to move to one side of the central frequency instead of splitting to both sides like the BOC signal.

As mentioned before, AltBOC modulation can be made up of either two or four components. If we only need to carry two signal components on the AltBOC signal, we can use the bipolar complex subcarrier for single sideband modulation (see Sect. 7.8.1 for more details). Define

$$s_{\text{Alt}}(t) = c_L \gamma_{sc}^*(t) + c_U \gamma_{sc}(t) \ , \tag{4.84}$$

where c_L and c_U are the codes of the lower and upper sideband respectively, and $\gamma_{sc}(t)$ is a complex subcarrier, i.e.

$$\gamma_{sc} = \text{sgn}\left(\cos\left(2\pi f_s t\right)\right) + \text{j}\,\text{sgn}\left(\sin\left(2\pi f_s t\right)\right) \ , \tag{4.85}$$

which can be seen as a bipolar version complex exponential function. The code c_U of the upper sideband is multiplied by the complex subcarriers so that the spectrum is moved to $+f_s$, while the spectrum of the lower sideband code c_L is moved to $-f_s$.

For the four-component case, the baseband signal expression can be written as

$$s_{\text{Alt}}(t) = \left(c_L^D + \text{j}c_L^P\right)\gamma_{sc}^*(t) + \left(c_U^D + \text{j}c_U^P\right)\gamma_{sc}(t) \ , \tag{4.86}$$

where c_L^D and c_L^P are two signal components of the lower sideband, and the phases are in quadrature with each other, c_U^D and c_U^P are two signal components of the upper sideband, and their phases are also in quadrature with each other. The above expression corresponds to the general case of AltBOC. However, it is not difficult to verify that the envelope of this composite signal is not constant. For high power amplifiers on satellites, the constant envelope characteristic of the signal is very important. By introducing appropriate intermodulation signals, we can ensure that the complex envelope corresponding to all states of the signal maintains a constant value. The modified constant envelope AltBOC signal is expressed as

$$\begin{aligned}\tilde{s}_{\mathrm{AltBOC}}(t) = & \left(c_L^D + \mathrm{j}c_L^P\right)\left[sc_d(t) - \mathrm{j}sc_d\left(t - \frac{T_s}{4}\right)\right] \\ & + \left(c_U^D + \mathrm{j}c_U^P\right)\left[sc_d(t) + \mathrm{j}sc_d\left(t - \frac{T_s}{4}\right)\right] \\ & + \left(\bar{c}_L^D + \mathrm{j}\bar{c}_L^P\right)\left[sc_p(t) - \mathrm{j}sc_p\left(t - \frac{T_s}{4}\right)\right] \\ & + \left(\bar{c}_U^D + \mathrm{j}\bar{c}_U^P\right)\left[sc_d(t) + \mathrm{j}sc_d\left(t - \frac{T_s}{4}\right)\right],\end{aligned} \tag{4.87}$$

where T_s is the subcarrier period, $\bar{c}_L^D = c_U^P c_U^D c_L^P$, $\bar{c}_L^P = c_U^D c_U^P c_L^D$, $\bar{c}_U^D = c_L^D c_U^P c_L^P$, and $\bar{c}_U^P = c_U^D c_U^D c_L^P$ are the intermodulation signals, and subcarriers are respectively

$$\begin{aligned}sc_d(t) = & \frac{\sqrt{2}}{4}\,\mathrm{sgn}\left[\cos\left(2\pi f_s t - \frac{\pi}{4}\right)\right] + \frac{1}{2}\,\mathrm{sgn}\left[\cos\left(2\pi f_s t\right)\right] \\ & + \frac{\sqrt{2}}{4}\,\mathrm{sgn}\left[\cos\left(2\pi f_s t + \frac{\pi}{4}\right)\right]\end{aligned} \tag{4.88}$$

and

$$\begin{aligned}sc_p(t) = & -\frac{\sqrt{2}}{4}\,\mathrm{sgn}\left[\cos\left(2\pi f_s t - \frac{\pi}{4}\right)\right] + \frac{1}{2}\,\mathrm{sgn}\left[\cos\left(2\pi f_s t\right)\right] \\ & - \frac{\sqrt{2}}{4}\,\mathrm{sgn}\left[\cos\left(2\pi f_s t + \frac{\pi}{4}\right)\right].\end{aligned} \tag{4.89}$$

Comparing (4.85) and (4.88), we can find that in the original non-constant envelope AltBOC signal, the real part of the subcarrier has cosine phase and the imaginary part has sine phase. However, the modified real part and imaginary part become a combination of sine and cosine subcarriers with advanced and delayed phases.

AltBOC is not only a spreading modulation but also a multiplexing scheme. This technique was first proposed to solve the problem of constant envelope multiplexing of signals on E5a and E5b, which are two sub-bands of the Galileo system on the E5 band. Since the AltBOC signal can be regarded as a single entity, it is possible for the receiver to process it as a wideband signal. However, because the PRN codes and modulated data messages of those signal components in the upper and lower sides of the AltBOC signal are completely independent, such processing is difficult. Therefore, for receivers, AltBOC is more regarded as a multiplexing technique, and the signal components on the upper and lower sides of the signal can be processed separately. In Chap. 7, we will discuss the features of AltBOC technique and the more general dual-frequency multiplexing technologies and the design mechanism behind them from the perspective of constant envelope multiplexing. For a specific implementation of the AltBOC signal used by the Galileo system, i.e. AltBOC(15, 10), a detailed introduction is given in [5].

For the non-constant envelope AltBOC modulation, assuming that the signal is wide-sense stationary, and the correlation property of PRN codes is ideal, it is not difficult to obtain from (4.86) that

$$G_{\mathrm{AltBOC}}(f) = 4\left[G_{\mathrm{BOC_s}}(f) + G_{\mathrm{BOC_c}}(f)\right]. \tag{4.90}$$

That is, the PSD of the non-constant envelope AltBOC is a superposition of the PSD of a sine phase BOC signal and that of a cosine phase BOC signal. In Sect. 4.3 we have given the power spectrum expressions for the sine phase BOC signal and the cosine phase BOC signal. Therefore, when the modulation order is even, we have

$$G_{\mathrm{AltBOC}}(f) = 8f_c\left[\frac{\sin\left(\frac{\pi f}{f_c}\right)}{\pi f\cos\left(\frac{\pi f}{2f_s}\right)}\right]^2\left[1-\cos\left(\frac{\pi f}{2f_s}\right)\right]. \tag{4.91}$$

For the odd-order non-constant envelope AltBOC, we have

$$G_{\mathrm{AltBOC}}(f) = 8f_c\left[\frac{\cos\left(\frac{\pi f}{f_c}\right)}{\pi f\cos\left(\frac{\pi f}{2f_s}\right)}\right]^2\left[1-\cos\left(\frac{\pi f}{2f_s}\right)\right]. \tag{4.92}$$

For the constant envelope AltBOC, the PSD is different from the non-constant envelope AltBOC due to the change of the subcarriers. When the modulation order is even, we have

$$\begin{aligned}\tilde{G}_{\mathrm{AltBOC}}(f) = &\frac{4f_c}{\pi^2 f^2}\frac{\sin^2\left(\dfrac{\pi f}{f_c}\right)}{\cos^2\left(\dfrac{\pi f}{2f_s}\right)}\\ &\times\left[\cos^2\left(\frac{\pi f}{2f_s}\right)-\cos\left(\frac{\pi f}{2f_s}\right)-2\cos\left(\frac{\pi f}{2f_s}\right)\cos\left(\frac{\pi f}{4f_s}\right)+2\right].\end{aligned} \tag{4.93}$$

When the modulation order is odd, we have

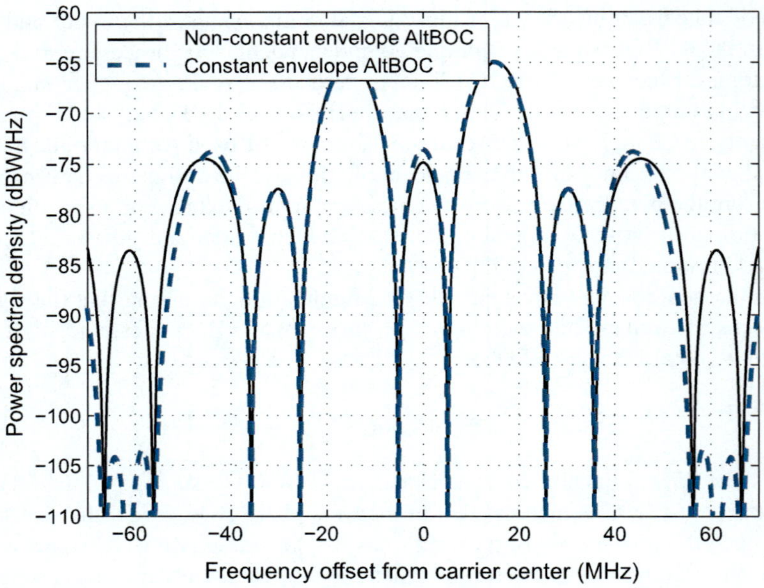

Fig. 4.39 Power spectral density of non-constant envelope AltBOC(15,10) and constant envelope AltBOC(15,10) signals

$$\tilde{G}_{\text{AltBOC}}(f) = \frac{4f_c}{\pi^2 f^2} \frac{\cos^2\left(\dfrac{\pi f}{f_c}\right)}{\cos^2\left(\dfrac{\pi f}{2f_s}\right)} \times \left[\cos^2\left(\frac{\pi f}{2f_s}\right) - \cos\left(\frac{\pi f}{2f_s}\right) - 2\cos\left(\frac{\pi f}{2f_s}\right)\cos\left(\frac{\pi f}{4f_s}\right) + 2\right]. \tag{4.94}$$

Figure 4.39 shows the PSDs of the non-constant envelope AltBOC(15,10) and that of the constant envelope AltBOC(15,10). As the figure shows, the PSDs of these two signals have substantially the same shape, and the main lobe shapes are almost the same, with only minor differences in several side lobes away from the center frequency. These differences in the side lobes mainly come from the intermodulation terms that are added to ensure a constant envelope. In fact, in receiver processing, most of the high-frequency intermodulation terms will be filtered out by the receiver's front-end filter, so it appears that the two AltBOC received signals have almost no difference in PSD. However, from the point of view of the on-board transmitter implementation, it is clear that the constant enveloping AltBOC has more advantages.

4.9.2 MSK Modulation

The BPSK-R modulation used in traditional GPS signals, the BOC, MBOC, and AltBOC modulations used in the next-generation GNSS signals all can be considered to be SCS based modulation, which is a family of discontinuous phase modulations. Phase jump in discontinuous phase modulation leads to slower power spectral roll-off. Although the L-band currently used by GNSS signals is crowded, its constraint on the out-of-band spectrum of the signal has not been very severe. Therefore, the spectrum broadening feature of SCS modulation has little effect on its applications in GNSS. However, in cases where the spectrum resources are strictly limited, and the out-of-band constraints are severe, such as in the design of C-band and S-band GNSS signals, SCS modulation may no longer be suitable.

In order to find solutions for future GNSS, spreading modulations with better spectral roll-off characteristics have received widespread attention in the field of satellite navigation signal design in recent years. Typical modulation techniques with high spectrum efficiency include minimum shift keying (MSK) modulation, Gaussian-filtered MSK (GMSK) modulation, square-root raised cosine (SRRC) waveform modulation, and prolate spheroidal wave functions (PSWF) waveform modulation etc.

MSK modulation is a constant envelope phase continuous modulation with good side lobe decay characteristics, and was first used in the wireless communication field [42]. If a spreading code $c_k(t)$ is subjected to MSK modulation, the mathematical expression of the corresponding MSK signal is defined as [43]

$$s_{\mathrm{MSK}}(t) = A\cos\left(2\pi f_0 t + \phi(t)\right), \tag{4.95}$$

where f_0 is the carrier frequency of the signal and

$$\phi(t) = 2\pi h \int_{-\infty}^{t} \sum_k c_k(t)\, p(t - kT_c)\,\mathrm{d}t \tag{4.96}$$

is the phase of the modulated carrier where T_c is the chip duration, h is the modulation index of the carrier phase (usually $h = 0.5$), and

$$p(t) = \begin{cases} (2T_c)^{-1} & 0 \le t \le T_c \\ 0 & \text{otherwise} \end{cases} \tag{4.97}$$

is the pulse waveform of the code sequence.

Following the naming rules of BPSK-R signals in GNSS community, we denote the spread spectrum signal modulated by MSK as MSK-BPSK(n) signal, where $n = f_c/f_0$, and $f_0 = 1.023$ MHz. The normalized PSD of MSK-BPSK(n) is [8]

$$G_{\mathrm{MSK}} = \frac{8f_c^3}{\pi^2} \frac{\cos^2\left(\frac{\pi f}{f_c}\right)}{\left(f_c^2 - 4f^2\right)^2}. \tag{4.98}$$

If the spreading code is modulated by the BOC subcarrier before MSK modulation, the resulting modulation waveform is called MSK-BOC [8]. Similar to the representation of the BOC signal, we denote the MSK-BOC signal of the spreading code rate f_c and the subcarrier rate f_s as MSK-BOC(f_s,f_c) or MSK-BOC(m,n), where $f_s = mf_0$, $f_c = nf_0$. The normalized PSD of MSK-BOC(f_s,f_c) is

$$G_{\mathrm{MSK}} = \begin{cases} \dfrac{2f_s^2 f_c}{\pi^2} \dfrac{\sin^2\left(\frac{\pi f}{f_c}\right)}{\left(f_s^2 - f^2\right)^2}, & \Phi = \dfrac{f_s}{f_c} \text{ is even,} \\ \dfrac{2f_s^2 f_c}{\pi^2} \dfrac{\cos^2\left(\frac{\pi f}{f_c}\right)}{\left(f_s^2 - f^2\right)^2}, & \Phi = \dfrac{f_s}{f_c} \text{ is odd.} \end{cases} \tag{4.99}$$

Figure 4.40 shows the normalized PSD of BPSK-R(1), BOC(1,1), MSK-BPSK(1), and MSK-BOC(1,1) signals. As seen in Fig. 4.40, when the spreading code rate is 1.023 MHz, the main lobe width of the MSK-BPSK(1) signal is significantly larger than that of the BPSK-R(1) signal, but its side lobes are significantly smaller than those of BPSK-R(1) signal. At the same time, the side lobes of the MSK-BPSK(1) and MSK-BOC(1,1) signals are more rapidly attenuated compared to BPSK-R(1) and BOC(1,1) signals, thus have better frequency attenuation characteristics.

As will be explained in Chap. 5, the RMS bandwidth is often used to evaluate the theoretical lower bound of the code tracking accuracy of GNSS signals. In theory, with the same receiving bandwidth, the larger the RMS bandwidth, the lower the limit of code tracking error, that is, the better the ranging accuracy. Figure 4.41 compares the RMS bandwidths of BPSK-R(1), BOC(1,1), MSK-BPSK(1), and MSK-BOC(1,1) signals with respect to the double-sided receiving bandwidth. It can be seen that the frequency attenuation of the MSK-BPSK signal and the MSK-BOC signal is more rapid, and their main lobe energy is relatively concentrated. Therefore, in cases of very limited receiving bandwidth, the RMS bandwidth of the MSK-BPSK signal is larger than that of the BPSK-R signal, and the RMS bandwidth of MSK-BOC signal is greater than that of the BOC signal. However, in cases of a large receiving bandwidth, since BPSK-R and BOC signals have higher side lobes, the RMS bandwidth of the BPSK-R and BOC signals is larger than that of MSK-BPSK and MSK-BOC signals. Meanwhile, the MSK-BPSK and MSK-BOC signals' side lobes contain very little signal energy due to rapid frequency decay. Therefore, when the receiving bandwidth is greater than the main lobe width, the RMS bandwidth of MSK-BPSK and MSK-BOC signals does not change significantly with the increase of the receiving bandwidth.

Figure 4.42a, b show the multipath error envelopes of BPSK-R(1), BOC(1,1), MSK-BPSK(1), and MSK-BOC(1,1) signals with a receiving bandwidth of 3 MHz and 10 MHz respectively, where the non-coherent early-minus-late discriminator is employed with a 0.1 chip E-L correlator. The multipath and direct signal amplitude ratio is −10 dB. Similar to the conclusion of RMS bandwidth comparison, in cases where the receiver bandwidth is close to the main lobe width of the signal, the MSK signals have a better multipath resistant performance than BPSK and BOC signals due to the concentration of the main lobe energy as shown in Fig. 4.42a. However, when the receiver bandwidth is greater than the signal main lobe bandwidth, the MSK signals' side lobes are rapidly attenuated, which results in less side lobe energy, and inferior multipath resistant performance compared to BPSK-R and BOC signals, as shown in Fig. 4.42b.

4.9.3 GMSK Modulation

The MSK signal has good out-of-band attenuation characteristics, and is also a constant envelope signal. However, it should be noted that the side lobes of the MSK signal still have a certain power. In some cases where the radiant power of the signal is strictly limited, the MSK modulation cannot meet power limitation requirements [44].

In order to further compress the side lobe energy of the MSK signal and reduce the out-of-band radiant power, Murota et al. [45] proposed the GMSK modulation technique in 1981. The principle of GMSK modulation is to add a Gaussian pre-filter

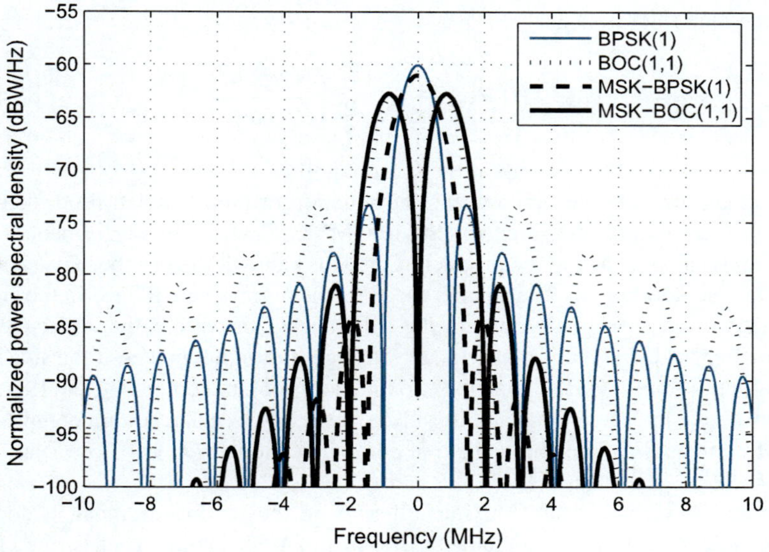

Fig. 4.40 Normalized PSD of BPSK-R(1), BOC(1,1), MSK-BPSK(1), and MSK-BOC(1,1)

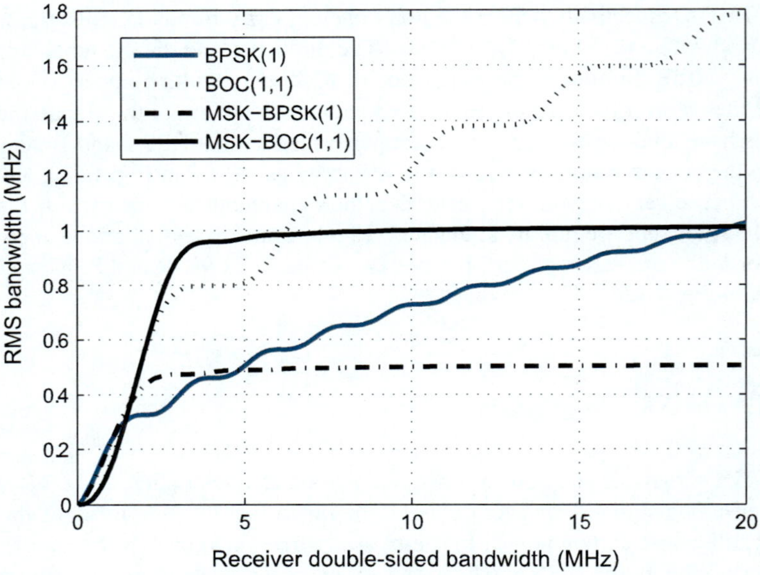

Fig. 4.41 Comparison of RMS bandwidth of BPSK-R(1), BOC(1,1), MSK-BPSK(1), and MSK-BOC(1,1)

before MSK modulation. Through the Gaussian filtering of the input baseband signal, the signal can achieve good out-of-band attenuation. The unit impulse response of the Gaussian filter is

$$g_{\mathrm{GMSK}}(t) = \sqrt{\frac{2\pi}{\ln 2}} B \exp\left(-\frac{2\pi^2 B^2 t^2}{\ln 2}\right), \tag{4.100}$$

where B is the 3dB bandwidth of the Gaussian filter.

It should be noted that due to the introduction of the Gaussian filter, there is a certain inter-symbol interference (ISI) in GMSK signals. The performance of the GMSK signal is related to the product of the 3dB bandwidth B of the Gaussian filter and the chip duration T_c. The smaller the BT_c, the more severe the ISI in the signals, and the weaker the signal power of the side lobes. Conversely, the larger the BT_c, the smaller the ISI in the signals, and the stronger the signal power of the side lobes. Therefore, for the GMSK signal, the two indicators of band efficiency and ISI cannot be satisfied at the same time, and it is often necessary to make certain compromises for different applications. In the field of communications, $BT_c = 0.3$ is usually used [46].

Since GMSK performs Gaussian filtering on the baseband signal, it does not affect the constant envelope characteristic of the MSK signal. That is, the GMSK signal is still a constant envelope signal, so it can still be used in GNSS. Due to the advantageous spectrum efficiency and constant envelope characteristics of GMSK

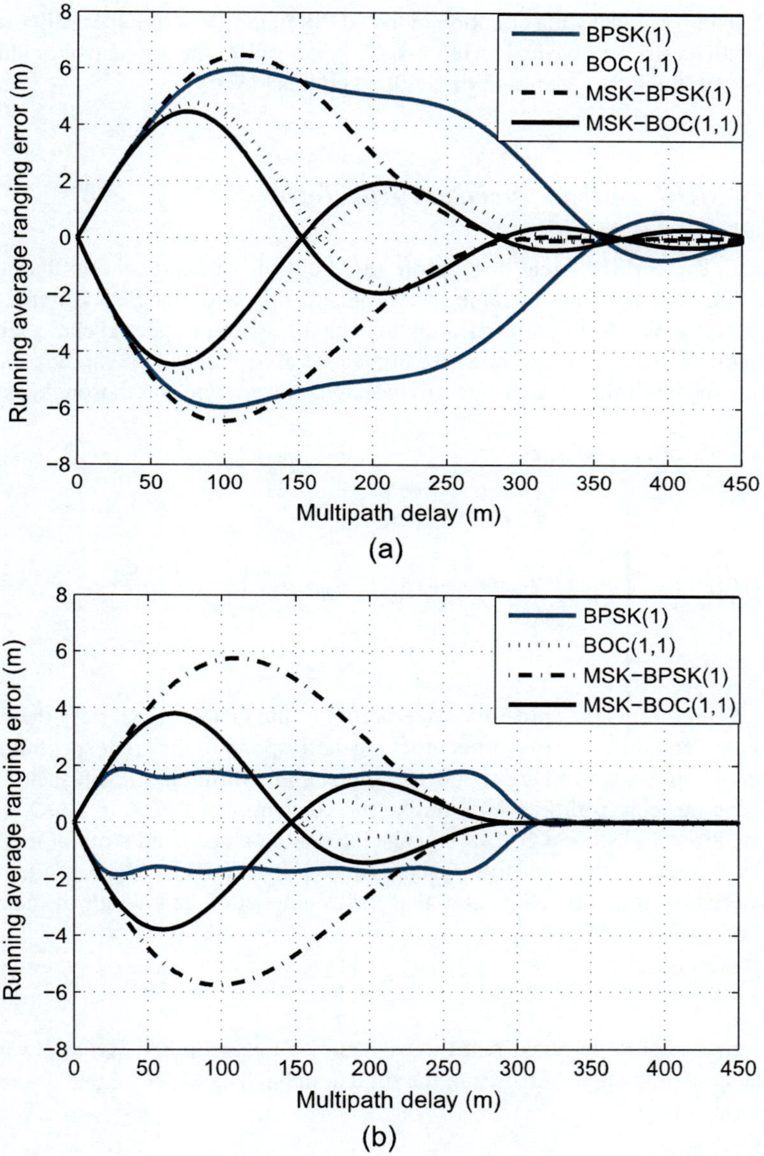

Fig. 4.42 Comparison of BPSK(1), BOC(1,1), MSK-BPSK(1), and MSK-BOC(1,1) multipath error envelopes under the receiving bandwidth of **a** 3MHz and **b** 10MHz

signal, it has received wide attention in the GNSS field [46, 47]. Especially in the C-band, with its strict spectrum constraints, GMSK is one of the alternative modulation methods for the next generation of Galileo systems [48, 49].

4.9.4 SRRC Spread Spectrum Waveform

Although the GMSK signal has a more rapid out-of-band attenuation than that of the MSK signal, there is still some energy outside the band. The SRRC signal that is widely used in wireless communication field has exceptional spectral characteristics. During the design of Galileo, the SRRC signal was used as one of the candidate signals [50], and the feasibility of applying it to the navigation system was thoroughly studied [2].

The PSD of the SRRC signal is

$$G_{\mathrm{SRRC}}(\alpha, f) = \begin{cases} T_c, & |f| \leq \dfrac{1-\alpha}{2T_c} \\ \dfrac{T_c}{2}\left\{1+\cos\left[\dfrac{\pi T_c}{\alpha}\left(|f| - \dfrac{1-\alpha}{2T_c}\right)\right]\right\}, & \dfrac{1-\alpha}{2T_c} < |f| \leq \dfrac{1+\alpha}{2T_c} \\ 0, & |f| > \dfrac{1+\alpha}{2T_c} \end{cases} \quad (4.101)$$

where T_c represents the spreading code period of the GNSS signal, α is the roll-off coefficient, and different roll-off coefficients correspond to different spectral forms. Figure 4.43 shows the PSD of the SRRC signal under different roll-off factors.

As can be seen from Fig. 4.43, unlike the conventional BPSK-R, BOC, and the aforementioned MSK and GMSK signals, the double-sided bandwidth of the SRRC signal is limited. At the same time, in cases of $\alpha = 0$, its PSD has the best attenuation characteristics. It can be calculated that the double-sided bandwidth of the SRRC signal is

$$B_{\mathrm{SRRC}} = \frac{1+\alpha}{T_c}. \quad (4.102)$$

However, it should be noted that the frequency domain limited signal always has infinite spreading characteristics in the time domain. The SRRC signal's spreading waveform is [51]

$$s_{\mathrm{SRRC}}(t) = \frac{4\alpha}{\pi\sqrt{T_c}} \frac{\cos\left[\dfrac{(1+\alpha)}{T_c}\pi t\right] + \dfrac{T_c}{4\alpha t}\sin\left[\dfrac{(1-\alpha)}{T_c}\pi t\right]}{1-\left(\dfrac{4\alpha t}{T_c}\right)^2}. \quad (4.103)$$

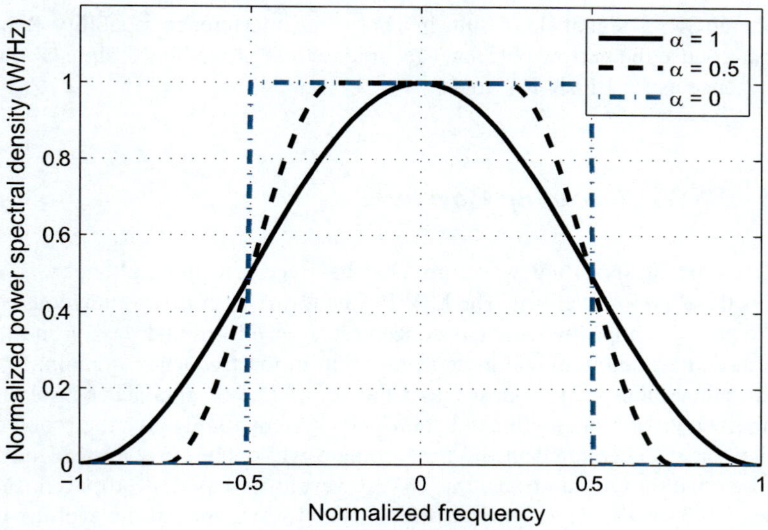

Fig. 4.43 The PSD of the SRRC signal

It is not difficult to verify that the SRRC signal time domain waveform has an infinite stretch characteristic, which causes ISI in this signal. Moreover, the smaller the roll-off coefficient α, the more severe the ISI in the signal.

Since the SRRC signal has the characteristics of optimal spectral efficiency, it can propagate the most signal energy in a limited bandwidth, and thus optimal performance can be obtained within a limited bandwidth. Many scholars and practitioners have noticed this feature of the SRRC signal. This signal was considered at the beginning of the Galileo signal design. However, since the SRRC signal originates from communication system design, and navigation and communication applications have different requirements for signals, SRRC signals that perform well in the field of communication are not applicable to the navigation system, and finally SRRC signals were not selected in Galileo system [2]. Its main defects focus on the following aspects.

The SRRC signal is strictly band-limited, and its energy is concentrated in the band, which severely limits its ranging performance. Generally speaking, for traditional GNSS signals, the larger the receiving bandwidth, the better the ranging performance. However, for the SRRC signal, the signal ranging performance does not improve as the receiving bandwidth grows. The same problem can be found in the time domain. The BPSK-R signal and the BOC signal have sharp autocorrelation peaks, while the autocorrelation peak of the SRRC signal is very rounded. We know that the sharpness of the autocorrelation peak of the signal directly determines its ranging performance and multipath resistance. The SRRC signal with its relatively smooth autocorrelation peak is inferior to the BPSK and BOC signals in the thermal noise and multipath environment. Moreover, the excessive concentration of the spectrum of the SRRC signal also results in a poor anti-interference capability. Receivers also cannot use the narrow correlator technique to improve SRRC signals' tracking performance as for BPSK-R signals.

4.9.5 PSWF Spreading Waveform

Another type of spreading waveform that has received much attention in recent years is the PSWF waveform. The PSWF waveform has both advantageous of time and frequency compactness. It has concentrated waveform distribution in the time domain, and a maximum energy concentration in the frequency domain. In 1961, Selpian and Pollak [52] proposed that the set of PSWF obtained by solving the Helmholtz equation in an ellipsoid coordinate system could solve the problem that functions are concentrated in the time domain while energy is maximized in the finite bandwidth. On this basis, the PSWF waveform was first proposed. Further, Parr et al. [53] proposed to apply the PSWF waveform to the pulse waveform design of ultra-wide band (UWB) communication. Antreich et al. [54–56] were the first to suggest PSWF waveform for the design of satellite navigation signals.

As its name suggests, the PSWF signal is obtained by solving the prolate spheroidal wave function set obtained by solving the Helmholtz equation in the ellipsoidal coordinate system. Halpern [57] shows that the condition under which the function $s(t)$ can be concentrated in the time domain $[-T_c/2, T_c/2]$ while its energy is maximally concentrated in the band $[-B, B]$ is

$$\int_{-T_c/2}^{T_c/2} s_{\mathrm{PSWF}}(x)\, h(t-x)\, \mathrm{d}x = \lambda s_{\mathrm{PSWF}}(t) \ , \tag{4.104}$$

where

$$h(t) = \frac{\sin Bt}{\pi t} \tag{4.105}$$

is the Lagrange multiplication factor.

Equation (4.104) is the integral equation of the prolate spheroidal wave function. The solution $s_{\mathrm{PSWF}}(t)$ is called the *prolate spheroidal wave function*. However, the closed-form solution of (4.104) is difficult to obtain directly. An approximate solution of this equation can be obtained by numerical solution [53]. That is, by performing N-point sampling in $[-T_c/2, T_c/2]$, (4.104) can be discretized, and the obtained discrete equation can be expressed as

$$\lambda s[n] = \sum_{m=-N/2}^{N/2} s[m]\, h[n-m] \ , \tag{4.106}$$

where $n = -N/2, -N/2+1, \ldots, N/2$.

Equation (4.106) can be expanded for all n and organized into a matrix form

$$\lambda \boldsymbol{s} = \boldsymbol{H}\boldsymbol{s}\,, \tag{4.107}$$

where

$$\boldsymbol{s} = \left[s\left[-N/2\right], s\left[-N/2+1\right], \ldots, s\left[0\right], \ldots, s\left[N/2\right]\right]^{\mathrm{T}}\,, \tag{4.108}$$

$$\boldsymbol{H} = \begin{pmatrix} h\left[0\right] & h\left[-1\right] & \cdots & h\left[-N\right] \\ h\left[1\right] & h\left[0\right] & \cdots & h\left[-N+1\right] \\ \vdots & \vdots & \vdots & \vdots \\ h\left[N/2\right] & h\left[N/2-1\right] & \cdots & h\left[-N/2\right] \\ \vdots & \vdots & \vdots & \vdots \\ h\left[N\right] & h\left[N-1\right] & \cdots & h\left[0\right] \end{pmatrix}, \tag{4.109}$$

λ represents the eigenvalue of the matrix $\boldsymbol{H}$, and $\boldsymbol{s}$ is the corresponding eigenvector, which is also the N-point discretized time-domain waveform of the prolate spheroidal wave function.

By eigenvalue decomposition of $\boldsymbol{H}$ and sorting these eigenvalues in descending order $\lambda_0 > \lambda_1 > \cdots > \lambda_m$, the eigenvectors corresponding to each eigenvalue are the N-point discrete time-domain waveforms of the prolate spheroidal wave functions of each order.

4.9.6 Summary of Bandwidth-Limited Spreading Modulation Techniques

Compared with the traditional BPSK-R and BOC signals, the above-mentioned MSK, GMSK, PSWF, and SRRC waveforms have better out-of-band attenuation characteristics, making them suitable for applications with strict spectrum constraints. However, admittedly, these signals have several common shortcomings when using them for navigation signals.

In cases where the receiving bandwidth is not greater than the bandwidth of the signal's main lobe, the bandwidth-limited spreading modulations tend to display better ranging performance and multipath resistance due to the concentration of their spectrum. However, once the receiver bandwidth is greater than the signal's main lobe bandwidth, their ranging performance is not as good as the traditional GNSS signal.

Furthermore, theoretical analysis shows that for GNSS signals, the more dispersed the spectrum, the stronger its ability to resist narrowband and wideband interference. Conversely, the more concentrated the spectrum, the weaker its ability to resist interference [8]. Since the spectrum of signals such as MSK and GMSK is relatively concentrated, their anti-interference ability is poor compared to conventional GNSS signals.

References

1. Kaplan ED, Hegarty CJ (2006) Understanding GPS: principles and applications. Artech House, Norwood, MA
2. De Gaudenzi R, Hoult N, Batchelor A, Burden G, Quinlan M (2000) Galileo signal validation development. Wiley, New York
3. Spilker JJ (1998) A family of split spectrum GPS civil signals. In: ION GPS-98, pp 1905–1914
4. Betz JW (2001–2002) Binary offset carrier modulations for radio navigation. Navig J Inst Navig 48(4):227–246
5. Lestarquit L, Artaud G, Issler J-L. AltBOC for dummies or everything you always wanted to know about AltBOC, pp 961–970
6. Pratt AR, Owen J (2001) Performance of GPS Galileo receivers using m-PSK BOC signals. In: Proceedings of the 2004 national technical meeting of the institute of navigation, pp 362–370
7. Hegarty CJ, Betz JW, Saidi A (2001) Binary coded symbol modulations for GNSS. In: Proceedings of the 60th annual meeting of the institute of navigation, pp 56–64
8. Avila-Rodriguez JA (2008) On generalized signal waveforms for satellite navigation. University FAF Munich, Munich, Germany
9. Hein GW, Avila-Rodriguez J-A, Ries L, Lestarquit L, Issler J-L, Godet J, Pratt T (2005) A candidate for the Galileo L1 OS optimized signal. In: Presented at the proceedings of the 18th international technical meeting of the satellite division of the institute of navigation (ION GNSS 2005), pp 833–845
10. Ries L, Issler J-L, Lestarquit L et al (2006) Spread-spectrum signal, EU Patent WO/2006/075018
11. Avila-Rodriguez J-A, Hein GW, Wallner S et al. The MBOC modulation: the final touch to the Galileo frequency and signal plan, pp 1515–1529
12. Hein GW, Avila-Rodriguez J-A, Wallner S, Pratt AR, Owen J, Issler J-L, Betz JW, Hegarty CJ, Lenahan LS, Rushanan JJ et al (2006) MBOC: the new optimized spreading modulation recommended for GALILEO L1 OS and GPS L1C. In: Proceedings of IEEE/ION PLANS, vol 2006, pp 884–892
13. Betz JW, Blanco MA, Cahn CR et al (2006) Description of the L1C signal. In: ION GNSS, pp 2080–2091
14. Lohan ES, Lakhzouri A, Renfors M (2006) Complex double-binary-offset-carrier modulation for a unitary characterization of Galileo and GPS signals. IEE Proc Radar Sonar Navig 153(5):403–408
15. Cote FD, Psaromiligkos IN, Gross WJ (2011) GNSS modulation: a unified statistical description. IEEE Trans Aerosp Electron Syst 47(3):1814–1836
16. Yao Z, Lu M (2011) Optimized modulation for Compass B1-C signal with multiple processing modes. In: ION GNSS 24th international technical meeting of the satellite division. Portland, USA, pp 1234–1242
17. Proakis JG (2001) Digital communications, 4th edn. McGraw-Hill Companies Inc., Boston
18. Pratt AR, Owen JI, Hein GW, Avila-Rodriguez JA (2006) Tracking complex modulation waveforms - how to avoid receiver bias. In: Proceedings of IEEE/ION PLANS 2001, pp 853–864
19. Yao Z, Lu M, Feng Z (2010) Quadrature multiplexed BOC modulation for interoperable GNSS signals. Electron Lett 46(17):1234–1236
20. Cobb HS (1997) GPS pseudolites: theory, design and applications. PhD thesis, Stanford University, Sept 1997
21. Helstrom CW (1968) Statistical theory of signal detection. International series of monographs in electronics and instrumentation
22. Van Dierendonck AJ, Fenton P, Ford T (1992) Theory and performance of narrow correlator spacing in a GPS receiver. Navigation 39(3):265–283
23. Fine P, Wilson W. Tracking algorithm for GPS offset carrier signals, pp 671–676
24. Fishman PM, Betz JW (2000) Predicting performance of direct acquisition for the M-code signal. In: Proceedings of the 2000 national technical meeting of the institute of navigation, pp 574–582

25. Martin N, Leblond V, Guillotel G, Heiries V (2001) BOC (x, y) signal acquisition techniques and performances. In: Proceedings of the 16th international technical meeting of the satellite division of the institute of navigation (ION GPS/GNSS 2003), pp 188–198
26. Burian A, Lohan ES, Renfors M (2006) BPSK-like methods for hybrid-search acquisition of Galileo signals. In: Communications, 2006. ICC'06. IEEE international conference, vol 11, pp 5211–5216
27. Hodgart MS, Blunt PD (2007) Dual estimate receiver of binary offset carrier modulated signals for global navigation satellite systems. Electron Lett 43(16):877–878
28. Hodgart MS, Simons E (2012) Improvements and additions to the double estimation technique. In: IEEE. Proceedings of 2012 6th ESA workshop on satellite navigation technologies and European workshop on GNSS signals and signal processing (NAVITEC). Noordwijk: IEEE, pp 1–7
29. Blunt P, Weiler R, Hodgart MS (2007) Demonstration of BOC (15; 2:5) acquisition and tracking with a prototype hardware receiver. In: The European institute of navigation. Proceedings of European navigation conference GNSS 2007. Geneva, Switzerland, The European Institute of Navigation, pp 1–11
30. Ren J, Jia W, Chen H et al (2012) Unambiguous tracking method for alternative binary offset carrier modulated signals based on dual estimate loop. IEEE Commun Lett 16(11):1737–1740
31. Borio D (2014) Double phase estimator: a new unambiguous BOC tracking algorithm. IET Radar Sonar Navig 8(7):729–741
32. Wendel J, Schubert FM, Hager S. A robust technique for unambiguous BOC tracking
33. Ward P (2003) A design technique to remove the correlation ambiguity in binary offset carrier (BOC) spread-spectrum signals. In: Proceedings of ION AM, Albuquerque, NM, pp 146–155
34. Julien O, Macabiau C et al (2007) ASPeCT: unambiguous Sine-BOC (n, n) acquisition/tracking technique for navigation applications. IEEE Trans Aerosp Electron Syst 43(1):150–162
35. Yao Z, Lu M (2011) Side-peaks cancellation analytic design framework with applications in unambiguous processing of BOC signals. In: Proceedings of ION ITM, San Diego, CA, pp 775–785
36. Yao Z, Lu M, Feng Z (2010) Unambiguous sine-phase binary offset carrier modulated signal acquisition technique. IEEE Trans Wirel Commun 9(2):577–580
37. Yao Z, Cui X, Lu M, Feng Z, Yang J (2010) Pseudo-correlation function-based unambiguous tracking technique for sine-BOC signals. IEEE Trans Aerosp Electron Syst 46(4):1782–1796
38. Yao Z (2012) Unambiguous processing techniques of binary offset carrier modulated signals. In: Jin S (ed) Global navigation satellite systems - signal, theory and applications (Chapter 3). InTech Press, Rijeka, Croatia, pp 53–76
39. Schubert FM, Wendel J, Sollner M, Kaindl M, Kohl R (2014) The Astrium correlator: unambiguous tracking of high-rate BOC signals. In: Proceedings of IEEE/ION PLANS 2014, Monterey, CA, May 2014, pp 589–601
40. Yao Z, Gao Y, Gao Y, Lu M (2017) Generalized theory of BOC signal unambiguous tracking with two-dimensional loops. IEEE Trans Aerosp Electron Syst 53(6)
41. Gao Y, Yao Z, Lu M (2018) Theoretical analysis of unambiguous 2-D tracking loop performance for band-limited BOC signals. GPS Solut 22(1)
42. Simon MK (2001) Bandwidth-efficient digital modulation with application to deep-space communications. JPL Publication 00–17:3–5
43. Avila-Rodriguez JA, Wallner S, Won JH et al (2008) Study on a Galileo signal and service plan for the C-Band. In: ION GNSS 21st international technical meeting of the satellite division. GA, Savannah, pp 2515–2529
44. Ungerboeek G (1987) Trellis-coded modulation with redundant signal sets. Introduction. IEEE Commun Mag 25(2):12–21
45. Murota K, Hirade K (1981) GMSK modulation for digital mobile radio telephony. IEEE Trans Commun 29(7):1044–1050
46. Fontanella D, Paonni M, Eissfeller B (2010) A novel evil waveforms threat model for new generation GNSS signals: theoretical analysis and performance. In: Satellite navigation technologies and European workshop on GNSS signals and signal processing (NAVITEC), 2010, 5th ESA workshop, IEEE

47. Emmanuelle A (2012) Signal design and theoretical bounds for time-of-arrival estimation in GNSS applications. PhD thesis at the University of Pisa
48. Avila-Rodriguez JA, Wallner S, Won JH, Eissfeller B, Schmitz-Peiffer A, Floch JJ, Colzi E, Gerner JL (2008) Study on a potential Galileo signal and service plan for the C-band. In: ION-GNSS 2008. Savannah, USA
49. Fontanella D, Paonni M, Eissfeller B (2008) C-band user terminal concepts and acquisition performance analysis for European GNSS evolution programme. In: ION-GNSS 2008. Savannah, USA
50. Hein GW, Godet J, Issler JL, Martin JC, Lucas-Rodriguez R, Pratt T (2001) The Galileo frequency structure and signal design. In: Proceedings of the international technical meeting of the institute of navigation, ION-GNSS (2001) Sept 2001. Salt Lake City, Utah, USA
51. Lee EA, Messerschmitt DB. Digital communications, 2nd edn. Kluwer Academic Publishers, Norwell, Massachusetts
52. Selpian D, Pollak O (1961) Prolate spheroidal wave functions, Fourier analysis and uncertainty. Bell Syst Technol J 40(1)
53. Parr B, Cho BL, Wallace K (2003) A novel ultra-wideband pulse design algorithm. IEEE Commun Lett 7(5):219–221
54. Antreich F, Nossek JA, Issler JL, Meurer M (2009) On methodologies for GNSS signal design. In: Proceedings of the 4th European workshop on GNSS signals and signal processing. Oberpfaffenhofen, Germany
55. Floch JJ, Antreich F, Meurer M, Issler JJ (2010) S-band signal design considering interoperability and spectral separation. In: Proceedings of the 23rd annual international technical meeting ION GNSS 2010. Portland, Oregon, USA
56. Antreich F, Floch JJ, Issler JL, Nossek JA (2011) On backward compatibility in GNSS signal design. In: Proceedings of the 5th European workshop on GNSS signals and signal processing. Toulouse, France
57. Halpern P (1979) Optimum finite duration Nyquist signals. IEEE Trans Commun 27(6):884–888

Chapter 5
Performance Evaluation Theory for Satellite Navigation Signals

Admit that complexity always increases, first from the model you fit to the data, thence to the model you use to think about and plan about the experiment and its analysis, and thence to the true situation.

— John Tukey

5.1 Introduction

The design of the GNSS navigation signal spreading modulation is an optimization problem under multiple constraints. The optimization of the spreading chip waveform must account for various factors in satellite payload, ground control, and user terminals. Furthermore, the optimization must comprehensively consider the requirement on ranging performance, the limitation of spectrum resources, constraints posed by compatibility and interoperability, and the implementation complexities of payload and user equipment. In order to evaluate the performance of the navigation signal spreading modulation and compare modulation modes more comprehensively and objectively, it is necessary to establish evaluation criteria for each key performance and provide the corresponding quantitative evaluation methods. Only then can the spreading modulation optimization, comparison and evaluation be carried out in a quantitative form under a reasonable evaluation system.

Signal performance analysis is crucial both to designers of navigation systems and receiver developers. Signal designers usually need to begin with requirements, decompose the indicators of the signal design, and evaluate the performance of candidates of the signal designs in order to check whether the candidate technique can meet the corresponding requirements. Receiver developers also need to evaluate whether the performance of the signal under certain receiving methods can meet the application requirements.

Z. Yao and M. Lu, *Next-Generation GNSS Signal Design*, Navigation: Science and Technology 6, https://doi.org/10.1007/978-981-15-5799-6_5

In general, there are four methods for assessing the performance of satellite navigation signals: theoretical evaluation, computer simulation, physical simulation, and verification by satellite/ground transmission and reception. Among them, theoretical evaluation mathematically models the generation, transmission, and reception of signals based on reasonable assumptions and derivations. It can calculate the characteristic quantity of a certain aspect of the signal performance. Therefore, this method has strong universality. Computer simulation is based on the principle of similitude. It simulates the generation, reception, and processing of satellite navigation signals by computer, and evaluates the receiving performance using statistical methods. The physical simulation method and the satellite/ground transmission-reception verification are based on a navigation signal simulator and the actual GNSS satellite respectively. Actual GNSS receivers are used to receive and process the navigation signals to complete the confirmation of the signal's performance.

These four assessment methods approach the actual situation successively with the increasing evaluation cost, and the universality of the evaluation results gradually decreases. Physical verification is necessary before the signal scheme or receiving algorithm is finalized. However, in the early stages of signal design and processing algorithm development, new ideas may emerge at any time, and rapid verification is required. Therefore, theoretical evaluation and computer simulation evaluation become the most important evaluation means.

The theoretical evaluation method is universally applicable and can quickly detect problems in the signal scheme or receiving algorithm design, which greatly reduces the risk of failing to pass the subsequent verifications, thereby shortening the design cycle. According to varying research concerns, theoretical analysis models can be established at different levels of complexity. With a deeper understanding of the system internal structure, the theoretical evaluation model can also be refined and made more accurate.

However, any model is just a simplification of the actual system. In terms of feasibility and practicality, it is impossible or perhaps even unnecessary to enumerate all the details of the actual system. A real signal or system can be considered to be simulated and described by a simplified model if its essential features are preserved. Further refinement is taken only when the accuracy of the existing model can no longer satisfy the analysis requirements. Therefore, in theoretical modeling, an appropriate compromise must be made between the simplification of the model and the accuracy of the analytical results. This is a principle followed by theoretical evaluation.

The estimation of the TOA of the spread-spectrum signal is the basis for realizing positioning, speed measurement, and timing. Therefore, the evaluation of the tracking performance is an important focus of the navigation signal performance analysis.

The main error sources in spreading code tracking are divided into three categories: code tracking jitter caused by thermal noise and interference, code tracking deviation caused by multipath effects, and dynamic stress errors caused by relative dynamics between the receiver and the satellite. Among them, the dynamic stress error is more related to the design of the receiver loop, while the code tracking jitter caused by thermal noise and interference, and the code tracking deviation caused by multipath effect are closely related to the time domain and frequency domain characteristics of the spread-spectrum signal.

Code tracking jitter is mainly measured by the variance of the code tracking error, which is called code tracking accuracy. Some of the early research on the theoretical performance analysis of the spreading code tracking loop was undertaken by Lindsey and Simon et al. [1, 2]. In [2], the theoretical evaluation formula of the spreading code tracking for a non-coherent delay locked loop (DLL) under thermal noise is given. However, the evaluation formula is based on the assumption of an ideal triangular ACF, and thus can only be applied to a spread-spectrum signal using a BPSK-R spread chip waveform and a receiver whose front-end bandwidth is much larger than the signal bandwidth.

With the popularization of GPS applications, designers began to seek ways of improving the tracking accuracy of the spreading code and proposed more accurate methods of analyzing code tracking performance. Van Dierendonck et al. [3] pointed out that narrow early-late correlator spacing of GPS receivers can improve the code tracking accuracy in a thermal noise environment and reduce the tracking derivation caused by multipath effects. They also developed an expression for the tracking error of the coherent/non-coherent E-L DLL under thermal noise with respect to the early-late correlator spacing. This expression is still based on the assumption of an ideal triangle ACF and does not consider the effects of front-end bandwidth. For GPS C/A codes, the transmission bandwidth is much larger than the spreading code rate, so the infinite bandwidth approximation is reasonable in most engineering applications. Therefore, for a long time, most of the code tracking error evaluation in the navigation field has been based on this formula. However, for the GPS P(Y) code signal and wideband and split spectrum signals in the new generation of GNSS, due to the limited bandwidth, the shape of the ACF deviates greatly from the ideal triangle. If we continue to use the evaluation formula in [3], the assessment results will significantly deviate from the actual situation.

In 2009, Betz et al. [4] introduced the influences of the spreading chip waveform, receiver front-end bandwidth, and non-white noise interference into the analysis of code tracking accuracy. Further, this work provided a series of more generalized code tracking accuracy expressions which integrated the influence of factors such as signal power spectrum, interference power spectrum, front-end bandwidth of the receiver, correlator spacing, code loop bandwidth, and pre-detection integration time. This analysis relates the code tracking accuracy to the spreading modulation characteristics of navigation signals and has been widely used in the design and evaluation of the next-generation GNSS signal.

However, the analysis in [4] is based on the assumption that the de-spreading signal waveform generated in the receiver exactly matches the spreading waveform of the signal transmitted by the system. In fact, with the continuous development of processing technology and the increasing complexity of next-generation navigation signals, mismatch de-spreading solutions are commonly used in some advanced navigation receivers in order to improve performance or reduce the complexity of processing. That is, the receiver uses a spreading chip waveform inconsistent with that of the transmitted signal. For example, the double-delta discriminator [5, 6], which is commonly used to resist multipath effects, can be regarded as a mismatch de-spreading technique in essence. The use of BPSK-like methods [7–9] in the unambiguous processing of BOC signals (also see Sect. 4.8.5), and the use of BOC(1,1) local codes to de-spread MBOC signals [10] are also successful applications of mismatch de-spreading techniques.

In order to support the signal performance evaluation under the emerging mismatch de-spreading solutions, a set of evaluation methods that both supports the various local receiving strategies and decouples from unrelated factors is proposed in [11]. Specifically, the mismatch of the code chip waveform of the local de-spreading signal is considered.

In this chapter, we will discuss the theory and methods of navigation signal performance analysis based on time and frequency-domain characteristics of the spreading modulation waveform. These methods have provided direct guidance for the spreading modulation optimization and selection of the next-generation satellite navigation signals. The content of this chapter is mainly based on the content of [4, 11, 12]. It is inevitable to make some assumptions and simplifications about the signal, the transmission environment, and the receiver characteristics during the analysis. These assumptions and simplifications are consistent with those made by most existing literature on code ranging performance analysis. Some of the most important assumptions are:

- The influences of carrier synchronization and dynamic stress on the code tracking loop are ignored;
- The nonlinear distortion introduced by the on-board signal generation and transmission link are ignored;
- The influence of the receiver's sampling and quantizing is ignored.

Although in practice, when noise or interference is severe, the tracking jitter of the carrier synchronization loop and the demodulation error will have a certain impact on the code tracking loop, and some research has shown that the effects of non-linear distortion [13], receiver sampling, and quantization [14] on the tracking performance cannot be ignored in some extreme scenarios. However, in most application scenarios, the simplified model is accurate enough to reflect the actual performance of the signal. Therefore, this chapter does not cover the signal performance analysis in specific harsh scenarios. Readers who are interested in those topics are recommended to read the corresponding literature.

5.2 Baseband Equivalent Expressions of Received Signals

5.2.1 Complex Envelope Representation of Signals

Apart from the general expression in (3.1), the spreading modulated signal $s(t)$ can also be expressed by a complex envelope [15], which can offer a high degree of convenience to the analysis. Using the Euler formula, we can rewrite $s(t)$ in (3.1) as

$$s(t) = A_s \operatorname{Re}\left\{D(t)\, g(t)\, \mathrm{e}^{\mathrm{j}(2\pi f_0 t+\theta)}\right\} = A_s \operatorname{Re}\left\{s_l(t)\, \mathrm{e}^{\mathrm{j}2\pi f_0 t}\right\}, \tag{5.1}$$

where $\operatorname{Re}\{x\}$ represents the real part of the complex number x, and $g(t)$ and $D(t)$ are defined by (3.2) and (3.48) respectively. In (5.1), $s_l(t) \triangleq D(t)\, g(t)\, \mathrm{e}^{\mathrm{j}\theta}$ is referred to as the *complex envelope* of $s(t)$.

Since when we study the spreading chip waveform, the RF carrier is seen as a vessel of the baseband signal that does not carry useful information itself, the complex envelope can separate the effect of the carrier from the whole signal, so that we can use the same analysis and design method to operate signals at different carrier center frequencies f_0.

After $s(t)$ is generated on-board, it passes through the transmission channel, causing a propagation delay of τ. Since the carrier tracking is assumed to be perfect, we can focus on the baseband complex envelope of the spread-spectrum signal. The data obtained by the receiver can be modeled as

$$r(t) = s_l(t) + w(t) = A_s g(t-\tau)\, \mathrm{e}^{\mathrm{j}\theta} + w(t), \tag{5.2}$$

which is the sum of a complex baseband signal $s_l(t)$ and noise-plus-interference $w(t)$.

5.2.2 PSD of the Signal and Noise-Plus-Interference

As described in Chap. 3, for a long period or infinite period spread-spectrum signal, its PSD $S(f)$ is approximately in the form of (3.51). In order to be more explicit in physical meaning, and to carry out subsequent analysis and comparison under a uniform condition, according to the conventions in Sect. 3.4.7, the PSD $S(f)$ of the signal is further decomposed into the product of the signal carrier power and the normalized PSD, i.e. $S(f) = C_s G_s(f)$, where $C_s = \int_{-\infty}^{+\infty} S(f)\mathrm{d}f$ is the signal carrier power and $G_s(f) = S(f)/C_s$ is the normalized signal PSD, which satisfies $\int_{-\infty}^{+\infty} G_s(f)\mathrm{d}f = 1$. In this way, when we compare and analyze the normalized PSDs of different signals, we can eliminate the performance difference caused by the difference in transmission power.

There are multiple sources of noise and interference in the actual signal reception, and white noise is the main consideration. However, in many cases, in addition to thermal noise, the signal is inevitably disturbed by intra-system and inter-system signals and other RF signals in the same frequency band. The power spectrum of the interfering signal is not necessarily white. If it is assumed that both white noise and interference signals are wide-sense stationary random processes with zero means, and they are independent of the signals and are statistically independent of each other, the PSD of noise-plus-interference can be written as

$$W(f) = N_0 + C_\iota G_\iota(f) \ , \tag{5.3}$$

where N_0 is the PSD of white noise, C_ι is the interference signal carrier power under the infinite bandwidth, and $G_\iota(f)$ is the normalized interference PSD, which satisfies $\int_{-\infty}^{+\infty} G_\iota(f)\mathrm{d}f = 1$.

In general, there are several typical non-white interference signals:

- For narrow-band interference with center frequency f_ι, the PSD can be modeled as $G_\iota(f) = \delta(f - f_c)$, where $\delta(\cdot)$ is the Dirac δ-function;
- If the interference has the same PSD as the desired signal, it is referred to as matching spectral interference, which is modeled as $G_\iota(f) = G_s(f)$;
- When the interference power spectrum is flat, the center frequency is f_ι, and the double-sided bandwidth is β_ι, it is called band-limited white noise interference, and its power spectrum can be expressed as

$$G_\iota(f) = \begin{cases} \dfrac{1}{\beta_\iota}, & f_\iota - \beta_\iota/2 \le f \le f_\iota + \beta_\iota/2 \\ 0, & \text{otherwise} . \end{cases} \tag{5.4}$$

In the baseband equivalent expression model, the combined effects of the on-board transmitter filter, the transmission channel, and the front end filter of the receiver can be modeled by a single filter with a frequency characteristic of $H_L(f)$. Therefore, $\tilde{S}(f)$, the PSD of the receiver front-end filter output signal and $\tilde{W}(f)$, the PSD of the noise plus interference are respectively

$$\tilde{S}(f) = C_s G_s(f) \left|H_L(f)\right|^2 \ , \tag{5.5}$$

and

$$\tilde{W}(f) = \left[N_0 + C_\iota G_\iota(f)\right] \left|H_L(f)\right|^2 \ . \tag{5.6}$$

5.3 Lower Bound of the Spreading Code Tracking Error

The actual accuracy of the spreading code tracking is the comprehensive result of the signal structure, the receiver processing algorithm, and the receiving environment that includes dynamics, shadowing, multipath, and interference and so on. The performance of the same signal structure with different receiving processing algorithms may vary greatly. However, no matter how advanced the receiving technologies are, the optimal performance they can achieve will be subject to the performance limits of the navigation signal itself. With the full development of receiving technology, the signal structure will become a bottleneck limiting performance improvement.

In Sects. 3.3 and 3.4, we have seen that for the spread-spectrum signals with the approximate ideal PRN code correlation, the various time domain and frequency domain characteristics can be determined uniquely by the spreading modulation, that is, the spreading chip waveform. In the design of spreading chip waveforms, one issue of concerns is whether there is a correspondence between the selected spreading chip waveform and the inherent ranging accuracy of the signal, and if there is such an explicit correspondence, how to optimize the spreading chip waveform to achieve the highest possible inherent ranging accuracy. We will answer this question in this subsection.

5.3.1 *Cramér–Rao Lower Bound of the Code Tracking Error*

The Cramér–Rao lower bound (CRLB) of the code tracking error can be used to characterize the performance limits of signal ranging errors. The propagation delay τ of the signal is estimated by processing the received signal-plus-noise $r(t)$, and the lower bound of the estimated error variance is [16]

$$\begin{aligned} \sigma_\tau^2 &= \mathbb{E}_r\left\{(\tau-\hat{\tau})^2\right\} \\ &\geq \sigma_{\mathrm{CRLB}}^2(\tau) \triangleq \left[1+\frac{\mathrm{d}\varepsilon(\tau)}{\mathrm{d}\tau}\right]\left[\mathbb{E}_r\left\{\left[\frac{\mathrm{d}\ln p(\boldsymbol{r}\,|\tau)}{\mathrm{d}\tau}\right]^2\right\}\right]^{-1}, \end{aligned} \tag{5.7}$$

where $\boldsymbol{r}$ is the vector of a series of observations of $r(t)$ at $t = i\Delta$ when $i = 0, 1, \ldots, N-1$, Δ is the observation interval, $\varepsilon(\tau) = \mathbb{E}_r\left\{\hat{\tau}\right\} - \tau$ is the deviation of the estimated expectation from the true value, and $p(\boldsymbol{r}\,|\tau)$ is the likelihood function of $\boldsymbol{r}$ for a given τ. In cases where the estimation of τ is unbiased, i.e. $\varepsilon(\tau) = 0$, for a tracking loop with a pre-detection integration time of T and an equivalent noise bandwidth of B_L Hz, the CRLB of the code tracking error is approximately equal to [4]

$$\sigma_{\mathrm{CRLB}}^2(\tau) \approx \frac{B_L(1-0.5B_LT)}{(2\pi)^2\displaystyle\int_{-\infty}^{+\infty} f^2\left[\tilde{S}(f)/\tilde{W}(f)\right]\mathrm{d}f}, \tag{5.8}$$

where $\tilde{S}(f)$ and $\tilde{W}(f)$ are given by (5.5) and (5.6) respectively.

Equation (5.8) shows that the CRLB of the code tracking error is inversely proportional to the integral value in the equation when the pre-detection integration time and the equivalent noise bandwidth are given. Obviously, a signal with a larger integral value in (5.8) can have a more accurate code tracking potential. In addition to the spectral information of the signal, the integral value is also related to the spectral characteristics of noise and interference. In fact, it can be seen from the square of the frequency in the integral that the high-frequency components in the signal spectrum contribute more to the tracking accuracy than the low-frequency components.

Under normal conditions, the receiver's front-end filter bandwidth is always smaller than the signal's transmission bandwidth, so $H_L(f)$ is mainly determined by the receiver's front-end filter bandwidth. Under the baseband equivalent expression model, the receiver's front-end filter can be approximated as an ideal low-pass filter with a double-sided bandwidth of β_r, and its frequency response is

$$H_L(f) = \begin{cases} 1, & |f| \leq \beta_r/2 \\ 0, & |f| > \beta_r/2 \end{cases}. \tag{5.9}$$

Now, (5.8) can be further organized into

$$\sigma^2_{\mathrm{CRLB}}(\tau) \approx \frac{B_L(1 - 0.5 B_L T)}{(2\pi)^2 \int_{-\beta_r/2}^{+\beta_r/2} f^2 G_s(f) \left[\left(\frac{C_s}{N_0} \right)^{-1} + \frac{C_\iota}{C_s} G_\iota(f) \right]^{-1} \mathrm{d}f}, \tag{5.10}$$

where C_s/N_0 is the CNR of the desired signal, and C_ι/C_s is the power ratio of the non-white-noise interference to the desired signal.

5.3.2 CRLB in a Gaussian White Noise Environment

When $w(t)$ contains only thermal noise and does not contain non-white interference, we have $W(f) = N_0$. At this time, (5.8) is simplified to

$$\sigma^2_{\mathrm{CRLB}}(\tau) \approx \frac{B_L(1 - 0.5 B_L T)}{(2\pi)^2 \frac{C_s}{N_0} \eta \beta^2_{\mathrm{RMS}}}, \tag{5.11}$$

where

$$\eta = \int_{-\beta_r/2}^{+\beta_r/2} G_s(f)\,\mathrm{d}f \tag{5.12}$$

is the power proportion of the signal remaining after front-end filtering, which characterizes the power loss caused by the receiver filtering, and

$$\beta_{\mathrm{RMS}} = \left(\int_{-\beta_r/2}^{+\beta_r/2} f^2 G_s(f) \eta^{-1} \mathrm{d}f \right)^{1/2} = \left(\frac{1}{T_c \eta} \int_{-\beta_r/2}^{+\beta_r/2} f^2 |P(f)|^2 \mathrm{d}f \right)^{1/2} \tag{5.13}$$

is referred to as the RMS bandwidth, or Gabor bandwidth [17], of the signal, where $P(f)$ is the Fourier transform of the chip waveform $p(t)$.

Equations (5.11) and (5.13) show that when the CNR, the pre-detection integration time, and the loop equivalent noise bandwidth are given, the larger the RMS bandwidth of the signal is, the higher the limit of the ranging accuracy. Increasing the RMS bandwidth of a spread-spectrum signal can be achieved in three ways:

- Increase β_r, mainly by increasing the transmission bandwidth and the receiver bandwidth;
- Decrease T_c, that is, increase the spreading code rate;
- Increase the high-frequency components in the signal spreading chip waveform $p(t)$ when T_c is constant.

The position of T_c in the denominator of (5.13) suggests that the higher the spreading code rate, the larger the RMS bandwidth of the signal. It is clear that wideband signals have better ranging performance than narrowband signals, as can be seen from the ranging performance comparison of GPS P(Y) code and C/A code. However, the spreading code rate of the signal cannot be increased infinitely considering the implementation complexity or spectrum compatibility. An increase in the spreading code rate means a wider signal bandwidth. Even if the satellite transmission bandwidth can meet the requirements, the receiver may not be compatible from the perspective of cost and power consumption. Moreover, the degree of competitiveness between systems in the use of the spectrum also makes it impossible to increase the bandwidth of the signal indefinitely. Currently, the spectrum available for use in the L-band and C-band is strictly limited.

Therefore, if we want the signal to have the highest potential ranging accuracy in the AWGN channel, the signal is actually designed to optimize the signal spreading chip waveform $p(t)$ to maximize β_{RMS} when the transmission and reception bandwidth of the signal is limited.

When the total bandwidth usage cannot exceed B_{IF} and all the power of the baseband signal is within the frequency band $[-B_{\mathrm{IF}}/2, B_{\mathrm{IF}}/2]$, we have

$$\int_{-B_{\mathrm{IF}}/2}^{+B_{\mathrm{IF}}/2} G_s(f)\,\mathrm{d}f = 1\,. \tag{5.14}$$

Therefore, the above optimization problem can be formulated as

$$\begin{cases} \arg\max\limits_{P(f)} \displaystyle\int_{-B_{\mathrm{IF}}/2}^{+B_{\mathrm{IF}}/2} f^2 \left|P(f)\right|^2 \mathrm{d}f\,, \\ \text{s.t.} \displaystyle\int_{-B_{\mathrm{IF}}/2}^{+B_{\mathrm{IF}}/2} \left|P(f)\right|^2 \mathrm{d}f = T_c\,. \end{cases} \tag{5.15}$$

It is not difficult to obtain the optimal solution for this optimization problem. Mathematically, this is analogous to a classic problem of distributing mass on a rod to maximize torque under a constraint of total mass. Obviously, the optimal solution can be obtained when the mass is concentrated at both ends of the rod. Therefore, the optimal solution of the optimization problem (5.15) is obtained when

$$\left|P_{\text{opt}}(f)\right|^2 = \frac{1}{2B_{\text{IF}}}\delta\left(f + \frac{B_{\text{IF}}}{2}\right) + \frac{1}{2B_{\text{IF}}}\delta\left(f - \frac{B_{\text{IF}}}{2}\right). \tag{5.16}$$

That is to say, using a sine wave with a frequency of $B_{\text{IF}}/2$ as the spreading chip waveform can allow the signal to reach the ranging accuracy limitation under the band-limited condition. Substituting (5.16) into the (5.11), we have the performance limit of a spread-spectrum signal in a thermal noise environment when the total bandwidth occupation is limited to B_{IF}, which is

$$\sigma^2_{\tau_{\min}} = \frac{B_L}{C_s/N_0} \cdot \frac{1}{\pi^2 B_{\text{IF}}^2}. \tag{5.17}$$

This conclusion is coincident with our intuitive results. In Chap. 4, we have analyzed many spreading modulation techniques. Clearly, split spectrum signals that have emerged in recent years have put this conclusion into practice. This type of signal pushes the spectrum towards both edges of the transmitting bandwidth by introducing subcarriers. As can be seen in Figs. 4.10 and 4.11, for higher-order BOC modulation signals, the shape of their power spectrum has neared the optimal solution given by (5.16).

The last thing to note in this subsection is that CRLB is the lower bound for unbiased estimates. In some cases, such as weak signal processing, the unequal assumption of the estimator may not be satisfied. At this time, the lower bound of accuracy obtained by using CRLB may have some deviation from the real situation so other metrics need to be considered. In fact, in addition to CRLB, there are several lower bounds for accuracy measurement, such as the Ziv–Zakai bound used for biased estimation [18, 19].

5.4 Signal Processing Model for Satellite Navigation Receivers

We cannot solely rely on CRLB to analyze the signal performance, as many aspects in the actual implementation of the receiver would be ignored. Many textbooks offer detailed explanations of the signal processing algorithms of the receiver. Since this chapter focuses on signal design and evaluation, the engineering implementation details of the receiver are not discussed in depth here. Because signal design is often carried out prior to the receiver development, in order to be able to predict the receiving performance of a signal, it is necessary to simplify the actual receiver in many cases. This subsection will provide a mathematical model of the signal processing aspects of the receiver.

The receiver front end filters and amplifies the satellite navigation signal received by the antenna, and then down-converts and digitalizes the signal. The baseband digital signal processing channels are mainly responsible for processing the signal component from each satellite. Each channel corresponds to the processing of one signal component. Signal processing can be roughly divided into three stages: acquisition, tracking, and data demodulation.

For the received signal, there are three important parameters to be estimated, namely the transmission delay τ, the Doppler shift f_d caused by the relative motion of the satellite and the user, and the phase difference θ between the carrier of the received signal and its local replica. The transmission delay τ is used to calculate the receiver position and clock bias, and f_d is used to calculate the user speed and the clock frequency drift. For receivers that need to use carrier phase measurement for precise positioning, the carrier phase deviation θ also needs to be estimated.

The estimation of these key parameters in a receiver has two steps. The first stage roughly evaluates τ and f_d and is usually referred to as signal acquisition. The second stage is to accurately estimate τ and θ on this basis, which is called signal tracking and can be further divided into code tracking and carrier tracking.

5.4.1 Pre-detection Integration

The acquisition and tracking algorithms do not directly use the original digital signal sample sequence, but first multiply it with the locally reproduced in-phase and quadra-phase carrier and spreading code, and then conduct a coherent integration of length T in the integrate and dump (I&D) filter, which is also referred to as pre-detection integration. This process can be described mathematically as

$$I_k = \frac{1}{T}\int_{kT}^{(k+1)T} r\left(t\right)\hat{g}\left(t-\hat{\tau}_k\right)\cos\left(2\pi\left(f_{\mathrm{IF}}+\hat{f}_d\right)t+\hat{\varphi}_0\right)\mathrm{d}t\,, \tag{5.18}$$

$$Q_k = \frac{1}{T}\int_{kT}^{(k+1)T} r\left(t\right)\hat{g}\left(t-\hat{\tau}_k\right)\sin\left(2\pi\left(f_{\mathrm{IF}}+\hat{f}_d\right)t+\hat{\varphi}_0\right)\mathrm{d}t\,, \tag{5.19}$$

where $\hat{\tau}$, $\hat{f}_d$ and $\hat{\varphi}_0$ are the local code phase offset, the local carrier Doppler estimation, and the local carrier initial phase respectively. Without loss of generality, the output of the in-phase and quadra-phase I&D filters of the data channel at $t=(k+1)\,T$ can be approximated as

$$\begin{cases} I_k \approx A\,\mathrm{sinc}\left(\pi\Delta f_k T\right)R\left(\Delta\tau_k\right)d_k\cos\left(\phi_k\right)+n_{I,k} \\ Q_k \approx A\,\mathrm{sinc}\left(\pi\Delta f_k T\right)R\left(\Delta\tau_k\right)d_k\sin\left(\phi_k\right)+n_{Q,k} \end{cases}, \tag{5.20}$$

where R is the correlation function of the spread-spectrum signal, d_k is the data sign, $\phi_k=\Delta\varphi_k+\pi\Delta f_k T$, $\Delta\tau_k=\tau_k-\hat{\tau}_k$, $\Delta f_k=f_{d,k}-\hat{f}_{d,k}$, and $\Delta\varphi_k=\varphi_k-\hat{\varphi}_k$ are the estimated errors of the spreading code phase, carrier Doppler, and carrier phase respectively, and $n_{I,k}$ and $n_{Q,k}$ are the output noise of the in-phase and quadra-phase branch I&D filtering. Since the key factor that really influences the performance is the power ratio of the signal component to the noise component in the output value but not the signal absolute amplitude A in (5.20), we let the noise power be normalized, i.e. $\text{Var}(n_{I,k}) = \text{Var}(n_{Q,k}) = 1$ for ease of analysis. At this time, the corresponding signal amplitude $A = \sqrt{2TC/N_0}$, where C/N_0 is the CNR.

5.4.2 Acquisition

A block diagram of a typical serial acquisition strategy is shown in Fig. 5.1. Acquisition can be seen as a search process in a two-dimensional space composed of local spreading code phase offset estimation $\hat{\tau}$ and the carrier Doppler estimation $\hat{f}_d$. After the pre-detection integration is performed on the received signal, the output of the I&D filter is sent to the signal detector to generate a detection statistic, and then a threshold decision is made. Due to the influence of noise, the threshold decision is a statistical process with a detection probability P_d and a false alarm probability P_fa.

In the acquisition of the BPSK-R signal, the following detection statistics are widely used [20–22]:

$$\Psi = \sum_{k=0}^{L-1} \left(I_k^2 + Q_k^2 \right) , \tag{5.21}$$

where L is the number of non-coherent accumulations. Without loss of generality, taking the data channel as an example, assuming that the code phase offset and the

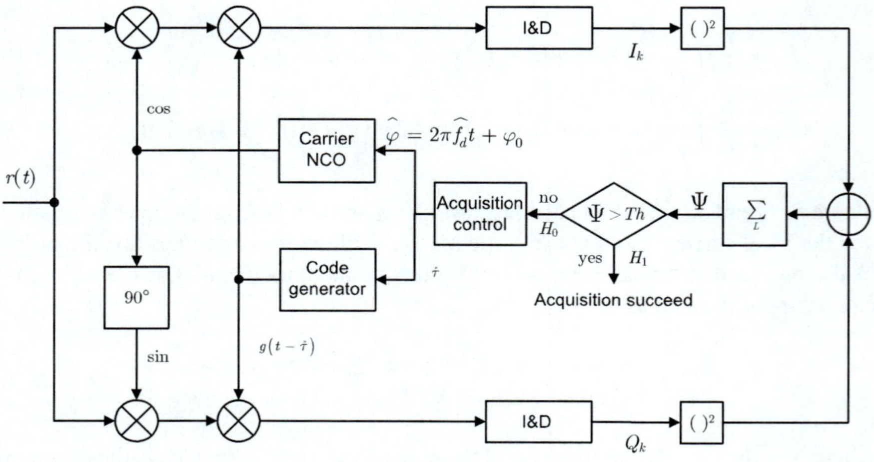

Fig. 5.1 Block diagram of the serial acquisition strategy

Doppler shift change slowly enough during the non-coherent accumulation process, then ignoring noise, (5.21) can be written as

$$\Psi = LTC/N_0 \,\mathrm{sinc}^2\left(\pi \Delta f T\right) R^2\left(\Delta\tau\right) . \tag{5.22}$$

For a BPSK-R signal with good correlation characteristics, when the estimation errors of both the spreading code phase and the Doppler shift are small enough, Ψ will have a significant peak. The acquisition process takes advantage of this feature.

When processing a BPSK-R signal, the search step of the code phase is typically 1/2 chip, and the Doppler search step is approximately $2/(3T)$ [23]. This size of search step can ensure that the attenuation of the detection statistic caused by resolution is within an acceptable range and does not miss the correlation peak. Of course, a higher search resolution will make the initial estimate of the tracking loop more accurate, but at the cost of an increased number of grids to search and a lower acquisition speed. Due to the slow speed of the serial acquisition process, multiple sets of correlators can be used simultaneously in real applications. Parallel acquisition techniques such as the fast Fourier transform (FFT)-based cyclic correlation method [24, 25] can also be used to increase acquisition speed. Refer to [26] for more details on the theory of BPSK-R signal acquisition.

In practice, the design of the acquisition algorithm is actually a compromise of detection performance, mean acquisition time, implementation complexity, and reliability. For decades, research into acquisition algorithms has been ongoing for a variety of different applications and scenarios [27–29]. New GNSS signals, such as the adoption of the BOC modulation signal, bring new challenges to the design of signal acquisition (see Sect. 4.8).

5.4.3 Code Tracking

From the perspective of whether or not to rely on carrier information, the code tracking loop can be divided into coherent tracking and non-coherent tracking loops. Since it is extremely difficult to generate a coherent local carrier before despreading, GNSS receivers generally use a non-coherent DLL [23] to track the code phase. Figure 5.2 is a block diagram of a non-coherent DLL that processes BPSK-R signals using the EMLP discriminator.

In the DLL, the received signal is multiplied with the carriers of the in-phase and quadrature branches, and correlated with the Early (E), Prompt (P), and Late (L) spread signals respectively. The spacing between the signal code phases of E and L is δ chips. Still assuming that the data channel is being processed, the I&D filter output is

$$\begin{cases} I_X = \sqrt{2TC/N_0}\,\mathrm{sinc}\left(\pi \Delta f T\right) R\left(\Delta\tau - \tau_X T_c\right) d \cos\left(\phi\right) + n_{I,X} \\ Q_X = \sqrt{2TC/N_0}\,\mathrm{sinc}\left(\pi \Delta f T\right) R\left(\Delta\tau - \tau_X T_c\right) d \sin\left(\phi\right) + n_{Q,X} \end{cases} , \tag{5.23}$$

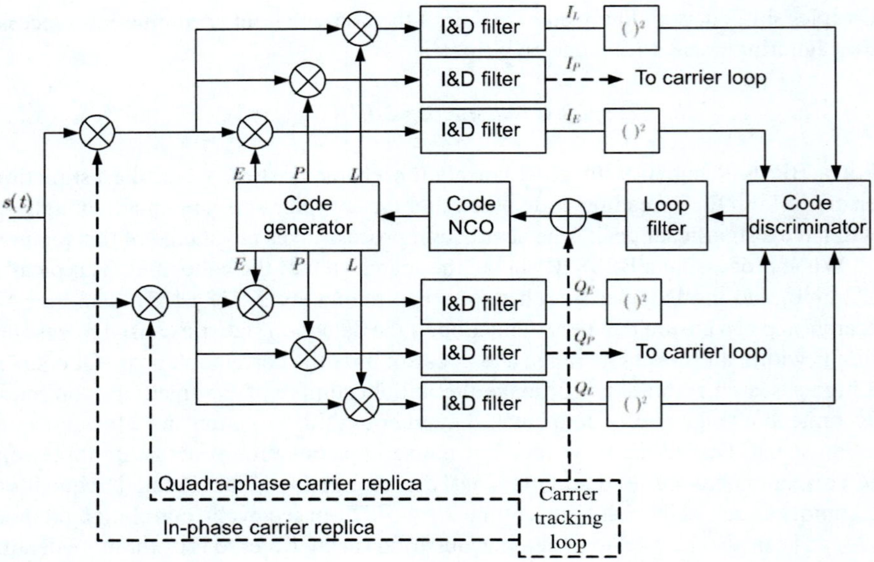

Fig. 5.2 Typical block diagram of a non-coherent delay-locked loop

where X corresponds to E, P, and L correlators, corresponds to the additional code phase delay, that is, $\tau_E = -\delta/2$, $\tau_P = 0$, and $\tau_L = \delta/2$. $n_{I,X}$ and $n_{Q,X}$ are the output noise of in-phase and quadra-phase correlators respectively. The representations of other symbols are the same as in (5.20). The outputs of the E and L correlators perform the following operations in the discriminator:

$$\varepsilon(\Delta\tau) = \left(I_E^2 + Q_E^2\right) - \left(I_L^2 + Q_L^2\right) . \tag{5.24}$$

Without considering noise and interference, the output of the discriminator is approximated as

$$\varepsilon(\Delta\tau) \approx 2TC/N_0 \operatorname{sinc}^2(\pi\Delta fT)\left[R^2\left(\Delta\tau + \frac{\delta T_c}{2}\right) - R^2\left(\Delta\tau - \frac{\delta T_c}{2}\right)\right] \tag{5.25}$$

$$\approx K_d \Delta\tau . \tag{5.26}$$

Although nonlinear operations are used in (5.24), $\varepsilon(\Delta\tau)$ can be linearized around $\Delta\tau = 0$. In (5.26), K_d is referred to as the *discriminator gain*. The range that approximately satisfies the linear characteristic is called the linear range of the discriminator.

5.4.4 *Carrier Tracking*

When the code tracking employs a non-coherent loop, carrier tracking can be performed simultaneously with or after code tracking. Figure 5.3 shows a general carrier tracking loop structure. The received signal is multiplied by the local replica carrier and the spreading code respectively, and the pre-detection integration operation is performed in the I&D filter. The carrier phase discriminator estimates the carrier phase error $\Delta\varphi$ using the correlation values. The result of the estimation is sent to the carrier NCO after passing through the loop filter, and the frequency of the local carrier is controlled to adjust the carrier phase.

The main indicators for measuring the performance of the tracking loop are the discriminator pull-in range, the quasi-linear range, and the closed loop thermal noise variance. In the absence of noise, when the true phase error is within the pull-in range, the discriminator's output value can correctly reflect the trend of the phase error, that is, the same sign as the phase error. When the error is in the quasi-linear range, the discriminator's output value is expected to be close to the true phase error.

In a conventional GNSS receiver, since there is data bit modulation on the carrier, a 180° phase flip occurs at the symbol conversion point. Therefore, the carrier tracking usually uses a Costas loop that is insensitive to data modulation. The discriminator used in the Costas loop comes in several forms. The phase error and characteristics of various discriminators are analyzed and summarized in [23, 30]. The most commonly used Costas discriminator is the dot-produce form:

$$\Delta\hat{\varphi} = \frac{I_P Q_P}{2TC/N_0}, \tag{5.27}$$

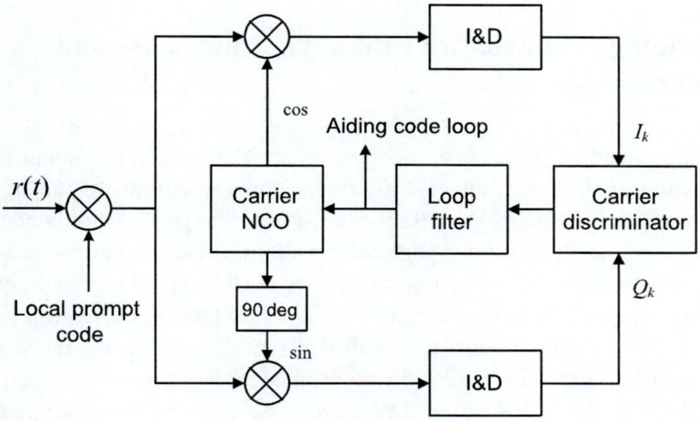

Fig. 5.3 Block diagram of a general GNSS receiver carrier tracking loop

where the product operation eliminates the sensitivity to the data bit symbols. Based on the assumptions of ideal code tracking, when the carrier frequency error and phase error are small, the expected value of the carrier discriminator output can be approximated as

$$\mathbb{E}\left[\Delta\varphi\right] = \frac{1}{2}\sin\left(2\Delta\varphi\right) \approx \Delta\varphi\,, \tag{5.28}$$

and the closed loop thermal noise variance is

$$\sigma^2_{\text{Costas}} = \frac{B_L}{C/N_0}\left(1 + \frac{1}{2TC/N_0}\right). \tag{5.29}$$

Similar to non-coherent DLLs, there is also a square loss in the Costas loop. Increasing the coherent integration length can reduce the square loss. As discussed in Sect. 2.10, for next-generation GNSS signals with pilot channels, the use of pilot channels for tracking can improve tracking performance.

In traditional GNSS receivers, most of the baseband signal processing is based on the matched receiving mode. That is, the local de-spreading waveform is the same as the spreading waveform of the received signal, i.e. $\hat{g}\left(t\right) = g\left(t\right)$. According to the matched filtering theory [31], the maximum receiver correlator's output SNR can be obtained under the matched de-spreading waveform condition. However, in fact, as we saw in the analysis of Sect. 5.1, the de-spreading waveform does not have to be the same as the spreading waveform. In the following sections, the signal performance in the more general mismatch de-spreading mode will be studied, and the corresponding signal performance evaluation indicators decoupled from the receiver implementation will be given.

5.5 Ranging Performance Under Thermal Noise and Interference

The ranging error CRLB given in Sect. 5.3 is an indicator independent of the signal processing method and reflects only the performance limits of the signal itself. Moreover, we can prove that the CRLB of the pseudorange measurement error actually corresponds to the tracking performance limit when the early-late interval of the coherent code tracking loop approaches zero [1, 4]. However, in the process of signal design, especially in the design of the next-generation navigation signals, we are more concerned with the inherent ability of signals in cases of general mismatch receiving modes. That is, the tracking performance limit of a specific spread spectrum signal under a given de-spreading waveform. This requirement is generally posed for the following purposes.

(1) For high-order BOC signals, it is unrealistic to adopt a matched receiving processing method. As described in Sect. 4.8, there are multiple peaks in the ACF of the BOC signal, and matched receiving will introduce ambiguity in terms of acquisition and tracking. When the BOC order is low, the bump-jumping method [32] (see also Sect. 4.8.5) can remove ambiguity. However, when the order of BOC increases, the bump-jumping technique is no longer applicable because the difference between the main peak and the side peak is reduced. In this case, we may need to use the BPSK-like method, which is a typical mismatch receiving technique. In order to evaluate the actual tracking ability of the high-order BOC signal, it is necessary to evaluate the ranging accuracy in this mismatch receiving mode.
(2) The same problem also exists for the AltBOC and ACE-BOC signals. The tracking accuracy of the BPSK-like method for these dual-frequency multiplexed signals still lacks theoretical analysis and evaluation.
(3) For the MBOC modulated signal, since the bandwidth of the entire signal exceeds 24 MHz, the receiver needs to use wide receiving bandwidth to receive most of the signal energy. Moreover, the time domain waveform of the MBOC signal is complicated. In order to generate a BOC(6,1) chip component in the receiver, the rate of the receiver code clock needs to increase six times compared to cases of a BOC(1,1) chip component. Furthermore, when considering multi-system interoperable receivers, a matched receiver needs to use three different processing structures simultaneously in order to process TMBOC, CBOC, and QMBOC signals, which undoubtedly increases the implementation complexity of the receiver. In some complexity limited applications, it is more likely to use a common unmatched receiving method to process all types of the MBOC signals. For example, the receiver can ignore the BOC(6, 1) component of the signal, approximates all MBOC signals as BOC(1,1) signals, and generates the BOC(1,1) signal locally to perform unmatched correlation with the input MBOC signal. In this way, the local waveform generation circuit can be greatly simplified, and the receiver can use a narrower receiving bandwidth and a lower sampling rate than in the matched receiving mode. A common channel structure can be used for processing different signals. This method of processing is undoubtedly very attractive for low-cost receivers.

Therefore, in this subsection, we will provide the evaluation indicators of code tracking error in mismatch de-spreading mode, and the corresponding signal ranging performance through theoretical derivation.

5.5.1 Spreading Chip Waveform Mismatch

We ignore the influence of the carrier tracking and use the signal baseband simplified model as described in (5.2). The simplified code tracking loop is shown in Fig. 5.4, where $r(t)$ first passes through a front-end low-pass filter with a double-sided band-

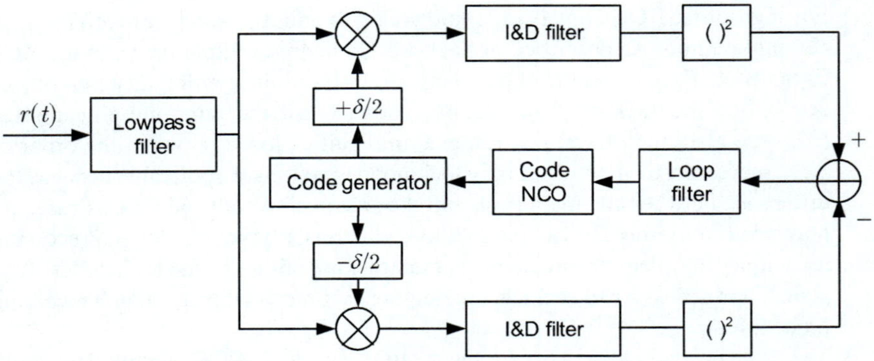

Fig. 5.4 Simplified model of the receiver code tracking loop signal processing

width of β_r, filtering out the out-of-band noise, and then multiplies the local signal replica, sending the result to the I&D filter with a coherent integration time of T. It is assumed that the low-pass filter has the ideal characteristics given by (5.9). The I&D filter along with the multiplier and the code generator is regarded as a linear filter, the transfer function of which is the Fourier transform of the local de-spreading waveform.

The original noise term $w(t)$ in (5.2) is further decomposed into a Gaussian white noise term $n(t)$ and a non-white interference term $\iota(t)$, and (5.2) is rewritten as

$$r(t) = A_s g(t - \tau_0)\,\mathrm{e}^{\mathrm{j}\theta} + n(t) + \iota(t)\,. \tag{5.30}$$

Traditional despreading follows the matched receiving principle that maximizes the SNR at the correlator output [31]. This requires the local despreading signal to be exactly the same as the received signal. However, in navigation signal processing, as mentioned before, maximizing SNR and matched receiving is no longer the only guideline to select local despreading signals. In practice, a variety of selections of despreading waveform have been applied in GNSS signal processing for the purpose of reducing receiver complexity [33], multipath mitigation [5] and unambiguous tracking of BOC signals [7, 9]. These receiving strategies, which are associated with the mismatch of local despreading code chip waveform, are termed as "unmatched receiving" hereafter.

In GNSS, since the PRN spreading sequence is known for both the transmitter and the receiver, hereafter we consider a general case where the despreading code chip waveform $\hat{p}(t)$ may be different from that of the received signal. Taking the sine-phase BOC(6,1) signal for example, the code chip waveform of the received signal is a bipolar pulse $\hat{p}(t) = \mathrm{sgn}\left[\sin\left(2\pi f_{sc} t\right)\right]\left[u(t) - u(t - T_c)\right]$, where $u(t)$ is the unit step signal, f_{sc} is the subcarrier frequency. When a BPSK-like unmatched receiving algorithm is applied, the code chip waveform of local despreading signal is a rectangular pulse $u(t) - u(t - T_c)$ at the central frequency of f_{sc} or $-f_{sc}$, when the intermediate frequency is assumed to be zero. From this aspect, the local despreading signal $\hat{g}(t)$ is

$$\hat{g}(t) = \sum_{i=-\infty}^{+\infty} (-1)^{c_i} \hat{p}(t - iT_c) . \quad (5.31)$$

5.5.2 Formulation and Statistical Properties of Correlator Output

Assuming that τ_0 is unchanged within the pre-detection integration time $(k-1)T < t < kT$, the output values of the E and L correlators can be written as

$$\Gamma_i = \frac{1}{T}\int_{(k-1)T}^{kT} r(t)\hat{g}(t - \hat{\tau}_0 + \tau_i)\mathrm{d}t \triangleq c_{g\hat{g}}(\tau_i) + c_{w\hat{g}}(\tau_i) , \quad (5.32)$$

where $\hat{g}(t - \hat{\tau}_0 + \tau_i)$ is the local signal, in which $i \in \{\mathrm{E}, \mathrm{L}\}$ represents the early and late branches, the replica code phase delay are $\tau_\mathrm{E} = \delta/2$ and $\tau_\mathrm{L} = -\delta/2$, respectively,

$$c_{g\hat{g}}(\tau_i) = \frac{1}{T}\int_{(k-1)T}^{kT} \mathrm{e}^{\mathrm{j}\theta} g(t - \tau_0)\hat{g}^*(t - \hat{\tau}_0 - \tau_i)\mathrm{d}t \quad (5.33)$$

and

$$c_{w\hat{g}}(\tau_i) = \frac{1}{T}\int_{(k-1)T}^{kT} w(t)\hat{g}^*(t - \hat{\tau}_0 - \tau_i)\mathrm{d}t \quad (5.34)$$

are respectively the useful signal component and the noise-and-interference component in the correlator output.

The output of the code discriminator is the square difference of the early and late coherent integration results

$$\varepsilon(\Delta\tau) = |\Gamma_E|^2 - |\Gamma_L|^2 , \quad (5.35)$$

which is the equivalent representation of the discriminator given by (5.24), where $\Delta\tau \triangleq \tau_0 - \hat{\tau}_0$ is the open-loop tracking error. When we need to consider the effects of filtering in the receiver, it is best to express the output of the correlator in frequency domain. Based on the Wiener–Khinchin theorem, (5.33) can be rewritten as

$$c_{g\hat{g}}(\tau_i) = \mathrm{e}^{\mathrm{j}\theta} C_s \int_{-\beta_r/2}^{+\beta_r/2} \mathcal{H}_{g\hat{g}}(f)\, \mathrm{e}^{-\mathrm{j}2\pi f(\Delta\tau - \tau_i)}\mathrm{d}f , \quad (5.36)$$

where $\mathcal{H}_{g\hat{g}}(f)$ is the normalized CSD of the local and received signals, the definition of which can be found in Sect. 3.4.6. Similarly, the statistical properties of the correlator output interference-plus-noise term (5.34) can also be represented by the PSDs of noise, interference, and the local de-spreading signal, which is

$$\mathrm{Var}\left(c_{w\hat{g}}\right)=\frac{C_s}{T}\int_{-\beta_r/2}^{+\beta_r/2}G_w\left(f\right)G_{\hat{g}}\left(f\right)\mathrm{d}f\,. \tag{5.37}$$

5.5.3 Statistical Characteristics of the Discriminator Output

The discriminator output obtained by (5.35) and (5.32) is

$$\begin{aligned}\varepsilon\left(\Delta\tau\right)=&\left|c_{g\hat{g}}(\delta/2)\right|^2-\left|c_{g\hat{g}}(-\delta/2)\right|^2+\left|c_{w\hat{g}}(\delta/2)\right|^2-\left|c_{w\hat{g}}(-\delta/2)\right|^2\\&+2\mathrm{Re}\{c_{g\hat{g}}(\delta/2)c_{w\hat{g}}^*(\delta/2)-c_{g\hat{g}}(-\delta/2)c_{w\hat{g}}^*(-\delta/2)\}\,.\end{aligned} \tag{5.38}$$

Substituting (5.36) and (5.37) into the above equation and conducting a series of trigonometric transformations, we can write the expectation of the discriminator output in the form of (5.26), where the discriminator gain is

$$K_d=8C_s^2\pi\int_{-\beta_r/2}^{\beta_r/2}\mathcal{H}_{g\hat{g}}(f)\cos(\pi f\delta)\mathrm{d}f\int_{-\beta_r/2}^{\beta_r/2}f\mathcal{H}_{g\hat{g}}(f)\sin(\pi f\delta)\mathrm{d}f\,, \tag{5.39}$$

and the variance of the discriminator output is

$$\begin{aligned}&\mathrm{Var}\{\varepsilon\left(\Delta\tau\right)\}=\\&\quad\frac{8}{T}\left(\int_{-\beta_r/2}^{\beta_r/2}\mathcal{H}_{g\hat{g}}(f)\cos(\pi f\delta)\mathrm{d}f\right)^2\int_{-\beta_r/2}^{\beta_r/2}G_w(f)G_{\hat{g}}\left(f\right)\sin^2(\pi f\delta)\mathrm{d}f\\&+\frac{2}{T^2}\left[\left(\int_{-\beta_r/2}^{\beta_r/2}G_w(f)G_{\hat{g}}\left(f\right)\mathrm{d}f\right)^2-\left(\int_{-\beta_r/2}^{\beta_r/2}G_w(f)G_{\hat{g}}\left(f\right)\cos(2\pi f\delta)\mathrm{d}f\right)^2\right].\end{aligned} \tag{5.40}$$

The variance of the estimated $\Delta\tau$ using the output of the discriminator is

$$\mathrm{Var}\{\Delta\tau\}\approx\frac{\mathrm{Var}\{\varepsilon\left(\Delta\tau\right)\}}{K_d^2}\,. \tag{5.41}$$

Then, by using the relationship between the open-loop error and closed-loop error

$$\sigma_{\mathrm{close}}^2=\mathrm{Var}\{\Delta\tau\}2B_LT(1-0.5B_LT)\,, \tag{5.42}$$

we can obtain the closed-loop code tracking error of the narrow EMLP loop, which is

$$\sigma^2_{\text{close}} = \frac{B_L(1-0.5B_LT)\int_{-\beta_r/2}^{\beta_r/2} G_w(f)G_{\hat{g}}(f)\sin^2(\pi f\delta)\mathrm{d}f}{(2\pi)^2 C_s\left(\int_{-\beta_r/2}^{\beta_r/2} f\mathcal{H}_{g\hat{g}}(f)\sin(\pi f\delta)\mathrm{d}f\right)^2} \times \left[1+\frac{\int_{-\beta_r/2}^{\beta_r/2} G_w(f)G_{\hat{g}}(f)\cos^2(\pi f\delta)\mathrm{d}f}{TC_s\left(\int_{-\beta_r/2}^{\beta_r/2} \mathcal{H}_{g\hat{g}}(f)\cos(\pi f\delta)\mathrm{d}f\right)^2}\right]. \tag{5.43}$$

5.5.4 Tracking Error Under Non-coherent Processing

When the early-late spacing δ of the tracking loop approaches zero, the lower bound of the code tracking error, in the case of waveform mismatch, can be obtained from the first-order Taylor series approximation of (5.43), which is

$$\sigma^2_{\text{NELP},\delta\to 0} = \frac{B_L(1-0.5B_LT)}{(2\pi)^2 C_s/N_0}\left(\frac{\int_{-\beta_r/2}^{\beta_r/2} f^2 G_{\hat{g}}(f)\,\mathrm{d}f}{\left(\int_{-\beta_r/2}^{\beta_r/2} f^2\mathcal{H}_{g\hat{g}}(f)\mathrm{d}f\right)^2}+\frac{C_\iota}{N_0}\frac{\int_{-\beta_r/2}^{\beta_r/2} f^2 G_\iota(f)G_{\hat{g}}(f)\,\mathrm{d}f}{\left(\int_{-\beta_r/2}^{\beta_r/2} f^2\mathcal{H}_{g\hat{g}}(f)\mathrm{d}f\right)^2}\right) \times \left(1+\frac{\int_{-\beta_r/2}^{\beta_r/2} G_{\hat{g}}(f)\,\mathrm{d}f}{T\frac{C_s}{N_0}\left(\int_{-\beta_r/2}^{\beta_r/2}\mathcal{H}_{g\hat{g}}(f)\mathrm{d}f\right)^2}+\frac{\int_{-\beta_r/2}^{\beta_r/2} G_\iota(f)G_{\hat{g}}(f)\,\mathrm{d}f}{T\frac{C_s}{C_\iota}\left(\int_{-\beta_r/2}^{\beta_r/2}\mathcal{H}_{g\hat{g}}(f)\mathrm{d}f\right)^2}\right), \tag{5.44}$$

where the term before the product sign is the lower bound of the coherent loop tracking error, and the other term is the square loss introduced by non-coherent processing. We find that both (5.43) and (5.44) are related to the receiver's specific configuration parameters such as the loop bandwidth, the coherent integration time, and the early-late spacing.

In the signal design and evaluation phase, in order to be able to uniformly compare signals using different spreading modulation methods, it is desired to seek some decoupled indicators, which can be as independent of the receiver's configuration parameters as possible, and only reflect the impact of band-limiting and local signal waveform mismatching on the ranging potential performance of different spreading modulation techniques. To this end, we define the following decoupled evaluation criteria. First, we define

$$\chi_{\iota\hat{s}} \triangleq \int_{-\beta_r/2}^{\beta_r/2} f^2 G_\iota(f)G_{\hat{g}}(f)\,\mathrm{d}f \tag{5.45}$$

as the anti-interference code tracking spectral separation factor, and define

$$\varpi \triangleq \frac{\int_{-\beta_r/2}^{\beta_r/2} f^2 \mathcal{H}_{s,\hat{s}}(f)\mathrm{d}f}{\sqrt{\int_{-\beta_r/2}^{\beta_r/2} f^2 \hat{G}_s(f)\mathrm{d}f}} \tag{5.46}$$

as the equivalent RMS bandwidth in case of unmatched receiving. Further, we define

$$\nu \triangleq \sqrt{\frac{\chi_{\iota\hat{s}}}{\varpi^2 \beta_{\mathrm{RMS},\hat{s}}^2}}$$

as the equivalent anti-interference ratio, and define

$$\zeta \triangleq \frac{1}{\frac{C_s}{N_0}\eta_n} + \frac{1}{\frac{C_s}{C_\iota}\eta_\iota} \tag{5.47}$$

as the squared loss factor for unmatched receiving, in which

$$\eta_n = \left(\int_{-\beta_r/2}^{\beta_r/2} \mathcal{H}_{s,\hat{s}}(f)\mathrm{d}f\right)^2 \Big/ \int_{-\beta_r/2}^{\beta_r/2} G_{\hat{g}}(f)\,\mathrm{d}f \tag{5.48}$$

$$\eta_\iota = \left(\int_{-\beta_r/2}^{\beta_r/2} \mathcal{H}_{s,\hat{s}}(f)\mathrm{d}f\right)^2 \Big/ \int_{-\beta_r/2}^{\beta_r/2} G_\iota(f) G_{\hat{g}}(f)\,\mathrm{d}f \tag{5.49}$$

correspond to the unmatched SNR attenuation under white noise and colored interference respectively. Thus, the lower bound of the code tracking error under waveform mismatch conditions can be expressed as

$$\sigma_{\mathrm{NELP,LB}}^2 = \frac{B_L(1-0.5B_LT)}{(2\pi)^2}\left(\frac{1}{\frac{C_s}{N_0}\varpi^2} + \frac{C_\iota}{C_s}\nu^2\right) \times \left(1 + \frac{\zeta}{T}\right). \tag{5.50}$$

We can see from (5.50) that the code tracking error of the signal under unmatched conditions is related to multiple factors such as signal-to-noise ratio, interference-to-signal ratio, receiver loop bandwidth, early-late spacing, and coherent integration time. However, in the above-given series of decoupled indicators $\chi_{\iota\hat{s}}$, ϖ, ν, η_n, and η_ι, only the characteristics of the interference type and the factors of receiving bandwidth and local despreading waveform, which are closely correlated to the essential properties of signal, are considered. Decoupled indicators are unrelated to the external factors of the interference strength, AWGN strength, and other receiver related factors. Thus, these decoupled criterion can better represent the inherent performance of the received signal under noise, interference, and waveform mismatching, with which we can conveniently compare the intrinsic performance differences of different signal modulations under specific despreading waveforms in terms of thermal noise resistance and anti-interference.

Under some special conditions, the form of the lower bound of the code tracking error can be simplified further. For example, when there is no non-white interference, (5.50) can be rewritten as

$$\sigma^2_{\mathrm{NELP,LB}} = \frac{B_L(1-0.5B_LT)}{(2\pi)^2\frac{C_s}{N_0}\varpi^2} \times \left(1+\frac{\zeta}{T}\right), \tag{5.51}$$

and when the pre-detection integration time T is long enough, and the CNR of the receiver's operating environment is sufficiently high, the effect of the squared loss ζ can be further approximately ignored.

Furthermore, it is easy to see from (5.46) that matched receiving is just a special case of the above unmatched receiving. When the local despreading waveform is the same as the spreading waveform used in received signal, i.e. $\hat{g}(t) = g(t)$, substituting $\mathcal{H}_{s,\hat{s}}(f)$ with $G_{\hat{g}}(f)$, the equivalent RMS bandwidth (5.46) is degraded to the normal RMS bandwidth defined by (5.13), while the unmatched SNR attenuation (5.48) is also degraded to the SNR attenuation (5.12) in the matched reception. At this time, the attenuation of the SNR is only caused by the band-limiting.

Example 5.1 (*Tracking performance of different MBOC signals*) In Sect. 4.7, we have introduced several different MBOC modulated signals. The total PSD of these signals satisfies the constraint of (4.48), but the time domain waveform and the PSD of each component are not the same. When the signal is received, each component is processed separately. The subtle differences of the component waveforms make a significant difference in their receiving performance. In this example, we use the performance evaluation method derived in this section to compare the tracking performances of different modulation techniques.

The modulation techniques used for comparison in this example include BPSK-R(1), BOC(1,1), CBOC$^+$(6,1,1/11), CBOC$^-$(6,1,1/11), QMBOC(6,1,4/33), and TMBOC(6,1,4/33). Among them, for both BPSK-R(1) and BOC(1,1) modulations, the processing mode is always matched receiving. As composite signals, MBOC signals such as CBOC$^+$(6,1,1/11), CBOC$^-$(6,1,1/11), QMBOC(6,1,4/33), and TMBOC(6,1,4/33) enable multiple receiving strategies, where different levels of accuracy can be obtained with corresponding receiver complexity. Matched receiving of MBOC signals enable the optimum received performance. However, an unmatched receiving mode, generally referred as BOC$_{11}$-like receiving, is more widely used, where only the narrowband BOC(1,1) component is processed in the receiver, so that the local replica generation in BOC$_{11}$-like receiving is less complex than in the matched method.

Figures 5.5 and 5.6 respectively show the equivalent RMS bandwidth with respect to the double-sided receiving bandwidths of the various modulated signals in cases of matched receiving and unmatched BOC$_{11}$-like receiving.

It can be seen from Figs. 5.5 and 5.6 that the tracking accuracy of the BOC(1,1) and MBOC signals is significantly better than that of the BPSK-R(1). We can also see that the tracking performance of different MBOC modulations is not the same. In Sect. 4.7.2, we qualitatively analyzed this phenomenon from the perspective of PSD

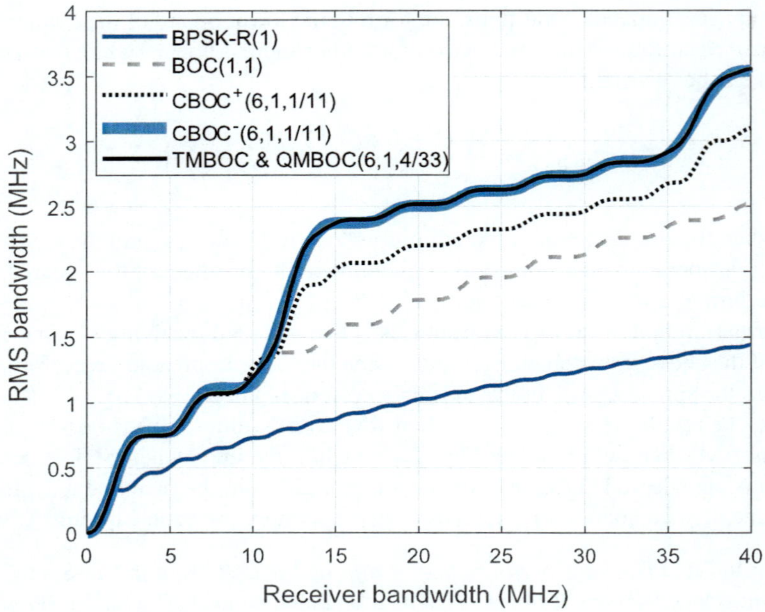

Fig. 5.5 Equivalent RMS bandwidths for different modulations under matched receiving mode

differences. Now the tracking performance difference of these modulation methods can be quantitatively presented through equivalent RMS bandwidth.

In Fig. 5.5, we can see that when the receiver bandwidth is less than 12 MHz, various MBOC modulations have approximately the same tracking accuracy as BOC(1,1). Moreover, comparing Figs. 5.5 and 5.6, we can see that when the receiver bandwidth is less than 12 MHz, for MBOC signals, the performance of matched receiving is not significantly different from that of the unmatched receiving. This phenomenon is an important reference for the development of low-end MBOC receivers. It means that if the front-end bandwidth of the receiver is less than 12 MHz, it is not necessary to generate MBOC local signals to use matched receiving mode [34]. For narrowband receivers to process MBOC signals, the BOC_{11}-like method is good enough.

When the receiving bandwidth is greater than 12 MHz, the tracking performance distinction between different modulations becomes more significant. CBOC^{-}(6,1,1/11), TMBOC(6,1,4/33) and QMBOC(6,1,4/33) have the best tracking accuracy, followed by CBOC^{+}(6,1,1/11) and BOC(1,1) modulation. BPSK-R(1) modulation has the lowest tracking accuracy. However, under a wideband unmatched receiving mode, only CBOC^{-}(6,1,1/11) performance is better than BOC(1,1), while the performance of the rest of MBOC modulations is worse than BOC(1,1). The performance degradation of TMBOC(6,1,4/33) is the most obvious. ■

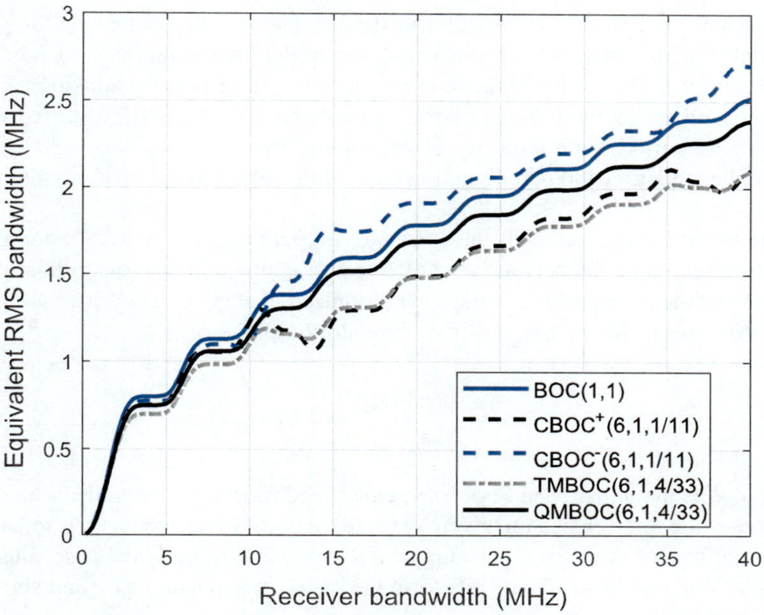

Fig. 5.6 Equivalent RMS bandwidths for different modulations under BOC_{11}-like receiving mode (only for MBOC signals)

Although MBOC modulated signals have a certain loss of tracking accuracy in the BOC11-like receiving mode, this mode can greatly simplify the circuit for generating the local waveform compared with that in the matched receiving mode, and the receiver can also use a narrower receiving bandwidth and a lower sampling rate than in matched receiving. The same channel structure can be used to process different MBOC signals. Therefore, this processing method is still very attractive for low-cost receivers.

5.6 The Performance of Acquisition, Carrier Tracking, and Data Demodulation

From the analysis in Sect. 5.4, we can see that the signal acquisition, carrier tracking, and data demodulation performance are all dependent on the signal-to-noise-plus-interference ratio (SNIR) at the output of the receiver's prompt branch correlator. In actual receivers, the acquisition performance is not only related to the type of spreading modulation used by the signal but is also coupled strongly with the receiver's coherent integration length and acquisition search strategy. Meanwhile, the demodulation performance is also concerned with factors such as the symbol rate of the modulated message and the channel coding that is used.

However, when other parameters in the receiver are the same, for the spread spectrum signals that adopt different spreading modulation techniques, a larger SNIR signal output at the prompt branch correlator can cause better acquisition, carrier tracking, and data demodulation performance. Therefore, the SNIR of the correlator output can be used as an indicator of decoupling from the receiver to evaluate the potential acquisition and demodulation capabilities of a given spreading modulated signal.

The statistical properties of the correlator output are given by (5.36) and (5.37). In order to evaluate the performance of carrier tracking and data demodulation, we assume that the code phase is already synchronized, that is, the TOA estimation error in (5.36) is zero, and the SNIR of the correlator output is

$$\rho_{\text{mis}} = \frac{\left|c_{g\hat{g}}(0)\right|^2}{\operatorname{Var}\left(c_{w\hat{g}}\right)} = T C_s \eta_{\text{mis}}\,, \tag{5.52}$$

where η_{mis} is the unmatched SNR attenuation, and its definitions under white noise and non-white interference are given by (5.48) and (5.49) respectively. It reflects the additional loss of the correlator output SNR compared to the ideal case caused by band-limiting and local signal waveform mismatch, when using a certain spreading modulated signal.

In the above analysis, we always specify that the received signal adopts the same spreading chip waveform at all times, i.e.

$$\Pr\left\{p(t) = p_1(t)\right\} \equiv 1\,. \tag{5.53}$$

However, when the transmitted signal employs different spreading waveforms with a given probability in different time periods, as modeled in Sect. 3.2, assuming the local despreading waveform keeps unchanged, the average power of the correlator output should be rewritten as

$$\left|c_{g\hat{g}}(0)\right|^2 = C_s\left(\sum_i \mathcal{P}_{p_i}\int_{-\beta_r/2}^{+\beta_r/2}\mathcal{H}_{s,\hat{s}}(f)\,\mathrm{d}f\right)^2. \tag{5.54}$$

At this time, the corresponding unmatched SNR is attenuated to

$$\eta_{\text{mis}} = \left(\sum_i \mathcal{P}_{p_i}\int_{-\beta_r/2}^{+\beta_r/2}\mathcal{H}_{s_i\hat{s}}(f)\,\mathrm{d}f\right)^2 \Big/ \int_{-\beta_r/2}^{\beta_r/2} G_\iota(f)G_{\hat{g}}(f)\,\mathrm{d}f\,. \tag{5.55}$$

Example 5.2 (*Correlator output SNIR loss of TMBOC modulation*) In the TMBOC(6, 1, 4/33) modulation used by the GPS L1C signal, the BOC(6, 1) spreading chip waveform is used for the 1st, 5th, 7th, and 30th chip in every 33 spreading chip waveforms, while the rest use the BOC(1,1) spreading chip waveform. Since the spreading code length of the L1C signal is 10,230 chips, the above pattern is repeated 310 times per code period [35].

In the BOC_{11}-like processing mode, using (5.55), the correlation output SNR loss of the TMBOC signal is

$$\eta_{SNR} = \frac{\left| \frac{29}{33} \int_{-\beta_r/2}^{+\beta_r/2} G_{11}(f)\, df + \frac{4}{33} \int_{-\beta_r/2}^{+\beta_r/2} \mathcal{H}_{11/61}(f)\, df \right|^2}{\int_{-\beta_r/2}^{+\beta_r/2} G_{11}(f)\, df}, \tag{5.56}$$

where $G_{11}(f)$ is the PSD of the BOC(1,1) component with expression

$$G_{11}(f) = f_c \left[\frac{\sin\left(\frac{\pi f}{f_c}\right) \sin\left(\frac{\pi f}{2 f_c}\right)}{\pi f \cos\left(\frac{\pi f}{2 f_s}\right)} \right]^2 \tag{5.57}$$

and $\mathcal{H}_{11/61}(f)$ is the CSD of BOC(1,1) and BOC(6,1). Since the spreading waveforms of BOC(1,1) and BOC(6,1) are from two different functions of the Rademacher functions [36], they are orthogonal to each other. Then we can prove that

$$\int_{-\beta_r/2}^{+\beta_r/2} \mathcal{H}_{11/61}(f)\, df = 0\,. \tag{5.58}$$

Therefore, the correlation output SNR loss of the TMBOC signal can be simplified as

$$\eta_{\text{mis}} = \left(\frac{29}{33}\right)^2 \int_{-\beta_r/2}^{+\beta_r/2} G_{11}(f)\, df\,. \tag{5.59}$$

Figure 5.7 plots the correlator output SNIR loss for the TMBOC$(6, 1, 4/33)$ modulated signal in both matched receiving and the BOC_{11}-like unmatched receiving modes. It can be seen from the figure that in the BOC_{11}-like mode, the performance of the TMBOC signal is severely degraded. When the receiving bandwidth is wide enough, the power loss due to band-limited filtering is negligible in matched receiving mode. However, as the receiving bandwidth increasing, the SNIR loss of the BOC_{11}-like mode approaches $-1.12\,\text{dB}$, corresponding to $(29/33)^2$ in the above equation. ■

Example 5.3 (*Correlator output SNIR loss of different MBOC signals*) Figures 5.8 and 5.9 show the correlator output SNIR with respect to the receiving bandwidth for various modulation types in the matched receiving mode and the BOC_{11}-like receiving mode respectively. Modulations involved in this comparison include BPSK-R(1), BOC(1,1), TMBOC(6,1,4/33), QMBOC(6,1,4/33), CBOC^{+}(6,1,1/11), and CBOC^{-}(6,1,1/11).

Comparing Figs. 5.8 and 5.9, there are some points worth noting: The correlation loss of the BPSK-R(1) modulated signal is significantly smaller than other modu-

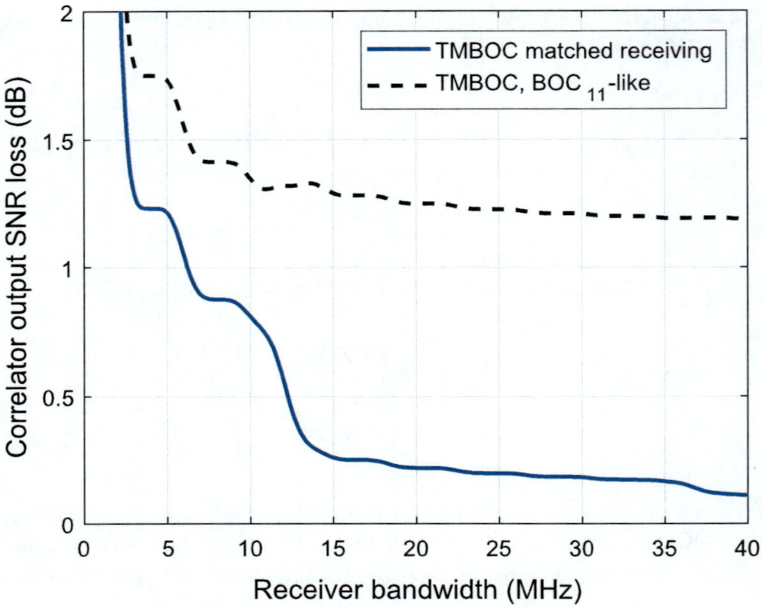

Fig. 5.7 Correlator output SNR loss of the TMBOC(6, 1, 4/33) signal in matched receiving and BOC_{11}-like unmatched receiving modes

lated signals. It can receive 95% of the signal power when the receiving bandwidth is only 4 MHz, and the SNR loss is only 0.2 dB. The performance of BOC(1,1) modulated signal comes second. The $CBOC^{+}$ signal has a lower SNR loss at the same bandwidth compared to other MBOC signals, with a 1 dB SNR loss under a 4 MHz bandwidth and approximately 0.6 dB SNR loss under a 10 MHz bandwidth. TMBOC and QMBOC offer the same performance under matched receiving, where the SNR loss is 1.2 dB in 4 MHz bandwidth. However, in the BOC_{11}-like receiving mode, TMBOC performance deteriorates severely. In cases of receiving bandwidths of 4 MHz and 10 MHz, the attenuation is 1.8 dB and 1.4 dB respectively. Even under a 40 MHz receiving bandwidth, the SNR loss is still 1.2 dB. For $CBOC^{+}$, $CBOC^{-}$, and QMBOC signals, when the receiving bandwidth is less than 10 MHz, the matched receiving performance is not significantly different from that of the unmatched receiving mode. ■

As we discussed in Example 5.3, although the BOC_{11}-like receiving mode of the MBOC signal has a larger loss in the correlation output SNR compared to matched receiving, this receiving mode has certain advantages for low-end and low-cost receivers that do not require high positioning performance. One of the advantages of the MBOC modulations is that it inherently has a variety of processing modes. Receiver designers can choose the appropriate receiving processing method accord-

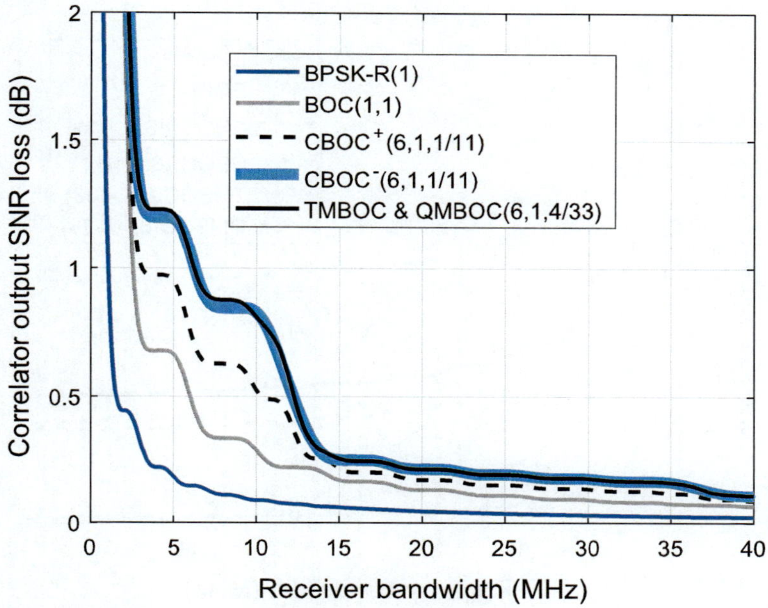

Fig. 5.8 Comparison of correlation output SNR losses for different modulation modes in matched receiving mode

ing to the application-oriented needs of the users, making a flexible trade-off between performance and complexity.

5.7 Multipath Resistant Performance

In the actual environment, multipath characteristics are complex and variable, so it is difficult to perform an accurate and generally applicable quantitative analysis of these effects. Even using computer simulations to perform an accurate and realistic analysis of a particular case, the conclusion cannot be universally applied. A common way of assessing anti-multipath performance is based on the computation of the multipath error envelope [37–39]. This assessment represents the influence of one single multipath signal with a given relative amplitude α that is constant for all considered geometric path delays, in which the received signal can be expressed as

$$s_m(t) = s(t) + \alpha s(t - \upsilon) \mathrm{e}^{-\mathrm{j}\phi}, \tag{5.60}$$

where υ is the relative time delay of the multipath, and ϕ is the received multipath reflection carrier phase.

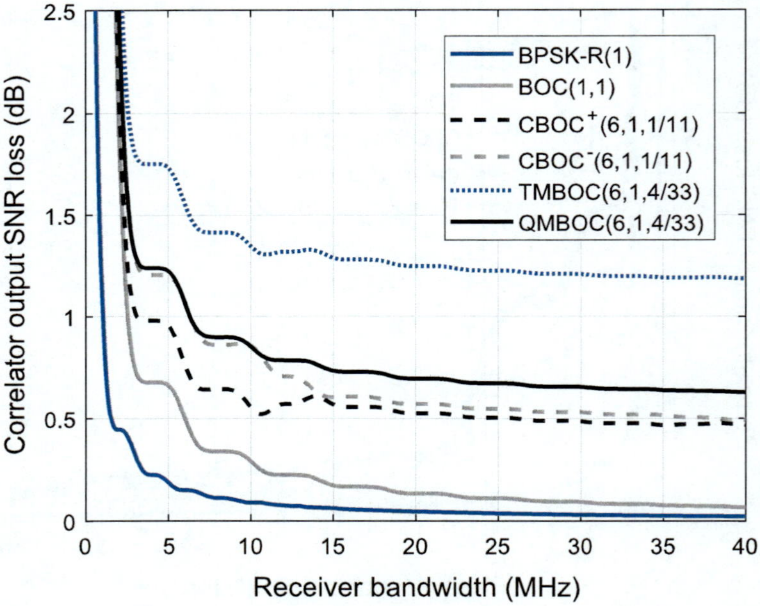

Fig. 5.9 Comparison of correlation output SNR losses for different modulation modes in BOC_{11}-like unmatched receiving mode

Here we take the case when the code tracking loop uses a non-coherent narrow-space EMLP discriminator as an example, where the output of the discriminator is shown in (5.25). Ignoring the amplitude term, we can simplify it as

$$\varepsilon(\tau) = R_{g\hat{g}}^2\left(\tau + \frac{\delta}{2}\right) - R_{g\hat{g}}^2\left(\tau - \frac{\delta}{2}\right). \tag{5.61}$$

If there is no multipath, duo to the symmetry of $R_{g\hat{g}}(\tau)$, $\varepsilon(\tau)$ can be zero at $\tau = 0$. The code tracking loop locks this zero-crossing point of the discriminator curve to track the peak of the correlation function. However, when a multipath signal exists, the shape of $R_{g\hat{g}}(\tau)$ will change due to its influence. For the received signal model of (5.60), the output of the correlator becomes

$$R'_{g\hat{g}}(\tau) = R_{g\hat{g}}(\tau) + \alpha R_{g\hat{g}}(\tau - \upsilon)\cos\phi. \tag{5.62}$$

Substituting (5.62) into (5.61), the expression of the discriminator output changes to

$$\begin{aligned}\varepsilon_m(\tau) =& R_{g\hat{g}}^2\left(\tau + \tfrac{\delta}{2}\right) - R_{g\hat{g}}^2\left(\tau - \tfrac{\delta}{2}\right) + \alpha^2\left[R_{g\hat{g}}^2\left(\tau + \tfrac{\delta}{2} - \upsilon\right) - R_{g\hat{g}}^2\left(\tau - \tfrac{\delta}{2} - \upsilon\right)\right] \\ &+ 2\alpha\cos\phi\left[R_{g\hat{g}}^2\left(\tau + \tfrac{\delta}{2}\right)R_{g\hat{g}}^2\left(\tau + \tfrac{\delta}{2} - \upsilon\right) - R_{g\hat{g}}^2\left(\tau - \tfrac{\delta}{2} - \upsilon\right)\right]. \end{aligned} \tag{5.63}$$

When the zero-crossing point is located at a certain position $\tau_e \neq 0$ by the influence of the multipath, the code tracking error caused by multipath will be τ_e. With the relative amplitude and delay of the multipath given, this error is a function of the signal CCF $R_{g\hat{g}}(\tau)$ and the early-late spacing δ. Under different receiving bandwidths, the shape of the signal correlation function may change, so the corresponding multipath error envelope will also change accordingly.

Under a given amplitude ratio of the multipath to the direct signal, the multipath error envelope curve can be calculated or simulated by traversing the delay of the multipath signal relative to the direct signal. The code tracking *multipath error envelope* with a given multipath delay can be defined as

$$\left(\max_{\phi} \varepsilon(\alpha, \upsilon, \phi), \min_{\phi} \varepsilon(\alpha, \upsilon, \phi)\right). \tag{5.64}$$

The multipath error envelope can describe the maximum code tracking errors with respect to different multipath delays caused by a single multipath with a given MDR. However, in most cases, we are more concerned with the average multipath performance of code tracking within a certain multipath delay range. In particular for BOC and MBOC modulated signals, the code tracking multipath error envelope fluctuates with delay. It is difficult to intuitively judge the multipath resistant capability of these signals only by the multipath error envelope.

To further evaluate the average multipath performance over the range of possible delays, we define the *running average range error envelope* as

$$\Gamma(\upsilon) = \frac{1}{\upsilon}\int_0^{\upsilon}\left\{\max\left[\left|\max_{\phi} \varepsilon(\alpha, u, \phi)\right|, \left|\min_{\phi} \varepsilon(\alpha, u, \phi)\right|\right]\right\} du. \tag{5.65}$$

The physical meaning of this index is the average error caused by the multipath whose delay is within $(0, \upsilon]$. For split-spectrum modulated signals whose multipath error envelope fluctuates with delay, the running average range error envelope can more intuitively reflect the resistance of such a signal to multipath within a certain delay.

For a given spreading waveform, different local de-spreading waveforms will generate different shapes of the CCF $R_{g\hat{g}}(\tau)$, thereby affecting the results of (5.62) and (5.63), thus changing the final multipath resistant performance.

Example 5.4 (*Multipath average range error envelope for MBOC signals*) Figures 5.10 and 5.11 show the multipath running average range error envelope of various MBOC modulated signals in matched and the BOC_{11}-like unmatched receiving modes for 40 MHz and 10 MHz bandwidth respectively, where the MDR is −5 dB. For comparison, the multipath performance of the BOC(1,1) signal matched receiving is also given on the same figure.

It can be seen that with a wide receiving bandwidth, CBOC^{-}, TMBOC, and QMBOC have the best multipath resistant capability in the matched receiving mode. When the BOC(1,1) local signal is used for the unmatched receiving of MBOC, the

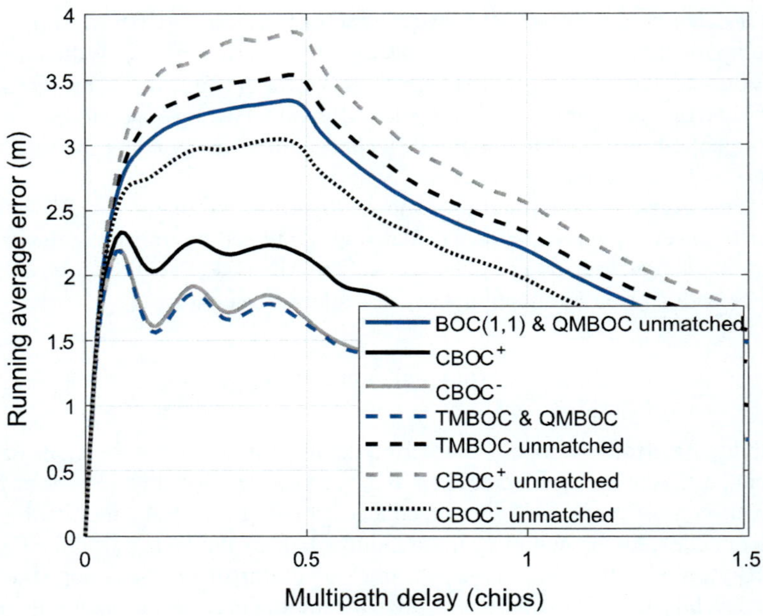

Fig. 5.10 Comparison of the multipath resistant performance of different modulation methods under 40 MHz bandwidth

degradation of the multipath resistant performance of CBOC$^-$ is the lowest, followed by QMBOC. TMBOC has the greatest performance degradation.

When the receiving bandwidth is narrow, there is no significant difference in the multipath performance of various MBOC modulations and processing methods. This phenomenon is consistent with accuracy analysis cases when the bandwidth is less than 12 MHz as shown in Example 5.1. It also shows that for a narrowband receiver, it is sufficient to treat the MBOC signal as a BOC(1,1) signal.

From Fig. 5.10, we can also see the significant performance improvement brought by processing the BOC(6,1) component in an MBOC signal for the wideband receiver. ■

5.8 Radio Frequency Compatibility

As seen in Sect. 2.4, with the development of GNSS, more and more navigation signals are sharing the limited navigation frequency band resources, making spectrum overlap of navigation signals inevitable. The problem of compatibility between signals has become the focus of attention in the next-generation GNSS construction.

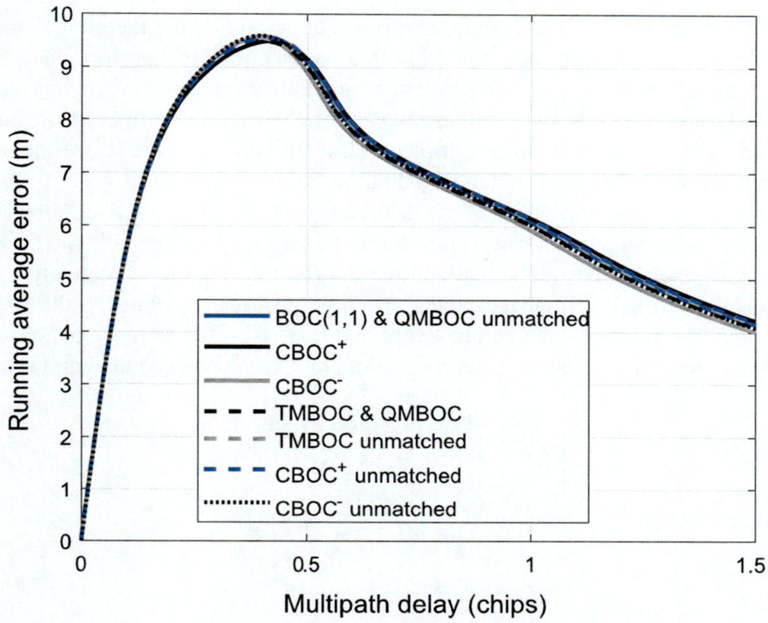

Fig. 5.11 Comparison of the multipath resistant performance of different modulation methods under 10 MHz bandwidth

In the field of satellite navigation, GNSS radio frequency compatibility (RFC) refers to the ability of a navigation satellite system to be used without causing unacceptable interference and harm to services of other navigation satellite systems. The problem of GNSS compatibility is mainly due to the mutual interference between the signals of various systems. We must analyze and define the compatibility indicator between signals in order to ensure that when multiple GNSSs coexist in the future, the performance degradation of each system due to signal spectrum interference can be within the expected range.

5.8.1 The Spectrum Separation Coefficient

As mentioned earlier, in the receiver, the performance of signal acquisition, carrier tracking, and data demodulation all depend on the SNIR of the prompt branch correlator output. Therefore, the RF compatibility between two DSSS signals can be evaluated by measuring their SNIR at the receiver prompt correlator output.

In the model that is widely used currently, it is assumed that the interference signal is a zero-mean, wide-sense stationary Gaussian random process. Only in this way can we ensure that the superposition of interference and thermal noise still satisfies the

Gaussian distribution at the output of the correlator, so that the theoretical analysis model of the acquisition, demodulation, and carrier tracking performance is still valid. Although a single DSSS interference signal does not fit well with the Gaussian random process hypothesis, the real interference is usually composed of multiple independent DSSS signals from multiple satellites. At this time, the above-mentioned assumptions can be approximately satisfied.

When we represent the PSD of the desired signal as $C_s G_s(f)$, and represent the PSD of the other signal that causes interference as $C_\iota G_\iota(f)$, where C_s and C_ι are the carrier power of the desired signal and the interference signal respectively, $G_s(f)$ and $G_\iota(f)$ are their normalized power spectral densities respectively, the SNIR of the prompt correlator output can be expressed by (5.52). Using (5.6) and assuming that the receiver is in matched receiving mode, (5.52) can be expanded and rewritten as

$$\rho = \frac{TC_s \int_{-\beta_r/2}^{\beta_r/2} G_s(f)\,\mathrm{d}f}{N_0 + C_\iota \chi}, \tag{5.66}$$

in which the term

$$\chi = \int_{-\beta_r/2}^{\beta_r/2} G_s(f) G_\iota(f)\mathrm{d}f \tag{5.67}$$

is referred to as the *spectral separation coefficient (SSC)* of the desired signal and the interference signal, which reflects the reduction of the SNIR of the desired signal due to the interfering signals.

From the form of the expression (5.67), we can also find that the SSC of the two signals reflects the degree of overlap of these two signals' spectra. The smaller the SSC value, the higher the degree of spectral separation between signal spectrum, and the smaller the mutual interference between signals.

Example 5.5 (*SSC of GPS, Galileo, and BDS in the upper L band*) Based on (5.67), Table 5.1 shows the SSC calculation results for GPS, Galileo, and BDS for signals in the L1 band with a receiver front-end bandwidth of 24 MHz. Among them, the signal structure parameters of each system used for calculation are based on the assumptions in Table 5.2.

From the data in Table 5.1, it can be found that the spectrum separation of the three authorization signals, GPS M code, Galileo E1 PRS, and BDS B1A, from other signals is obvious. This is because their power spectrum main lobes are far from the center frequency. In addition, the interoperable MBOC signals of these three GNSSs have good spectral separation with the GPS C/A codes. ■

5.8.2 The Code Tracking Spectral Sensitivity Coefficient

The SSC reflects the influence of the interference signal on the SNIR of the desired signal's prompt correlator output, which directly corresponds to the performance

Table 5.1 SSC in the L1 band (receiver front-end bandwidth is 24 MHz)

SSC(dB/Hz)			Desired signal							
			GPS				Galileo		BDS	
			C/A	P(Y)	M	L1C	PRS	OS	B1C	B1A
Interference signal	GPS	C/A	−61.80	−69.91	−87.11	−68.10	−108.7	−68.15	−68.10	−90.96
		P(Y)	−69.91	−71.25	−79.90	−70.39	−100.6	−70.45	−70.40	−87.63
		M	−87.11	−79.9	−71.69	−81.95	−88.88	−82.00	−81.95	−83.51
		L1C	−68.10	−70.39	−81.95	−65.35	−101.4	−65.40	−65.35	−86.26
	Galileo	E1	−108.7	−100.6	−88.88	−101.4	−100.9	−101.5	−101.4	−97.04
		OS	−68.15	−70.45	−82.00	−65.40	−101.5	−65.46	−65.41	−86.32
	BDS	B1C	−68.10	−70.40	−81.95	−65.35	−101.4	−65.41	−65.36	−86.27
		B1A	−90.96	−87.63	−83.51	−86.26	−97.04	−86.32	−86.27	−91.91

Table 5.2 Signal structure parameters of each system used to calculate results in Table 5.1

System	Service type	Carrier frequency (MHz)	Modulation type
GPS	C/A	1575.42	BPSK-R(1)
	P(Y)	1575.42	BPSK-R(10)
	M	1575.42	BOC(10.5)
	L1C	1575.42	MBOC(6,1,1/11)
Galileo	PRS	1575.42	$\text{BOC}_\text{c}(15,2.5)$
	OS	1575.42	MBOC(6,1,1/11)
BDS	B1C	1575.42	MBOC(6,1,1/11)
	B1A	1575.42	$\text{BOC}_\text{s}(14,2)$

impact on the desired signal's acquisition, carrier tracking, and data demodulation. The *code tracking spectral sensitivity coefficient (CT-SSC)* [40] can be used to evaluate the influence of the interference signal on the code tracking performance of the desired signal.

For simplicity, ignoring the squared loss term and dividing the effects of white noise and non-white interference into two separate parts, we can rewrite (5.43) as

$$\sigma_{\text{close}}^2 = \frac{B_L(1-0.5B_LT)\int_{-\beta_r/2}^{\beta_r/2} G_s(f)\sin^2(\pi f\delta)\mathrm{d}f}{(2\pi)^2 C_s\left(\int_{-\beta_r/2}^{\beta_r/2} f G_s(f)\sin(\pi f\delta)\mathrm{d}f\right)^2} \times \left[N_0 + \frac{C_\iota\int_{-\beta_r/2}^{\beta_r/2} G_\iota(f)G_s(f)\sin^2(\pi f\delta)\mathrm{d}f}{\int_{-\beta_r/2}^{\beta_r/2} G_s(f)\sin^2(\pi f\delta)\mathrm{d}f}\right]. \tag{5.68}$$

From the above equation, we can extract a term that plays a similar role as χ in the SNIR expression of the prompt branch correlator output, which is called CT-SSC:

$$\kappa_{ELS} \triangleq \frac{\int_{-\beta_r/2}^{\beta_r/2} G_w(f) G_s(f) \sin^2(\pi f \delta) \mathrm{d}f}{\int_{-\beta_r/2}^{\beta_r/2} G_s(f) \sin^2(\pi f \delta) \mathrm{d}f}. \tag{5.69}$$

Comparing the expression forms of χ and κ_{ELS}, it can be found that κ_{ELS} is related to not only the PSD of the desired signal and the interference signal and the front-end bandwidth of the receiver, but also the early-late spacing of the tracking loop. If the calculation of the SSC is visually regarded as calculating the overlap area of the PSDs of the two signals, then the CT-SSC calculation involves the overlap area of three components, including the PSDs of the desired signal and the interference signal, and $\sin^2(\pi f \delta)$ which is related to δ.

In the example below, we discuss CT-SSC and SSC with some specific signal modulations.

Example 5.6 (*CT-SSC analysis of several signals*) Figure 5.12 shows the curve of CT-SSC with respect to the early-late spacing δ when BPSK-R(1) is selected as the desired signal, and the interference signals are BOC(1,1), BPSK-R(1), and BPSK-R(10) signals respectively. Figure 5.13 selects BOC(1,1) as the desired signal, and the other conditions remain unchanged. For comparison, the value of SSCs are also drawn in same figures. Because they are independent of δ, they are straight lines in these figures.

Both of these two figures show that when δ takes different values, the value of CT-SSC has a significant change, and the variation range can reach more than 10 dB. For the interference of BOC(1,1) on BPSK-R(1), the CT-SSC maximum occurs near $\delta_{\max} = 0.6$ chips.

This phenomenon can be explained intuitively by Fig. 5.14, which shows the PSDs of BPSK-R(1) and BOC(1,1) and the curve of $\sin^2(\pi f \delta_{\max})$. We can find that $\sin^2(\pi f \delta_{\max})$ has a large overlap with the PSD of BOC(1,1) and a small overlap with the PSD of BPSK-R(1). This results in a small value for the denominator in (5.69), while the numerator has a large value. ■

We can find from this example that in cases where δ is not very large, and the desired signal and the interference signal have the same modulation, the value of CT-SSC is always lower than the value of SSC. Therefore, the degree of degradation of the code tracking performance caused by the interference signal is not as serious as that of the SNIR.

However, for a more general selection of desired signals and interference signals, the above rules are not always true. For example, when the PSD main lobe of the interference signal is far from that of the desired signal, the deterioration of the code tracking performance may exceed that of the SNIR. This phenomenon is also mentioned in [41, 42].

The characteristic difference between SSC and CT-SSC once again reflects the particularity of navigation signal design. For a navigation signal in which the ranging performance is the main evaluation criterion, the influence of the interference on the SNR does not necessarily directly correspond to the influence on the tracking performance. A more detailed discussion of CT-SSC can be found in [40].

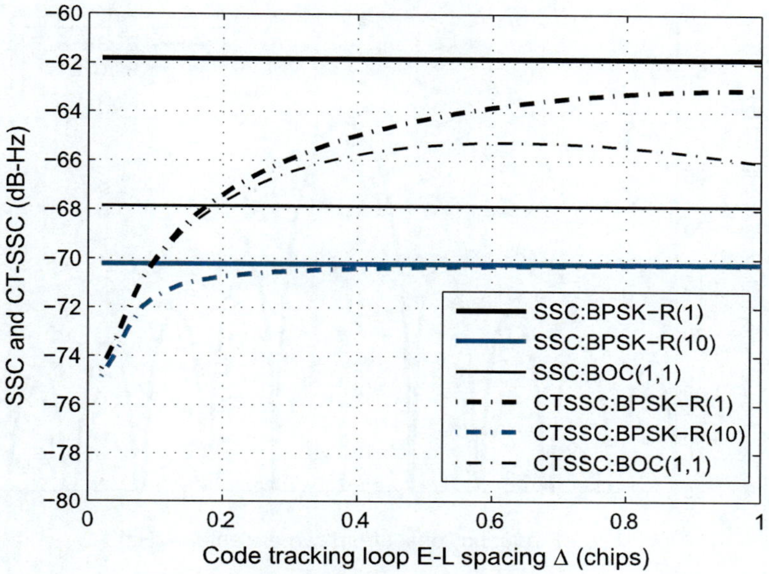

Fig. 5.12 CT-SSC and SSC when the desired signal is BPSK-R(1)

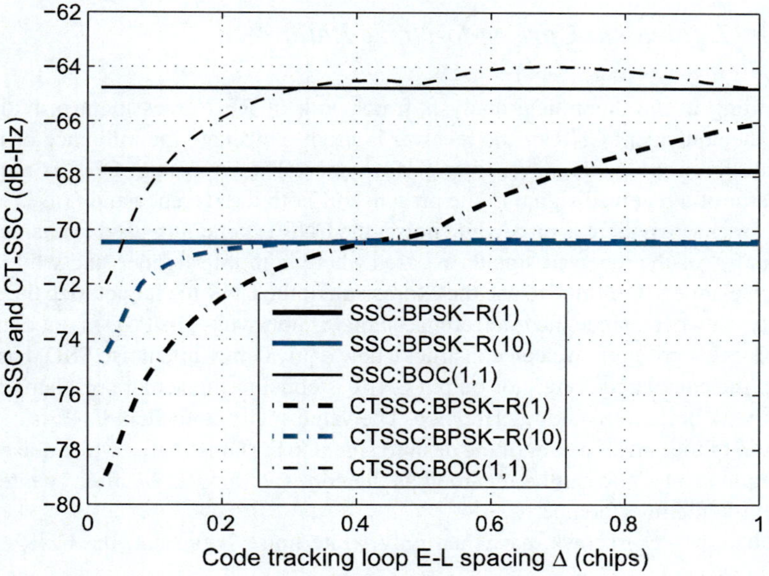

Fig. 5.13 CT-SSC and SSC when the desired signal is BOC(1,1)

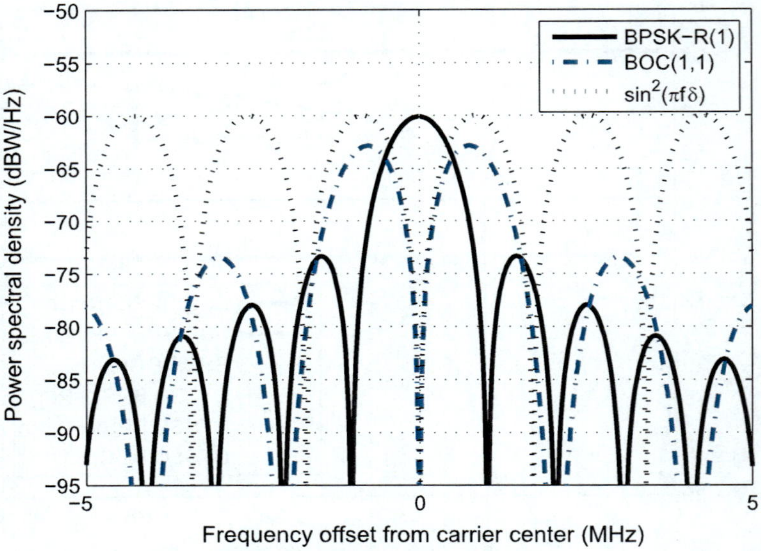

Fig. 5.14 Power spectral density of BPSK-R(1), BOC(1,1) and $\sin^2(\pi f \delta_{\max})$

5.8.3 Equivalent Carrier-to-Noise Ratio

According to the theoretical analysis framework of RFC assessment provided by ITU, the equivalent CNR of the receiver is used to evaluate the influence of different interference signals on receiver performance. The equivalent CNR is a quality indicator of a received signal in the presence of both interference and noise.

In cases when there is only white noise, the CNR is generally used to characterize the quality of the received signal. In cases where both interference and white noise exist, we hope to continue to use the expression of the CNR to characterize the signal quality. However, since the interference signal is non-white, its PSD is not constant. To solve this problem, we can construct a new equivalent white noise PSD, based on which the calculated correlator output SNIR is equal to the actual correlator output SNIR with both noise and interference. The value of the equivalent CNR is equal to the ratio of the carrier power of the desired signal to the constructed white noise PSD. The equivalent CNR can be regarded as the corresponding CNR after "whitening" the non-white interference.

Equation (5.66) shows that when only white noise is present, the CNR can be written as

$$\frac{C_s}{N_0} = \frac{\rho_c}{2T \int_{-\beta_r/2}^{\beta_r/2} G_s(f)\,\mathrm{d}f}. \tag{5.70}$$

When noise and interference are present at the same time, if we still use the right side of the above formula for calculation, the equivalent CNR

$$
\begin{aligned}
\left(\frac{C_s}{N_0}\right)_{\text{eff}} &= \frac{\rho_c}{2T\int_{-\beta_r/2}^{\beta_r/2} G_s(f)\,\mathrm{d}f} \\
&= \frac{C_s}{N_0 + C_\iota \int_{-\beta_r/2}^{\beta_r/2} G_s(f)G_\iota(f)\mathrm{d}f} \\
&= \frac{C_s}{N_0 + I_0}
\end{aligned}
\tag{5.71}
$$

can be obtained after some simplification, where $I_0 \triangleq C_\iota \chi$ is the equivalent noise PSD and χ is the SSC of the desired signal and the interference signal defined by (5.67).

The above discussion simply deals with the effect of one signal on another. When the actual GNSS is operating, the receiving process of a desired signal is actually subject to the radio frequency interference of all visible satellites in all GNSSs. Therefore, if we want to evaluate the RFC between navigation systems more accurately, we need a much more complicated computational model. Since the satellite signals are independent, the influence of each interference signal on the desired signal can be considered as separate, and the total impact can be regarded as the accumulation of these respective influences.

In the theoretical analysis framework of radio frequency compatibility evaluation given by ITU, I_0 is further divided into intra-system interference equivalent noise PSD I_{intra} caused by all signals except the desired signal itself in the system, inter-system interference equivalent noise PSD I_{inter} caused by other GNSS signals in the same frequency band, and extra-system interference equivalent noise PSD I_{ext} caused by radio signals of non-GNSS systems, i.e. $I_0 = I_{\text{intra}} + I_{\text{inter}} + I_{\text{ext}}$. Further, I_{intra} can be decomposed into

$$
I_{\text{intra}} = \sum_{m=1}^{M_{\text{intra}}} \sum_{n=1}^{N_{\text{intra},n}} C_{m,n}^{\text{intra}} \chi_{m,n}^{x}, \tag{5.72}
$$

where M_{intra} is the number of visible satellites in the system at a given time and place, $N_{\text{intra},n}$ is the number of signals transmitted by each satellite in the system except the desired signal, $C_{m,n}^{\text{intra}}$ is the received carrier power of the nth interference signal of the mth satellite in the system at a given time and place, and $\chi_{m,n}^{x}$ represents the SSC of the nth interference signal of the mth satellite in the system and the desired signal. It is similar for the equivalent noise PSD I_{inter} of the inter-system interference, and the terms will not be described here again.

In order to be able to calculate each $C_{m,n}^{\text{intra}}$ for a given time and place, we need to consider not only the center frequency and PSD of each signal, but also a series of factors regarding the Earth-satellite link including the satellite constellation, space loss, processing loss, satellite transmit antenna pattern, and characteristics of the receiver antenna. If we want to obtain the RFC results for various locations around the world over a long-enough period of time, we need to cover all possibilities simultaneously.

Completing a comprehensive inter-system compatibility analysis requires a large number of input conditions and a considerable amount of computation. Some scholars have performed RFC simulation analysis between several major GNSSs in specific scenarios [43]. The research on the further improvement of the RFC evaluation formula is also ongoing. For example, Hegarty [14] considers the further effects of sampling and quantification, and Avila-Rodriguez [44] offers a detailed analysis of the non-ideal performance of the signal for SSC calculations. Due to the limitations of space, it is not possible to expand them one by one. Readers who are interested can refer to the relevant literature.

In fact, we can see from (5.72) that the core of RFC analysis is based on SSC and CT-SSC. The series of parameters on the satellite link mainly affect the weight coefficients of each SSC. In the initial stage of signal design, if one is only considering the impact of the spreading chip waveform design of the signal on RF compatibility, it is sufficient to use SSC and CT-SSC as evaluation indicators.

References

1. Lindsey WC, Simon MK (1973) Telecommunication systems engineering. Courier Corporation, North Chelmsford
2. Simon MK (1977) Noncoherent pseudo-noise code tracking performance of spread-spectrum receivers. IEEE Trans Commun 25(3):327–345
3. Dierendonck AJ, Fenton P, Ford T (1992) Theory and performance of narrow correlator spacing in a GPS receiver. Navigation 39(3):265–283
4. Betz JW, Kolodziejski KR (2009) Generalized theory of code tracking with an early-late discriminator part I: lower bound and coherent processing. IEEE Trans Aerosp Electron Syst 45(4):1538–1556
5. Nunes F, Sousa F, Leitao J (2007) Gating functions for multipath mitigation in GNSS BOC signals. IEEE Trans Aerosp Electron Syst 43(3):941–964
6. Irsigler M, Eissfeller B, Comparison of multipath mitigation techniques with consideration of future signal structures
7. Martin N, Leblond V, Guillotel G et al (2003) BOC (x, y) signal acquisition techniques and performances. In: ION GPS, Portland, OR 2003, pp 188–198
8. Lohan ES, Burian A, Renfors M (2008) Low-complexity unambiguous acquisition methods for BOC-modulated CDMA signals. Int J Satell Commun Netw 26:503–522
9. Fishman P, Betz JW (2000) Predicting performance of direct acquisition for the M-code signal. In: Proceedings of the 2000 national technical meeting of the institute of navigation, pp 574–582
10. Hoult N, Aguado LE, Xia P (2008) MBOC and BOC (1, 1) performance comparison. J Navig 61(4):613–627
11. Zhang J, Yao Z, Lu M (2016) Generalized theory and decoupled evaluation criteria for unmatched despreading of modernized GNSS signals. Sensors 16(6):1128 July
12. Yao Z, Lu M (2011) Lower bound on spreading code tracking error under unmatched despreading mode. Electron Lett 47(15):878–879
13. Soellner M, Kohl R, Luetke W, Erhard P (2001) The impact of linear and non-linear signal distortions on Galileo code tracking accuracy. In: Proceedings of the 15th international technical meeting of the satellite division of the institute of navigation (ION GPS 2002), pp 1270–1285

14. Hegarty CJ (2011) Analytical model for GNSS receiver implementation losses. Navigation 58(1):29–44
15. Zhang W, Miller MJ (1992) Baseband equivalents in digital communication system simulation. IEEE Trans Educ 35(4):376–382
16. Mengali U, D'Andrea AN (1997) Synchronization techniques for digital receivers. Springer Science & Business Media, Berlin
17. Gabor D (1946) Theory of communication. Part 1: the analysis of information. J Inst Electr Eng-Part III: Radio Commun Eng 93(26):429–441
18. Ziv J, Zakai M (1969) Some lower bounds on signal parameter estimation. IEEE Trans Inf Theory 15(3):386–391
19. Chazan D, Zakai M, Ziv J (1975) Improved lower bounds on signal parameter estimation. IEEE Trans Inf Theory 21(1):90–93
20. Helstrom CW (1968) Statistical theory of signal detection: international series of monographs in electronics and instrumentation
21. Van Trees HL (1968) Detection estimation and modulation theory, part I. Wiley, Hoboken
22. Van Dierendonck AJ (1996) GPS receivers. In: Global positioning system: theory and applications, vol 1, pp 329–407
23. Ward PW, Betz JW, Hegarty CJ (2006) Satellite signal acquisition, tracking, and data demodulation. In: Understanding GPS principles and applications, vol 5, pp 174–175
24. Van Nee DJR, Coenen A (1991) New fast GPS code-acquisition technique using FFT. Electron Lett 27(2):158–160
25. Tsui JBY (2005) Fundamentals of global positioning system receivers: a software approach, 2nd edn. Wiley Interscience, New Jersey
26. Simon M et al (1994) Spreading spectrum communications handbook, revised. McGraw-Hill, New York, pp 751–900
27. Lin D, Tsui J (2000) Comparison of acquisition methods for software GPS receiver. In: Proceedings of ION GPS, Salt Lake City
28. Akopian D (2005) Fast FFT based GPS satellite acquisition methods. IEE Proc Radar Sonar Navig 152(4):277–286
29. Psiaki M (2001) Block acquisition of weak GPS signals in a software receiver. In: Proceedings of ION GPS, Salt Lake City, USA
30. Julien O (2005) Carrier-phase tracking of future data/pilot signals. In: Proceedings of ION GNSS, Long Beach, CA, pp 113–124
31. Proakis J (2001) Digital communications, 4th edn. McGraw-Hill, Boston
32. Fine P, Wilson W (1999) Tracking algorithm for GPS offset carrier signals. In: Proceedings of the 1999 national technical meeting of the institute of navigation, pp 671–676
33. Yang C, Miller M, Nguyen T, Akos D (2006) Generalized frequency-domain correlator for software GPS receiver: preliminary test results and analysis. In Proceedings of the 19th international technical meeting of the satellite division of the institute of navigation (ION GNSS 2006), Fort Worth, TX, USA, 26–29 September 2006, pp 2346–2630
34. Yao Z, Lu M (2011) Optimized modulation for Compass B1-C signal with multiple processing modes. In: ION GNSS 24th international technical meeting of the satellite division, Portland, USA, pp 1234–1242
35. Betz JW, Blanco MA, Cahn CR et al (2006) Description of the L1C signal. In: ION GNSS. 2006, pp 2080–2091
36. Beauchamp KG (1975) Walsh functions and their applications. Academic, Cambridge
37. Kaplan ED, Hegarty CJ (2006) Understanding GPS: principles and applications. Artech House, Norwood
38. Braasch MS (1996) Multipath effects. In: Global positioning system: theory and applications, vol 1. American Institute of Aeronautics and Astronautics, Washington, D. C.
39. Hagerman L (1973) Effects of multipath on coherent and noncoherent PRN ranging receiver. Development Planning Division, The Aerospace Corporation
40. Soualle F, Burger T (2007) Radio frequency compatibility criterion for code tracking performance. In: Proceedings of the 20th international technical meeting of the satellite division of the institute of navigation (ION GNSS 2007), Fort Worth, TX, September 2007, pp 1201–1210

41. Betz JW (2000) Effect of narrowband interference on GPS code tracking accuracy. In: Proceedings of the 2000 national technical meeting of the institute of navigation, Anaheim, CA, January 2000, pp 16–27
42. Lopez-Almansa J-M, Pablos PA (2000) Derivation of an analytical model for estimation of measurement error induced by an arbitrary disturbance on a coherent delay lock loop. Navigation 47(4):267–278
43. Liu W, Zhai CR, Zhan XQ, Zhang YH (2011) Assessment and analysis of radio frequency compatibility among several global navigation satellite systems. IET Radar Sonar Navig (IET) 5(2):128–136
44. Avila-Rodriguez JA (2008) On generalized signal waveforms for satellite navigation. University FAF Munich, Munich, Germany

Chapter 6
Fundamental Theory of Constant Envelope Multiplexing for Spread-Spectrum Signals

The true face of Lushan Mountain is lost to my sight,
for it is right in this mountain that I reside.

— Su Shi

6.1 Introduction

From the satellite to the ground, the satellite navigation signal has to travel through tens of thousands of kilometers, which results in severe path loss and fading. In order to maintain sufficient received power at the ground receiver, the onboard transmitter has to have as high a power efficiency as possible, since the transmission power of the satellite is limited. If different service signals in one frequency band use separate transmission antennas and amplifier links, this will pose great challenges to the antenna design [1] and the control of total power, cost, volume, and weight of the payload. Therefore, it is desirable to multiplex a plurality of signals onto one carrier and share one transmission link. In addition, in order to allow the onboard high power amplifier to achieve a high power efficiency by operating in its saturation nonlinear region, it is necessary to ensure the constant envelope characteristics of the multiplexed signal.

In the original GPS and GLONASS, as well as the BDS-2 system, there are only two signals, a civil signal and a military signal, to be multiplexed in each frequency band. Therefore, the QPSK technique, which is the simplest way to implement the CEM of two bipolar DSSS signals, can be used. However, in the new generation of GNSSs, with further refined services and the requirement of backward compatibility, more signals are required to be broadcast [2]. Most of these signals employ data-plus-pilot channel structures, which further increases the actual number of signal

Z. Yao and M. Lu, *Next-Generation GNSS Signal Design*, Navigation: Science and Technology 6, https://doi.org/10.1007/978-981-15-5799-6_6

components. For example, GPS III needs to multiplex at least four bipolar signal components in the L1 band [3]. Galileo system faces challenges with multiplexing non-bipolar components and dual-frequency multiplexing in the E1 band and E5 band respectively [4]. In BDS-3, not only are there more components to be multiplexed but, due to the changes in central frequencies for the regional and global evolution phases, there are also more complicated multi-frequency multiplexing requirements during the transition period [5].

The past 15 years have witnessed significant progress in multiplexing techniques. Some multiplexing techniques proposed for other radio systems are re-evaluated for potential use in GNSS, and many novel and advanced CEM techniques are being proposed. In spite of the important advances in solving specific multiplexing issues during the construction of GPS III and Galileo, the research field of signal multiplexing is still subject to active studies. Especially in the recent five years, significant progress in multicarrier CEM techniques has been made [6–12], and the generalized signal multiplexing design philosophy is becoming more mature and complete [13–17]. However, for the future evolutional of GNSSs, many challenges and much work still remain [18].

This chapter and the next chapter focus on the constant envelope multiplexing techniques for satellite navigation signals. Since this is a relatively new field of research, there is not much in-depth systematic analysis of this technology in the existing literature, other than listing the contents, which cannot provide a full picture of the development situation. In these two chapters, we intend to provide a higher standpoint of the discussion and analysis of emerging CEM technology, by a comprehensive tutorial on the fundamental theories, state-of-the-art development, applications, as well as the remaining challenges of GNSS signal multiplexing techniques so that readers will thoroughly understand the nature, difference, and internal relationships of these technologies.

Therefore, before moving on to specific multiplexing implementation techniques, we first explain the mathematical modeling of CEM for general satellite navigation signals. Then, based on the concept of the baseband complex envelope, we provide the mathematical and geometric description of the constant envelope as well as the concept of the phase map, constellation diagram, and multiplexing efficiency.

With a discussion of two seemingly different patterns of CEM design, the waveform domain processing (WDP) and the phase domain processing (PDP), we introduce the design equations and analysis methods of general constant envelope multiplexing techniques. In spite of the significant difference between starting points of these two design patterns, their consistency in essence will be revealed at the end of this chapter. Existing research [19] has shown that some specific CEM techniques have more than one representation, such as Interplex [20] and coherent adaptive subcarrier modulation (CASM) [21]. However, by introducing signal value space and basis signals [16, 17], and subsuming WDP and the PDP into an integrated framework [18], the relationship and mutual transformation of different CEM representation will be discussed more generally in this chapter. We will show that every

CEM technique has at least three different expressive forms, with their own features in realization or analysis, and they can be converted into each other. We believe that with this elaborate text structure, this chapter will give the reader a global overview of CEM, highlighting connections and differences between the assorted techniques.

6.2 High Power Amplifier

In satellite navigation systems, the propagation loss of signals from satellites to ground receivers is very large. In order to maintain sufficient receiving power, a HPA such as the travelling wave tube amplifier (TWTA) or the solid state power amplifier (SSPA) which operates in full saturation mode is required [22].

However, typical HPA amplification usually introduces a nonlinear transformation into the transmitted signal. The HPA input and output process can generally be built into a nonlinear memoryless model, which is characterized by two nonlinear memoryless functions called amplitude-to-amplitude (AM/AM) and amplitude-to-phase (AM/PM) features. The AM/AM is the nonlinear conversion characteristic of the input signal amplitude to the output signal amplitude, whereas the AM/PM is the nonlinear conversion characteristic of the input signal amplitude to the output signal phase.

If we write the HPA input signal as

$$x(t) = r(t)\cos\left[\omega_0 t + \phi(t)\right] , \tag{6.1}$$

where ω_0 is the carrier angular frequency, $r(t)$ is the modulation envelope and $\phi(t)$ is phase angle of the signal, then the output signal of HPA can be expressed as

$$y(t) = F_A\left[r(t)\right]\cos\left\{\omega_0 t + \phi(t) + F_\Phi\left[r(t)\right]\right\} , \tag{6.2}$$

where $F_A[r]$ is the AM/AM characteristic function of HPA, and $F_\Phi[r]$ is the AM/PM characteristic function of HPA.

As can be seen from the above model, the nonlinear AM/AM and AM/PM characteristics of the HPA are functions of the input signal amplitude $r(t)$. For a constant envelope signal, it passing through a non-linear amplifier is equivalent to passing through a system with a fixed amplitude gain and a fixed phase offset, so that the performance of the signal receiving is hardly affected by the nonlinear amplifier. Therefore, when using a high power amplifier, the input signal should be as close as possible to a constant envelope signal.

6.3 Constant Envelope Signal

The signal transmitted by the transmitter can be expressed as

$$\begin{aligned} s_{\mathrm{RF}}(t) &= \mathrm{Re}\left\{\sqrt{2P}s(t)\exp\left[\mathrm{j}(\omega_c t+\varphi)\right]\right\} \\ &= I(t)\cos(\omega_c t+\varphi) - Q(t)\sin(\omega_c t+\varphi), \end{aligned} \tag{6.3}$$

where

$$s(t) = I(t) + \mathrm{j}Q(t) \tag{6.4}$$

is the complex baseband signal, $I(t)$ is its real part, which is modulated on the in-phase component of the transmitted signal, $Q(t)$ is its imaginary part, which is modulated on the quadrature-phase component, P is the transmission power, ω_c is the carrier angular frequency, and φ is the initial phase of the carrier.

The complex baseband signal $s(t)$, apart from the form of the real and imaginary parts shown in (6.4), can also be written as a module-and-phase form

$$s(t) = A(t)\,\mathrm{e}^{\mathrm{j}\phi(t)}, \tag{6.5}$$

where

$$A(t) = |s(t)| = \sqrt{I^2(t) + Q^2(t)} \tag{6.6}$$

and

$$\phi(t) = \arg\{s(t)\} = \mathrm{atan2}(Q(t), I(t)) \tag{6.7}$$

are referred to as the amplitude envelope and instantaneous phase of $s(t)$ respectively. Based on the values of real and imaginary parts, the signal baseband complex envelope can be mapped to different points on a complex plane. The result of the mapping on this complex plane is called *constellation diagram.*

If the envelope of the signal is a constant amount that does not change over time, i.e.

$$|s(t)| = \sqrt{I^2(t) + Q^2(t)} \equiv A, \tag{6.8}$$

the signal is then referred to as a constant envelope signal.

6.4 Constant Envelope Multiplexing

Since the 1960s, scholars have studied techniques for modulating multiple signals onto the same carrier by using subcarrier modulation while maintaining the constant envelope characteristics of the synthesized signal [23–25]. The satellite navigation system multiplexes multiple signals to a satellite link transmission through a signal multiplexing module to provide diverse services. For different service performance requirement and the degree of interference with each other, the power ratio and relative phase relationship of these signals may not be the same. We want to combine these signals while maintaining a preset power ratio and phase relationship.

6.4.1 Signal Model

Considering a general case of combining N independent DSSS signal components located at several adjacent subcarrier frequencies, a generalized mixed RF signal can be expressed as

$$s_{\mathrm{RF}}(t) = \sum_{i=1}^{N} \mathrm{Re}\left\{\sqrt{P_i} s_i(t)\, \mathrm{e}^{\mathrm{j}(2\pi f_i t + \phi_i)}\right\}, \tag{6.9}$$

where P_i, f_i, ϕ_i are the nominal power, carrier frequency, and initial phase of component i, respectively, and $s_i(t)$ is the complex envelope of component i, which follows the definition of (3.2) and can be constructed as

$$s_i(t) = \sum_{n=-\infty}^{+\infty} c_n^{(i)} p_i\left(t - nT_c^{(i)}\right), \tag{6.10}$$

where $c_n^{(i)}$ is the navigation data modulated by the corresponding bipolar spreading code, with chip rate $f_c^{(i)} = 1/T_c^{(i)}$ and chip waveform $p_i(t)$, which is non-zero only in interval $t \in [0, T_c^{(i)})$.

In this chapter, we limit $p_i(t)$ to the SCS waveform (See Sect. 3.3.5), which can be further divided into $M^{(i)}$ segments with equal-length $T_s^{(i)} = T_c^{(i)}/M^{(i)}$, each of which is assigned a deterministic value, i.e.

$$p_i(t) = \sum_{k=0}^{M^{(i)}-1} p_k^{(i)} \psi_{T_s^{(i)}}\left(t - kT_s^{(i)}\right), \tag{6.11}$$

where $\psi_T(t)$ is unit amplitude rectangular pulse function with T duration, and $p_k^{(i)}$ is the waveform value in kth segment. Here, we follow the notaion of the SCS signal waveform, using the shape vector

$$\boldsymbol{p}_i = \left(p_0^{(i)}, p_1^{(i)}, \ldots, p_{M^{(i)}-1}^{(i)}\right)^{\mathrm{T}} \tag{6.12}$$

to characterize the waveform shape. For BPSK-R modulation, $p_i = 1$. For BOC(n, n) modulation, $\boldsymbol{p}_i = (1, -1)^{\mathrm{T}}$. By substituting (6.11) into (6.10), we can rewrite the ith signal component $s_i(t)$ as

$$s_i(t) = \sum_{n=-\infty}^{+\infty} \sum_{k=0}^{M^{(i)}-1} c_n^{(i)} p_k^{(i)} \psi_{T_s^{(i)}} \left(t - \left(M^{(i)} n + k \right) T_s^{(i)} \right). \tag{6.13}$$

When combining these N components into an integrated signal to enable the transmission of them through a single chain, it is equivalent to constructing an analytic signal $s_{\mathrm{MUX}}(t)$, and modulating it onto a single carrier with frequency f_0, while minimizing the difference to the separate transmission from the receiving perspective. This relation is presented as $s_{\mathrm{RF,MUX}}(t) = \mathrm{Re}\left\{ s_{\mathrm{MUX}}(t)\, \mathrm{e}^{\mathrm{j}2\pi f_0 t} \right\} \approx s_{\mathrm{RF}}(t)$, which is equivalent to

$$s_{\mathrm{MUX}}(t) \approx \sum_{i=1}^{N} \sqrt{P_i} \mathrm{e}^{\mathrm{j}\phi_i} s_i(t)\, \mathrm{e}^{\mathrm{j}2\pi (f_i - f_0)t}. \tag{6.14}$$

The sub chip rate $f_s^{(i)} = 1/T_s^{(i)}$ of these N components may not be the same. However, in a digitized transmitter, the generation of these N components is usually driven by clock frequencies derived from a common system clock. Therefore, one can find the least common multiple f_s of all the sub chip rates $f_s^{(i)}$ and obtain the shortest common segment length $T_s = 1/f_s$. Furthermore, for the digital implementation, the subcarrier $\mathrm{e}^{\mathrm{j}2\pi(f_i - f_0)t}$ is often replaced by its sample-and-hold version $\mathrm{e}^{\mathrm{j}2\pi \Delta f_i \ell}$ with T_s interval, where $\Delta f_i = (f_i - f_0) T_s$. By choosing the proper value of f_i and f_0, one can ensure that $\Delta T_i = 1/\Delta f_i = f_s/(f_i - f_0)$ is an integer. Actually for all of the deployed GNSS signals, the carrier frequency, the spreading chip rate, as well as the subcarrier rate are all multiples of 1.023 MHz. Therefore, this condition can be satisfied easily.

Thus, by substituting (6.13) into (6.14), we can express signal (6.14) as

$$\begin{aligned} s_{\mathrm{MUX}}(t) &= \sum_{\ell=-\infty}^{+\infty} \sum_{i=1}^{N} \sqrt{P_i} \mathrm{e}^{\mathrm{j}\phi_i} \tilde{c}_\ell^{(i)} \psi_{T_s}(t - \ell T_s) \\ &= \sum_{i=1}^{N} \sqrt{P_i} \mathrm{e}^{\mathrm{j}\phi_i} s_{\mathrm{IF}}^{(i)}(t)\,, \end{aligned} \tag{6.15}$$

where $\tilde{c}_\ell^{(i)} \triangleq c_n^{(i)} p_k^{(i)} \mathrm{e}^{\mathrm{j}(2\pi \Delta f_i \ell)}$ with $n = \lfloor \ell T_s / T_c^{(i)} \rfloor$, $k = \lfloor (\ell T_s - n T_c^{(i)}) / T_s^{(i)} \rfloor$, and

$$s_{\mathrm{IF}}^{(i)}(t) = \sum_{\ell=-\infty}^{+\infty} \tilde{c}_\ell^{(i)} \psi_{T_s}(t - \ell T_s) \tag{6.16}$$

is the digital IF carrier modulated by $s_i(t)$. It can be observed from (6.15) that in every duration of common segment $t_\ell \in [\ell T_s, (\ell+1) T_s)$, the value of the integrated signal $s_{\mathrm{MUX}}(t_\ell) = \sum_{i=1}^{N} \sqrt{P_i} \mathrm{e}^{\mathrm{j}\phi_i} \tilde{c}_\ell^{(i)}$ is a fixed complex number.

As previously mentioned, for transmitters, $s_{\mathrm{MUX}}(t)$ is desired to have a constant envelope to enable the HPA to operate in full-saturation mode to maximize transmission efficiency. That is, the integrated signal is desired to have a constant amplitude

A, so that it can have the same form as a phase shift keying (PSK) signal

$$s_{\mathrm{CE}}(t) = \sum_{\ell=-\infty}^{+\infty} A\mathrm{e}^{\mathrm{j}\theta[\ell]}\psi_{T_s}(t-\ell T_s) , \tag{6.17}$$

where $\theta[\ell]$ is the phase in the duration $t \in [\ell T_s, (\ell+1)T_s)$. Nevertheless, for an arbitrary given set of P_i and ϕ_i for $i = 1, 2, \ldots, N$, when $N > 2$, because the value of $\tilde{c}_\ell^{(i)}$ changes with time index ℓ, the envelope of $s_{\mathrm{MUX}}(t)$ is not guaranteed to be constant.

Most of the existing spreading chip waveforms in the satellite navigation signals are bipolar, i.e., $p_k^{(i)} \equiv \pm 1$ in (6.11). However, multilevel waveforms have been used in some spreading modulations such as CBOC and AltBOC (See Sects. 4.7.2 and 4.9.1). Considering a more general case, assume that as ℓ changes, each $p_k^{(i)}$ has up to $M^{(i)}$ possible values, and $\mathrm{e}^{\mathrm{j}2\pi\Delta f_i \ell}$ has up to ΔT_i different values. On the assumption that $s_i(t)$ for $i = 1, 2, \ldots, N$ are mutually independent, $\tilde{c}_\ell^{(i)}$ has a maximum of $\digamma = 2^N \prod_{i=1}^{N} M^{(i)}\Delta T_i$ possible complex values, which leads to the temporal fluctuation of the integrated signal's envelope.

Example 6.1 (*Two bipolar signals multiplexing on GPS L1 frequency*) A typical case is the multiplexing of two bipolar baseband signals by using QPSK technique. In the first-generation GPS, the C/A code and P(Y) code signals in L1 band are multiplexed in this way. The P(Y) code and the C/A code are phase-quadratured, and the amplitudes of the two are different by 3 dB. The composite baseband complex envelope signal can be expressed as

$$s_{\mathrm{L1}}(t) = As_{\mathrm{P(Y)}}(t) + \mathrm{j}\sqrt{2}As_{\mathrm{C/A}}(t) , \tag{6.18}$$

whose amplitude is $\sqrt{3}A$, which is a constant value.

At any time, depending on the value of the C/A code and the P(Y) code, $s_{\mathrm{L1}}(t)$ has a total of four possible phase angles. The constellation points of $s_{\mathrm{L1}}(t)$ are distributed on a circle, as shown in Fig. 6.1. The radius of this circle is the amplitude of the transmitted signal. ■

Example 6.2 (*Linear combination of three binary signals*) When three or more bipolar signals are combined, the direct superposition cannot achieve the constant envelope. In this example, we linearly combine three bipolar signals, where $s_1(t)$ and $s_2(t)$ are placed in the in-phase branch of carrier, while $s_3(t)$ is placed in the quadra-phase branch, i.e. the baseband composite signal is

$$s_{\mathrm{MUX}}(t) = \sqrt{P_1}s_1(t) + \sqrt{P_2}s_2(t) + \mathrm{j}\sqrt{P_3}s_3(t) , \tag{6.19}$$

where P_1, P_2, and P_3 are the powers of $s_1(t)$, $s_2(t)$, and $s_3(t)$ respectively.

In this case, the envelope of the composite signal is

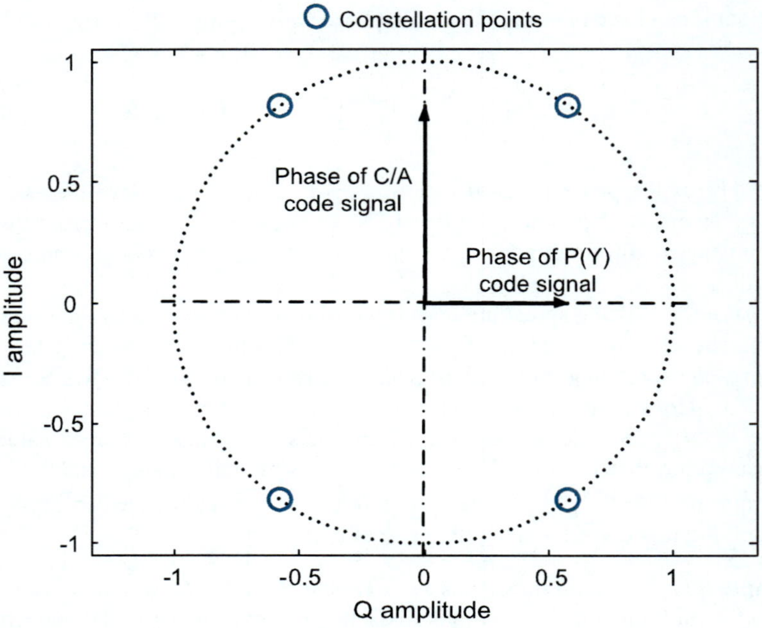

Fig. 6.1 Constellation diagram of QPSK multiplexing on the GPS L1 band

$$A(t) = \sqrt{P + 2\sqrt{P_1 P_2} s_1(t) s_2(t)}, \tag{6.20}$$

where $P = P_1 + P_2 + P_3$. Obviously, the envelope of this composite signal is not constant. It varies with the value of $s_1(t)$ and $s_2(t)$. Depending on the value at a certain time, the envelope of the signal may have two different values: $\sqrt{P + 2\sqrt{P_1 P_2}}$ or $\sqrt{P - 2\sqrt{P_1 P_2}}$. The modulation constellation of this case is depicted in Fig. 6.2 with $P_1 : P_2 : P_3 = 1 : 2 : 3$. ■

We can prove that the linear addition and combination of the N independent random signals does not have a constant envelope when $N > 2$. In order to achieve a constant envelope of the multiplexed signal, a new signal $I_{IM}(t)$ needs to be added to the direct linearly-added signal.

6.4.2 *Inter-modulation Term*

In order to keep the envelope constant, it is required to reconstruct $s_{MUX}(t)$. An intuitive idea is adding an additional component $I_{IM}(t)$ to the direct superposition signal to bring the phase points to a circle on complex plane. That is

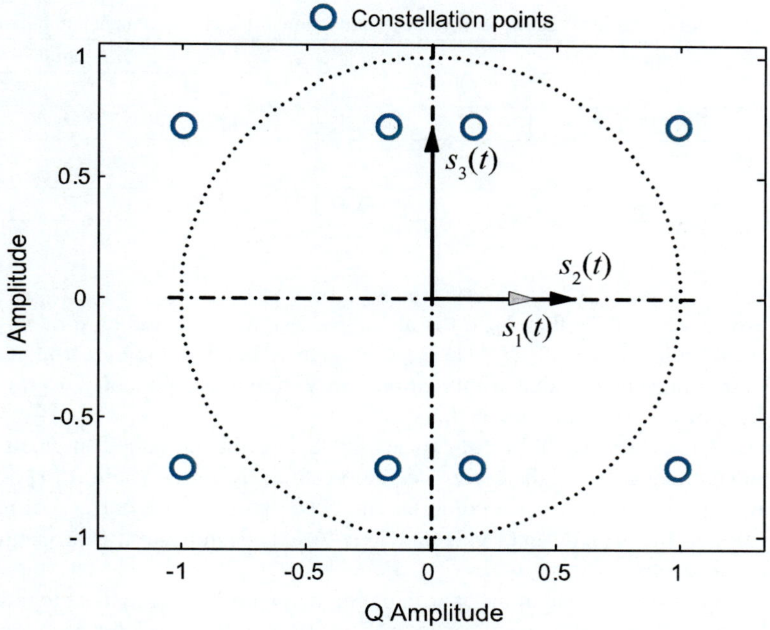

Fig. 6.2 Constellation diagram of three binary signals linear combination

$$s_{\mathrm{CE}}(t) = s_{\mathrm{MUX}}(t) + I_{\mathrm{IM}}(t). \tag{6.21}$$

Earlier research [15, 20, 21] has shown that some specific phase-modulation-based single carrier CEM signals for bipolar DSSS components can be expressed as the sum of the signal components and their cross-products. The study in [14] further proves that any single carrier bipolar components multiplexed signals expressed in a phase look-up table (LUT) has such an equivalent form. However, in this section, no prior restraint is imposed on the specific pattern of $I_{\mathrm{IM}}(t)$. We will discuss the general form of $I_{\mathrm{IM}}(t)$ under the constraint of CEM of multi-carrier multilevel components.

Evidently, $I_{\mathrm{IM}}(t)$ cannot be independent of N original components, otherwise the whole signal will be equivalent to the direct superposition of $N+1$ independent signals, in which case the constancy of the envelope is still a failure. Actually, in every time interval, the value of $I_{\mathrm{IM}}(t)$ should be determined by the values of the other N components. Therefore, this signal does not transmit any useful information in the general case. Following the notation given in [21], this term is called the *inter-modulation (IM)* term hereafter.

Using (6.17) and (6.15), we can express $I_{\mathrm{IM}}(t)$ as

$$I_{\mathrm{IM}}(t) = \sum_{\ell=-\infty}^{+\infty} \left(A\mathrm{e}^{\mathrm{j}\theta[\ell]} - \sum_{i=1}^{N} \sqrt{P_i}\mathrm{e}^{\mathrm{j}\phi_i} \tilde{c}_\ell^{(i)} \right) \psi_{T_s}(t - \ell T_s), \tag{6.22}$$

Table 6.1 Look-up table of the IM term value as a function of signal component combinations

$s_{\mathrm{IF}}^{(1)}$	$s_{\mathrm{IF}}^{(2)}$	…	$s_{\mathrm{IF}}^{(N)}$	I_{IM}	Probability
$\tilde{c}_1[1]$	$\tilde{c}_2[1]$	…	$\tilde{c}_N[1]$	$I_{\mathrm{m}}[1]$	$\mathcal{P}_1$
$\tilde{c}_1[2]$	$\tilde{c}_2[2]$	…	$\tilde{c}_N[2]$	$I_{\mathrm{m}}[2]$	$\mathcal{P}_2$
⋮	⋮	…	⋮	⋮	⋮
$\tilde{c}_1[F]$	$\tilde{c}_2[F]$	…	$\tilde{c}_N[F]$	$I_{\mathrm{m}}[F]$	$\mathcal{P}_F$

from which we can see that $I_{\mathrm{IM}}(t)$ is also a SCS signal with sub chip rate f_s, and in every duration $t_\ell \in [\ell T_s, (\ell+1)T_s)$, its value is fixed. Looping through all of ℓ, corresponding to F value combinations of N signal components, $I_{\mathrm{IM}}(t)$ has a maximum of F possible complex values.

An LUT in the form of Table 6.1 can be used to describe the value combinations of N signal components and the corresponding value of IM term. where $\tilde{c}_i[n]$ denotes the value of $\tilde{c}_\ell^{(i)}$ under the nth combination, accompanied by a corresponding IM term value $I_{\mathrm{m}}[n]$, which can be calculated via (6.22). $\mathcal{P}_n$ indicates the probability of this combination occurring, with $\sum_{n=1}^{F} \mathcal{P}_n = 1$.

Note that for N bipolar signals participating in the multiplexing that are independent of each other and the occurrence probability of $+1$ and -1 for each signal is the same, the $F = 2^N$ signal value combinations can be considered to have equal occurrence probability of $1/F$. However, for some uses of interleaved multiplexing techniques, such as interlaced majority voting multiplexing, modeling them into different combinations with different probabilities can simplify the analysis.

If one only uses envelope constancy as the design constraint, from (6.22) we can see that, with arbitrary selection of A and θ, there is an infinite number of $I_{\mathrm{IM}}(t)$ that can be chosen. Nevertheless, in constructing the IM term, some principles should be followed: Firstly, the influence on receiving performance caused by the IM term should be as small as possible. Secondly, the proportion of the IM term in the whole composite signal should be as low as possible, since the IM term introduces a waste of power into the transmitted signal. In the following three sections, these constraints on $I_{\mathrm{IM}}(t)$ are discussed respectively.

6.5 Transparency Constraint

To ensure forward and backward compatibility, an ideal multiplexing technique for satellite navigation system should be as *transparent* as possible to different receiver designs. In a DSSS signal receiver, since most of the key processes such as acquisition and tracking are a function of correlator output, the transparency constraint only needs to be guaranteed for this output. That is, the relative power and phase relationship of each input component signal measured at the corresponding matched correlator output should remain approximately unchanged before and after multiplexing.

6.5.1 Orthogonality Constraint

Considering a DSSS receiver. For the receiving channel of $s_{\mathrm{IF}}^{(i)}(t)$, when $s_{\mathrm{IF}}^{(i)*}(t)$ is used as the local replica, the output of the prompt correlator is

$$\begin{aligned}\tilde{R}_i &= \frac{1}{T}\int_0^T s_{\mathrm{CE}}(t)\, s_{\mathrm{IF}}^{(i)*}(t)\, \mathrm{d}t \\ &= \sum_{j=1}^{N} \sqrt{P_i} \mathrm{e}^{\mathrm{j}\phi_i} R_{ji} + \frac{1}{T}\int_0^T I_{\mathrm{IM}}(t)\, \tilde{s}_{\mathrm{IF}}^{(i)*}(t)\, \mathrm{d}t\,, \end{aligned} \tag{6.23}$$

where $R_{ji} = \frac{1}{T}\int_0^T s_{\mathrm{IF}}^{(j)}(t)\, \tilde{s}_{\mathrm{IF}}^{(i)*}(t)\, \mathrm{d}t$. On the assumption of ideal spreading codes, R_{ji} is equal to 1 if $i = j$, otherwise is equal to zero. Then (6.23) can be simplified as

$$\tilde{R}_i = \sqrt{P_i} \mathrm{e}^{\mathrm{j}\phi_i} + \frac{1}{T}\int_0^T I_{\mathrm{IM}}(t)\, s_{\mathrm{IF}}^{(i)*}(t)\, \mathrm{d}t. \tag{6.24}$$

The requirement that the IM term should not influence the correlation characteristic of every signal component is equivalent to making

$$\frac{1}{T}\int_0^T I_{\mathrm{IM}}(t)\, s_{\mathrm{IF}}^{(i)*}(t)\, \mathrm{d}t = 0 \tag{6.25}$$

hold for all $i = 1, 2, \ldots, N$.

Note that both $I_{\mathrm{IM}}(t)$ and $s_{\mathrm{IF}}^{(i)}(t)$ are SCS signals with sub chip rate f_s. In every duration of the common segment, their product remain unchanged. When $Tf_s \gg F$, it is rational to assume that every value combination of N signal components has occurred in the integral interval so that $s_{\mathrm{IF}}^{(i)}(t)$ and $I_{\mathrm{IM}}(t)$ can go through all of their possible values listed in Table 6.1. Thus, we can replace the time average in (6.25) with the statistical average, which yields

$$\mathbb{E}\left\{ I_{\mathrm{IM}}(t)\, s_{\mathrm{IF}}^{(i)*}(t) \right\} = \boldsymbol{c}_i^{\mathrm{H}} \boldsymbol{P} \boldsymbol{\lambda} = 0, \text{ for } i = 1, 2, \ldots, N\,, \tag{6.26}$$

where

$$\boldsymbol{c}_i = (\tilde{c}_i[1], \tilde{c}_i[2], \ldots, \tilde{c}_i[F])^{\mathrm{T}} \tag{6.27}$$

is referred to as the *signal value vector* of ith signal component, whose entries correspond to the column of $s_{\mathrm{IF}}^{(i)}$ in Table 6.1. The vector

$$\boldsymbol{\lambda} = (I_{\mathrm{m}}[1], I_{\mathrm{m}}[2], \ldots, I_{\mathrm{m}}[F])^{\mathrm{T}} \tag{6.28}$$

is referred to as the *IM value vector*, the entries of which correspond to the column of I_{IM} in Table 6.1, and

$$\boldsymbol{P} = \begin{bmatrix} \mathcal{P}_1 & & & \\ & \mathcal{P}_2 & & \\ & & \ddots & \\ & & & \mathcal{P}_F \end{bmatrix} \tag{6.29}$$

is referred to as the *probability matrix*. When each signal value combination occurs with equal probability, $\boldsymbol{P} = (1/F)\,\boldsymbol{I}$, and (6.26) can be simplified as

$$\boldsymbol{c}_i^{\mathrm{H}}\boldsymbol{\lambda} = 0, \text{ for } i = 1, 2, \ldots, N. \tag{6.30}$$

Through the above analysis we can see that the constraint of minimal impact on correlator outputs can be equivalent to the restriction of the value subspace of $\boldsymbol{\lambda}$. It should have $\boldsymbol{\lambda} \in \mathcal{W}^{\perp}$, where $\mathcal{W} = \text{span}\,\{\boldsymbol{c}_1, \boldsymbol{c}_2, \ldots, \boldsymbol{c}_N\}$, i.e., the value subspace of $\boldsymbol{\lambda}$ must be the orthogonal complement of the space expanded by $\boldsymbol{c}_1, \boldsymbol{c}_2, \ldots, \boldsymbol{c}_N$. Further, it is easy to verify that those $\boldsymbol{c}_i$ are orthogonal to each other. Therefore, they can be a set of basis of $\mathcal{W}$. It also can be proved that $\dim(\mathcal{W}) = N$, and $\dim\left(\mathcal{W}^{\perp}\right) = F - N$. If $\left\{\mathring{\boldsymbol{c}}_1, \mathring{\boldsymbol{c}}_2, \ldots, \mathring{\boldsymbol{c}}_{F-N}\right\}$ are a set of basis of $\mathcal{W}^{\perp}$, then $\boldsymbol{\lambda}$ is expressible in term of linear combination of them, that is

$$\boldsymbol{\lambda} = \left(\mathring{\boldsymbol{c}}_1, \mathring{\boldsymbol{c}}_2, \ldots, \mathring{\boldsymbol{c}}_{F-N}\right)\boldsymbol{w}_{\mathrm{m}} \triangleq \mathring{\boldsymbol{C}}\boldsymbol{w}_{\mathrm{m}}, \tag{6.31}$$

where $\boldsymbol{w}_{\mathrm{m}} = (w_{\mathrm{m},1}, w_{\mathrm{m},2}, \ldots, w_{\mathrm{m},F-N})^{\mathrm{T}} \in \mathbb{C}^{F-N}$ is referred to as the *IM coefficient vector*, and the entry $w_{\mathrm{m},i}$ corresponds to the weight of $\mathring{\boldsymbol{c}}_i$ in $\boldsymbol{\lambda}$.

In the analysis above, we convert time domain expressions of $s_{\mathrm{IF}}^{(i)}(t)$ and $I_{\mathrm{IM}}(t)$ into vectors of their possible values, which bring some facilities for our subsequent analysis. Similarly, the direct superposition signal $s_{\mathrm{MUX}}(t)$ and the envelope constantified signal $s_{\mathrm{CE}}(t)$ can also be converted into their possible value vectors.

When expressed in terms of vectors, the design constraints of the power and phase of the signal components can be written as

$$\begin{aligned}\boldsymbol{w}_{\mathrm{s}} &= \left(\tilde{R}_1, \tilde{R}_2, \ldots, \tilde{R}_N\right)^{\mathrm{T}} \\ &= \left(\sqrt{P_1}\mathrm{e}^{\mathrm{j}\phi_1}, \sqrt{P_2}\mathrm{e}^{\mathrm{j}\phi_2}, \ldots, \sqrt{P_N}\mathrm{e}^{\mathrm{j}\phi_N}\right)^{\mathrm{T}}\end{aligned} \tag{6.32}$$

referred to as the *signal coefficient vector*. Then, based on (6.15), the value vector of $s_{\mathrm{MUX}}(t)$ can be expressed as

$$\boldsymbol{s}_{\mathrm{MUX}} = \boldsymbol{C}_0\boldsymbol{w}_{\mathrm{s}}, \tag{6.33}$$

where $\boldsymbol{C}_0 \triangleq (\boldsymbol{c}_1, \boldsymbol{c}_2, \ldots, \boldsymbol{c}_N)$.

Note that $\boldsymbol{s}_{\mathrm{MUX}}$ is a F dimensional vector, of which the ith entry corresponds to the value of $s_{\mathrm{MUX}}(t)$ under the ith value combination of signal components. Further, based on (6.21) we can obtain the value vector of $s_{\mathrm{CE}}(t)$, which is

$$s_{\mathrm{CE}} = s_{\mathrm{MUX}} + \boldsymbol{\lambda} = \boldsymbol{C}_0 \boldsymbol{w}_{\mathrm{s}} + \mathring{\boldsymbol{C}} \boldsymbol{w}_{\mathrm{m}}, \tag{6.34}$$

where s_{CE} is also a F dimensional vector, and its ith entry corresponds to the value of $s_{\mathrm{CE}}(t)$ under the ith signal components value combination.

Defining the *augmented signal value matrix*

$$\boldsymbol{C} = (\boldsymbol{C}_0, \mathring{\boldsymbol{C}}) \tag{6.35}$$

and the *augmented coefficient vector* $\boldsymbol{w} = (\boldsymbol{w}_{\mathrm{s}}^{\mathrm{T}}, \boldsymbol{w}_{\mathrm{m}}^{\mathrm{T}})^{\mathrm{T}}$, we can simplify (6.34) into a concise form

$$s_{\mathrm{CE}} = \boldsymbol{C}\boldsymbol{w}. \tag{6.36}$$

It can be proved that

$$\boldsymbol{C}\boldsymbol{C}^{\mathrm{H}} = F \cdot \boldsymbol{I}. \tag{6.37}$$

For the general case, $\mathring{\boldsymbol{c}}_1, \mathring{\boldsymbol{c}}_2, \ldots, \mathring{\boldsymbol{c}}_{F-N}$ can be obtained through orthonormal basis expansion based on $\boldsymbol{c}_1, \boldsymbol{c}_2, \ldots, \boldsymbol{c}_N$. For instance, in [16], Gram-Schmidt (G-S) orthogonalization is employed. Nevertheless, in the following subsection, we can see that for the multiplexing of bipolar signals, there are some straightforward ways to construct the basis vectors of $\mathcal{W}^{\perp}$.

6.5.2 Basis Vector Construction in Bipolar Case

If all of the N components to be multiplexed are bipolar, such as BPSK-R, BOC, as well as BCS signals, and they are located on the same central frequency, then all of $\tilde{c}_i[n] \in \{\pm 1\}$, and $F = 2^N$. In such case, without using G-S orthogonalization, the bases of $\mathcal{W}^{\perp}$ can be obtained directly via Hadamard products of $\boldsymbol{c}_1, \boldsymbol{c}_2, \ldots, \boldsymbol{c}_N$.

Define the kth order *inter-modulation vector (IMV)* as the Hadamard product (element-wise product) of certain k signal value vectors, i.e., $\boldsymbol{v}^{(i_1 i_2 \cdots i_k)} = \boldsymbol{c}_{i_1} \circ \boldsymbol{c}_{i_2} \circ \cdots \circ \boldsymbol{c}_{i_k}$, where $\circ$ denotes the Hadamard product, and $i_1, i_2, \ldots, i_k$ are different from one another. Particularly, define $\mathbf{1} = (1, 1, \ldots, 1)^{\mathrm{T}}$ as the zero order IMV. It is easy to verify that the total number of kth order IMVs is ${}_N C_k$, which means the number of k-combinations from a set of N elements. Therefore, except ${}_N C_1$ first order IMVs which correspond to $\boldsymbol{c}_i$ themselves, the total number of nontrivial IMVs is ${}_N C_0 + \sum_{k=2}^{N} {}_N C_k = F - N$. It is interesting to note that both the zero order and the kth order IMVs ($k > 1$) are orthogonal to $\boldsymbol{c}_1, \boldsymbol{c}_2, \ldots, \boldsymbol{c}_N$ and to each other. Hence these IMVs can be employed as a set of basis of $\mathcal{W}^{\perp}$ directly. More specifically, the augmented signal value matrix can be constructed in this way such as

$$\boldsymbol{C}^{(N)} = \Bigg(\underbrace{\boldsymbol{c}_1, \ldots, \boldsymbol{c}_N}_{\text{signal components}}, \underbrace{\boldsymbol{v}^{(1,2)}, \ldots, \boldsymbol{v}^{(N-1,N)}}_{\text{2nd IMV}}, \ldots, \underbrace{\boldsymbol{v}^{(i_1 i_2 \cdots i_k)}}_{k\text{-th IMV}}, \ldots, \underbrace{\boldsymbol{v}^{(1,2,\ldots,N)}}_{N\text{-th IMV}}, \underbrace{\mathbf{1}}_{\text{0th IMV}} \Bigg), \tag{6.38}$$

where the superscript (N) denotes the number of signal components. With $N = 4$ for example, $\boldsymbol{C}^{(4)}$ has the form of

$$\begin{aligned} \boldsymbol{C}^{(4)} &= \big(\boldsymbol{c}_1, \boldsymbol{c}_2, \boldsymbol{c}_3, \boldsymbol{c}_4, \boldsymbol{v}^{(1,2)}, \boldsymbol{v}^{(1,3)}, \boldsymbol{v}^{(1,4)}, \boldsymbol{v}^{(2,3)}, \boldsymbol{v}^{(2,4)}, \boldsymbol{v}^{(3,4)}, \\ &\quad \boldsymbol{v}^{(1,2,3)}, \boldsymbol{v}^{(1,2,4)}, \boldsymbol{v}^{(1,3,4)}, \boldsymbol{v}^{(2,3,4)}, \boldsymbol{v}^{(1,2,3,4)}, \mathbf{1}\big) \\ &= \begin{pmatrix} - & - & - & - & 1 & 1 & 1 & 1 & 1 & 1 & - & - & - & - & 1 & 1 \\ - & - & - & 1 & 1 & 1 & - & 1 & - & - & - & 1 & 1 & 1 & - & 1 \\ - & - & 1 & - & 1 & - & 1 & - & 1 & - & 1 & - & 1 & 1 & - & 1 \\ - & - & 1 & 1 & 1 & - & - & - & - & 1 & 1 & 1 & - & - & 1 & 1 \\ - & 1 & - & - & - & 1 & 1 & - & - & 1 & 1 & 1 & - & 1 & - & 1 \\ - & 1 & - & 1 & - & 1 & - & - & 1 & - & 1 & - & 1 & - & 1 & 1 \\ - & 1 & 1 & - & - & - & 1 & 1 & - & - & - & 1 & 1 & - & 1 & 1 \\ - & 1 & 1 & 1 & - & - & - & 1 & 1 & 1 & - & - & - & 1 & - & 1 \\ 1 & - & - & - & - & - & - & 1 & 1 & 1 & 1 & 1 & 1 & - & - & 1 \\ 1 & - & - & 1 & - & - & 1 & 1 & - & - & 1 & - & - & 1 & 1 & 1 \\ 1 & - & 1 & - & - & 1 & - & - & 1 & - & - & 1 & - & 1 & 1 & 1 \\ 1 & - & 1 & 1 & - & 1 & 1 & - & - & 1 & - & - & 1 & - & - & 1 \\ 1 & 1 & - & - & 1 & - & - & - & - & 1 & - & - & 1 & 1 & 1 & 1 \\ 1 & 1 & - & 1 & 1 & - & 1 & - & 1 & - & - & 1 & - & - & - & 1 \\ 1 & 1 & 1 & - & 1 & 1 & - & 1 & - & - & 1 & - & - & - & - & 1 \\ 1 & 1 & 1 & 1 & 1 & 1 & 1 & 1 & 1 & 1 & 1 & 1 & 1 & 1 & 1 & 1 \end{pmatrix}, \end{aligned} \tag{6.39}$$

where $-$ denotes -1.

6.6 Envelope Constancy Constraint

In addition to the transparency constraint, $s_{\text{CE}}(t)$ should satisfy the constant envelope constraint, which is equivalent to making all of the entries $s_{\text{CE},k}$ $(k = 1, \ldots, F)$ of $\boldsymbol{s}_{\text{CE}}$ have the same modulus, i.e.,

$$\left|s_{\text{CE},1}\right| = \left|s_{\text{CE},2}\right| = \cdots = \left|s_{\text{CE},F}\right| . \tag{6.40}$$

If $s_{\mathrm{CE}}(t)$ is represented in the phase-modulated form (6.17), this constraint is satisfied inherently. However, if $s_{\mathrm{CE}}(t)$ is represented in the linear combination form (6.21), this constraint has to be given specifically.

In some optimization-based CEM design approaches, the constraint expression (6.40) is complicated for choosing a search direction in numerical optimization. For this reason, in [16], a cost function is introduced to measure the nonconstancy of $\boldsymbol{s}_{\mathrm{CE}}$. It first divides all $s_{\mathrm{CE},k}$ into two evenly sets, denoted as $\mathcal{G}^-$ and $\mathcal{G}^+$. The partition rule is

$$\max_{s_{\mathrm{CE},k}\in\mathcal{G}^-} \left|s_{\mathrm{CE},k}\right|^2 \leq \min_{s_{\mathrm{CE},k}\in\mathcal{G}^+} \left|s_{\mathrm{CE},k}\right|^2. \tag{6.41}$$

Namely, $\mathcal{G}^+$ contains the half of $s_{\mathrm{CE},k}$ with the larger modulus, while $\mathcal{G}^-$ contains the half with the smaller modulus. Then, the cost function is defined as the difference in the sum of $\left|s_{\mathrm{CE},k}\right|^2$ between $\mathcal{G}^-$ and $\mathcal{G}^+$, that is

$$\tilde{f}(\boldsymbol{w}_{\mathrm{m}}) \triangleq \boldsymbol{s}_{\mathrm{CE}}^{\mathrm{H}} \boldsymbol{G} \boldsymbol{s}_{\mathrm{CE}} \tag{6.42}$$

where $\boldsymbol{G} = \mathrm{diag}\left(g_1, g_2, \ldots, g_F\right)$ is a diagonal matrix with

$$g_k = \begin{cases} 1, & s_{\mathrm{CE},k} \in \mathcal{G}^+, \\ -1, & s_{\mathrm{CE},k} \in \mathcal{G}^-. \end{cases} \tag{6.43}$$

According to the definition, $\tilde{f}(\boldsymbol{w}_{\mathrm{m}})$ is a non-negative real scalar function indicating how $\boldsymbol{s}_{\mathrm{CE}}$ is close to being a constant envelope, and it is easy to prove that $\tilde{f}(\boldsymbol{w}_{\mathrm{m}}) = 0$ if and only if $\left|s_{\mathrm{CE},1}\right| = \left|s_{\mathrm{CE},2}\right| = \cdots = \left|s_{\mathrm{CE},F}\right|$. As a result, $\tilde{f}(\boldsymbol{w}_{\mathrm{m}}) = 0$ can be used to represent the constant-envelope constraint.

6.7 Efficiency of Constant Envelope Multiplexing

Multiplexing efficiency is an important indicator of constant envelope multiplexing. As mentioned earlier, in the process of composition, if the directly superposed signal cannot achieve a constant envelope, an additional IM term needs to be introduced. The IM term exists only to maintain a constant envelope. From a receiver perspective, it results in the waste of transmitting power. During the constant envelope transmission process, the percentage of the IM term power in the total power of the composite signal is defined as multiplexing loss, and the percentage of the power of useful signals is defined as the *multiplexing efficiency*.

If the composite signal does not have an explicit form of component linear combination, for example, in (6.17), the proportion of the useful signal and the IM signal is not directly seen from the signal expression, it is not easy to evaluate its multiplexing efficiency. In general, the satellite navigation signal CEM design defines the multiplexing efficiency as the ratio between the sum of all component signals power measured at the correlator output and the power of the integrated signal, i.e.,

$$\eta = \frac{1}{A^2}\sum_{i=1}^{N}\left|\tilde{R}_i\right|^2 = \frac{1}{A^2}\sum_{i=1}^{N} P_i. \tag{6.44}$$

Note that $\|\boldsymbol{w}_\mathrm{s}\|^2 = \sum_{i=1}^{N} P_i$, and by using (6.36) and (6.37) we can obtain

$$A^2 = F^{-1} \cdot \|\boldsymbol{s}_\mathrm{CE}\|^2 = \|\boldsymbol{w}_\mathrm{s}\|^2 + \|\boldsymbol{w}_\mathrm{m}\|^2. \tag{6.45}$$

Then the multiplexing efficiency (6.44) can be written as

$$\eta = 1 - \frac{\|\boldsymbol{w}_\mathrm{m}\|^2}{\|\boldsymbol{w}_\mathrm{s}\|^2 + \|\boldsymbol{w}_\mathrm{m}\|^2}. \tag{6.46}$$

With the given constraints of P_i for $i = 1, 2, \ldots, N$, maximizing η is equivalent to minimizing $\|\boldsymbol{w}_\mathrm{m}\|^2$.

6.8 Design Methodologies of CEM

In summary, the problem for GNSS signal multiplexing to solve is how to achieve an effect at correlator outputs similar to that of transmitting multiple DSSS signals when only one constant envelope signal is transmitted. Mathematically, this is equivalent to constructing a mapping f from multiple functions to a single function, i.e. $f : \{s_1(t), s_2(t), \ldots, s_N(t)\} \longmapsto s_\mathrm{MUX}(t)$, under the constraints of envelope constancy, receiving transparency, and high-efficiency.

In the most recent 15 years, many novel and advanced CEM techniques have been proposed. In the next chapter, we will discuss them in detail. Although the various methodologies differ, the essential differences between these assorted methods are the mapping objects and solving approaches. There exist two general design patterns: waveform domain processing (WDP) and phase domain processing (PDP), both of which we will explore.

6.8.1 Waveform Domain Processing

Since at any moment, the value of the IM term should be completely determined by the value combination of the component signals, the main concept of WDP is specifying a mapping rule

$$f_\mathrm{WDP} : \{s_1, s_2, \ldots, s_N\} \longmapsto I_\mathrm{IM} \tag{6.47}$$

from the value combination of $s_1, s_2, \ldots, s_N$ to the value of I_IM, or its equivalent vector form $\boldsymbol{\lambda}$. Based on the discussion in Sect. 6.5, the construction of mapping rule

$f_{\rm WDP}$ in the WDP view is equivalent to finding a proper weight vector $\boldsymbol{w}_{\rm s}$ of a set of bases under a given power and phase design constraint $\boldsymbol{w}_{\rm s}$ to make each complex entry's modulus of

$$s_{\rm CE}(\boldsymbol{w}) = \boldsymbol{C}_0\boldsymbol{w}_{\rm s} + \mathring{\boldsymbol{C}}\boldsymbol{w}_{\rm m} = \boldsymbol{C}\boldsymbol{w} \tag{6.48}$$

equal.

Different WDP-based methods, such as [9, 10, 16, 26], obtain the solution of the above problem differently, and the ways to present $I_{\rm IM}$ and $\boldsymbol{\lambda}$ may also be different. The mapping rule $f_{\rm WDP}$ can be obtained by quasi-structured approaches or by numerical optimization.

6.8.1.1 Quasi-structured WDP Approaches

In most of the existing WDP based CEM techniques, such as [9, 10, 26], the construction of $\boldsymbol{\lambda}$ is based on structured or quasi-structured approaches, in which the form of most of the auxiliary components is presupposed to be a weighted sum of input signals' third order cross-products $s_{k_1}s_{k_2}s_{k_3}$, where k_1, k_2, and k_3 are not equal to each other. The weight coefficients can be obtained by analytical or numerical optimization.

Zhang et al. [14] extends the third-order cross-product to a more general case. They defined the kth order inter-modulation term as the product of any signals in the whole set of component signal, and proved that for bipolar signals, those zeroth-order and $2 \sim N$th order IM terms form exactly a set of orthogonal bases of $\mathcal{W}^{\perp}$, i.e., $\mathring{\boldsymbol{C}}$ can be directly constructed by using (6.38). Such a result not only guarantees the feasibility of some of the quasi-structured WDP design methods that use third order IM terms $\boldsymbol{v}^{(k_1,k_2,k_3)}$ as bases, but also reveals that those quasi-structured approaches with incomplete bases have limited flexibility in the power and phase relationship between components.

In recent studies [16], it has been further proved that the cross-products form is not a necessary assumption of the bases of $\mathcal{W}^{\perp}$. $I_{\rm IM}(t)$ can have a more general form, which further widens the application range of WDP based CEM to multilevel and multi-carrier cases, as discussed in Sects. 7.8 and 7.11.

6.8.1.2 Optimization-Based WDP Approaches

As mentioned in Sect. 6.7, the IM term introduced to maintain the constant envelope occupies some of the transmitted power, without contributing to the received power of the desired component signal. Therefore, in the WDP view, if all of the bases are orthonormal, in the maximum multiplexing efficiency sense, the optimal $\boldsymbol{w}_{\rm m}$ is the one with the minimal norm in the feasible region. Furthermore, the WDP-based CEM design can be equivalent to solving the following nonlinear programming problem

$$\begin{cases} \underset{\boldsymbol{w}_{\mathrm{m}} \in \mathbb{C}^{F-N}}{\arg\min} \ \|\boldsymbol{w}_{\mathrm{m}}\|^2 \,, \\ \text{s.t.} \quad \tilde{f}\left(\boldsymbol{w}_{\mathrm{m}}\right) = 0. \end{cases} \tag{6.49}$$

where $\tilde{f}(\cdot)$ is defined in (6.42).

Yao et al. [16] studied the solution of this general optimization problem, and derived a numerical optimization based method termed constant envelope multiplexing via intermodulation construction (CEMIC). Unlike quasi-structured methods, CEMIC can be applied to any number of signals with arbitrary power and phase relationships and can achieve optimal power efficiency (see Example 6.3). We will describe this method in more detail in Sect. 7.7.

6.8.2 Phase Domain Processing

Since the constant envelope composite signal $s_{\mathrm{CE}}(t)$ inherently has a phase-modulated form (6.17), the main concept of PDP is specifying a mapping rule

$$f_{\mathrm{PDP}} : \{s_1, s_2, \ldots, s_N\} \longmapsto \theta \tag{6.50}$$

from the value combination of the component signals to the phase of the composite signal.

When the value combination of $s_1, s_2, \ldots, s_N$ corresponds to the kth row of the LUT as in Table 6.1, the composite signal has the phase angle θ_k thus the value $s_{\mathrm{CE}}[k] = A\mathrm{e}^{\mathrm{j}\theta_k}$. For the correlator channel of the component $s_{\mathrm{IF}}^{(i)}(t)$, the output of the prompt correlator is

$$\tilde{R}_i = \frac{1}{T}\int_T \sum_{\ell=-\infty}^{+\infty} A\mathrm{e}^{\mathrm{j}\theta[\ell]}\tilde{c}_{\ell}^{(i)}\psi_{T_s}(t - \ell T_s)\,\mathrm{d}t\,. \tag{6.51}$$

If T is long enough to traverse the combination of each pair of values of $\tilde{c}_i[k]$ and θ_k, the time average in the above equation can be replaced by the statistical average, which yields

$$\tilde{R}_i \approx A\mathbb{E}\left\{\mathrm{e}^{\mathrm{j}\theta_k}\tilde{c}_i[k]\right\} = A\boldsymbol{c}_i^{\mathrm{T}}\boldsymbol{P}\mathbf{Exp}(\boldsymbol{\Theta})\,, i = 1, 2, \cdots N\,, \tag{6.52}$$

where $\boldsymbol{\Theta} = \left[\theta_1, \theta_2, \ldots, \theta_F\right]^{\mathrm{T}}$, and $\mathbf{Exp}(\boldsymbol{\Theta}) \triangleq \left[\mathrm{e}^{\mathrm{j}\theta_1}, \mathrm{e}^{\mathrm{j}\theta_2}, \ldots, \mathrm{e}^{\mathrm{j}\theta_F}\right]^{\mathrm{T}}$. The signal value vector $\boldsymbol{c}_i$ and the probability matrix $\boldsymbol{P}$ are defined in (6.27) and (6.29) respectively. By using the signal coefficient vector $\boldsymbol{w}_{\mathrm{s}}$ defined in (6.32), we can rewrite (6.52) into a concise form

$$\boldsymbol{w}_{\mathrm{s}} = A\boldsymbol{C}_0^{\mathrm{T}}\boldsymbol{P}\mathbf{Exp}(\boldsymbol{\Theta})\,. \tag{6.53}$$

When each phase state value combination is of equal probability, the above equation can be simplified as

$$\mathbf{w}_{\mathrm{s}} = \frac{A}{F}\mathbf{C}_0^{\mathrm{T}}\mathbf{Exp}\left(\boldsymbol{\Theta}\right) . \tag{6.54}$$

It can be seen from the above analysis that, in the PDP view, the design of the constant envelope scheme is equivalent to finding a solution $\{A, \boldsymbol{\Theta}\}$ of (6.54) for a given design constraint $\mathbf{w}_{\mathrm{s}}$. If the solution to the above equation can be found, it means that the above design requirement can be met, and we can obtain the value vector of $s_{\mathrm{CE}}(t)$ by

$$s_{\mathrm{CE}}\left(\boldsymbol{\Theta}\right) = A\mathbf{Exp}\left(\boldsymbol{\Theta}\right) . \tag{6.55}$$

If the solution to the Eq. (6.54) cannot be found, it means that the above design requirements cannot be achieved.

Different PDP-based methods, such as [6, 7, 13, 20, 21, 27], obtain the solution of the underdetermined and nonlinear complex Eq. (6.54) differently, and the ways to present $\boldsymbol{\Theta}$ may also be different. The phase mapping rule can be obtained in an analytical way (e.g., Interplex [20]) or by numerical optimization (e.g., phase-optimized constant-envelope transmission (POCET) method [13]). In some methods, the phase-mapping rule may be an explicit expression, while in others it can only be presented in a more general look-up table form. Furthermore, the mapping relation from component values to phase angle can be one-to-one (e.g., POCET) or many-to-one (e.g., MV multiplexing [27]).

6.8.2.1 Structured and Quasi-structured PDP Approaches

For unique cases of N and $\mathbf{w}_{\mathrm{s}}$, some CEM techniques can directly construct particular solutions of (6.54). This group of methods are collectively referred to as structured PDP design methods. For example, in MV multiplexing (See Sect. 7.6), the composite signal takes the value of the majority of the component signals, which is equivalent to

$$\theta = \pi \left\{1 - \mathrm{sgn}\left(\sum_{i=1}^{N} s_i\right)\right\} /2 \tag{6.56}$$

mapping multiple combination values to one of two phase angles: 0 and π. Despite its simplicity, MV multiplexing has poor applicability to the number, power distribution, and phase relationships of component signals, because of its over-rigid mapping form. In [27], an interlacing operation is employed that improves the flexibility of MV multiplexing on power distribution to some extent.

Some quasi-structured PDP design methods, predetermining the specific form of the phase mapping rules, can obtain a degree of flexibility in the power ratio and phase relationship of components by tuning undetermined coefficients. For example, in the Interplex technique [20], the phase angle is preset to the form

$$\theta = a_1 s_1 + \sum_{i=2}^{N} a_i s_1 s_i \tag{6.57}$$

in which adjusting indices a_i can change the power ratio of the components in the composite signal. However, as with other structured and quasi-structured PDP design methods, the imposed prior restraint on the combination form artificially reduces the feasible domain of (6.54), which not only limits the flexibility of the phase relationship between components, but also affects power optimization efficiency in many cases (see Example 6.3).

6.8.2.2 Optimization-Based PDP Approaches

In order to further improve the flexibility and applicability, and guarantee the optimality of the solution, there have been some optimization-based PDP design methods proposed in recent years. By (6.44), for a given design constraint $\boldsymbol{w}_s$, the optimal solution for maximizing power efficiency is the one with the smallest A. Conforming to the PDP design pattern, the CEM design problem is equivalent to solving the following nonlinear programming problem

$$\begin{cases} \min\limits_{\boldsymbol{\Theta}} A(\boldsymbol{\Theta}), \\ \text{s.t. } \boldsymbol{w}_s = A\boldsymbol{C}_0^{\mathrm{T}} \mathbf{Exp}(\boldsymbol{\Theta}). \end{cases} \tag{6.58}$$

By introducing the penalty function, the POCET method [13] converts the above constrained optimization into an approximately equivalent but unconstrained search of the objective function, minimizing

$$\begin{aligned} \min F(\boldsymbol{\Theta}) = A^2 + \mu_a \sum_{i \in \mathcal{V}} \left(\left| \tilde{R}_i(\boldsymbol{\Theta}) \right|^2 - P_i \right) \\ + \mu_b \sum_{i,j \in \mathcal{U}} \mathrm{Im} \left\{ \tilde{R}_i(\boldsymbol{\Theta}) \tilde{R}_j^*(\boldsymbol{\Theta}) \exp\left(-\mathrm{j}\left(\phi_i - \phi_j\right)\right) \right\} \end{aligned} \tag{6.59}$$

where the penalty factors μ_a and μ_b are positive, and $\mathcal{V}$ and $\mathcal{U}$ are the sets of component indices which have power and phase constraints, respectively (See Sect. 7.4 for more detailed description). Strictly speaking, this is not exactly equivalent to solving (6.58). However, as the penalty factors are increased, the solution of this approximation problem is forced to approach the solution of (6.58).

The use of the scalar form of constraints and the penalty function has an additional advantage over directly solving (6.58) in that not all angle and amplitude relationships between any two signal components must be declared. For those components without power and phase restrictions, the optimization routine will lead to the power and phase relationship that has the best combining efficiency. The BDS B3 signal design is a classic example of using this approach. By relaxing the phase relation between the open service signals and the authorized service signals, a relative phase angle of $\pi/4$ is found by optimization, under which a high multiplexing efficiency 85.36% is obtained for four signal components of equal power.

Example 6.3 (*CEM of four bipolar signals*) Consider the CEM of four bipolar DSSS signals as an example, where s_1, s_2, and s_3 are located in the in-phase (I) portion of the carrier with equal power when s_4 is in the quadrature-phase (Q) portion of the carrier, with α times power of s_1. The modulation constellations of the direct superposition signal s_{MUX} in the case $\alpha = 3$ are illustrated as open markers in Fig. 6.3. It can be seen that s_{MUX} does not have a constant envelope. Four representative design methods, MV+QPSK, Interplex, POCET, and CEMIC, are employed, and their power efficiencies in this scenario with respect of α are shown in Fig. 6.4 for comparison.

As a WDP based method, CEMIC can output the explicit expression of the composite signal: When $\alpha = 3$,

$$s_{\text{CE}} = s_1 + s_2 + s_3 + \sqrt{3}\text{j} \cdot s_4 - 0.51 \cdot s_1 s_2 s_3 - 0.28 j \cdot (s_1 s_2 s_4 + s_1 s_3 s_4 + s_2 s_3 s_4)\ . \tag{6.60}$$

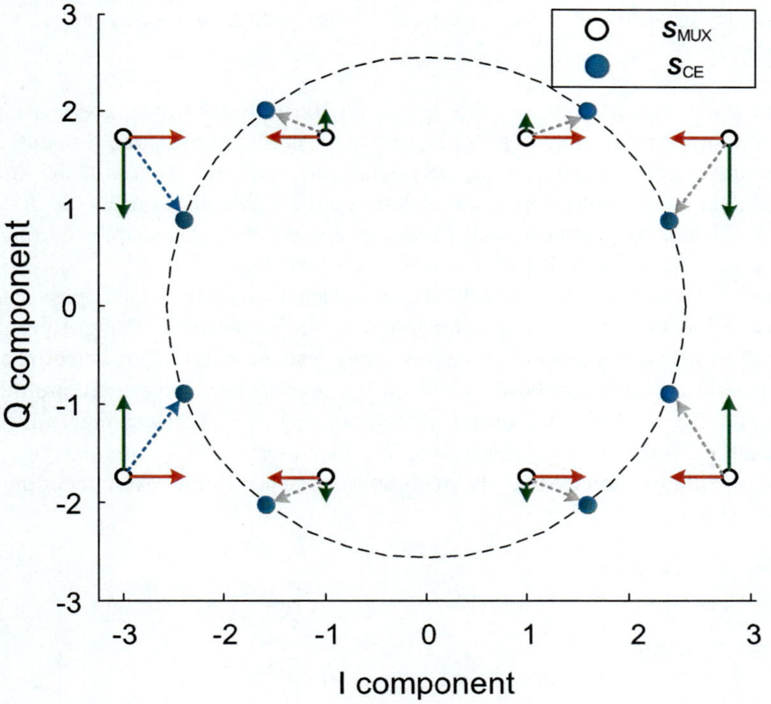

Fig. 6.3 The modulation constellations of direct superposition signal and CEM signal constructed by CEMIC technique with $\alpha = 3$

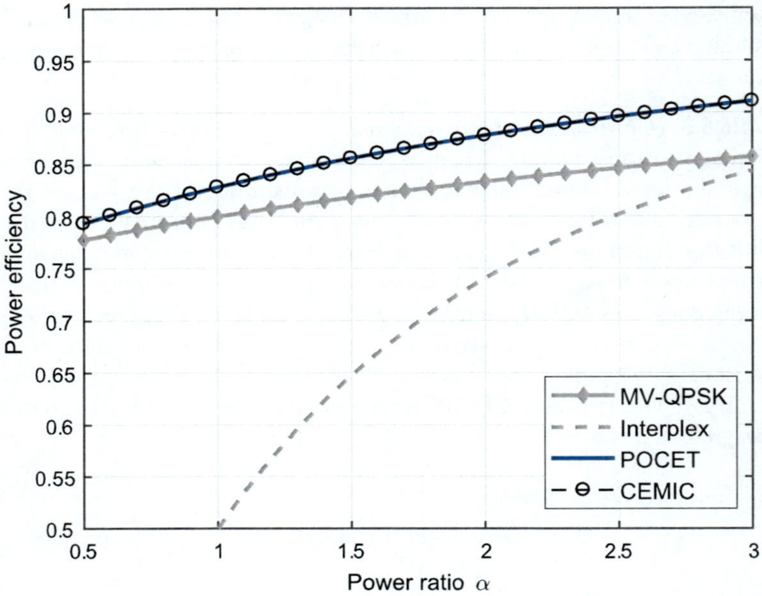

Fig. 6.4 Power efficiencies of four representative design methods with respect of α

The modulation constellations of this signal are illustrated as solid markers in Fig. 6.3. The role the IM term plays in the I component is moving the phase points ± 0.51 horizontally (indicated by red arrows), while the combined action of the three IM terms in the Q component makes the phase points move ± 0.28 or $\pm 0.28 \times 3$ vertically (indicated by green arrows). The overall effect of the IM term is to bring all of the phase points to a circle (indicated by gray arrows).

From Fig. 6.4, we can see that the structured and quasi-structured approaches have relatively low power efficiency compared to optimization-based design approaches. Note that different values of N and $\mathbf{w}_{\mathrm{s}}$ may lead to different power efficiencies. For general optimization based CEM design approaches, since there are multiple degrees of freedom, no obvious general relationship exists between power allocation and multiplexing efficiency. However, in either PDP or WDP, the design process based on optimization ensures the maximum efficiency of the result under the given conditions. ■

6.8.3 Summary

Both WDP and PDP design patterns have advantages and disadvantages, based on which, new CEM design methods are emerging. Through the comparison of (6.49) and (6.58), one can see that PDP optimizes the phase of a signal, so the solution meets the envelope constancy constraint inherently, while the transparency constraint is used to limit the feasible region. However, WDP is just the opposite: It optimizes the value of the auxiliary component, so the transparency constraint can be met inherently so long as $\boldsymbol{\lambda} \in \mathcal{W}^{\perp}$, but the envelope constancy constraint has to be given specifically.

Compared with (6.49), the problem (6.58) is easier to solve. Therefore, the early CEM design methods are mostly based on PDP. However, WDP is more flexible and extendable. Since WDP directly designs the weight of each IM term in the auxiliary component, additional constraints can be added during the optimization process to flexibly control the proportion of every IM term in the composite signal. As discussed in [16], this enables adjusting the spectrum distribution of the multiplexed signal, imposing the cancellation of unwanted IM terms, or relaxing the constant envelope constraint to realize a quasi-constant envelope which may have a higher power efficiency.

6.9 Representation and Implementation of CEM Signals

A specific CEM scheme can correspond to a phase LUT. If the phase LUTs of the two CEM schemes are the same, we can say that the two multiplexing schemes are mathematically equivalent.

However, for the multiplexing scheme represented by the same phase LUT, there may be different physical implementation methods. Moreover, with the increasing demand for flexibility and on-board reconfigurability of the next-generation navigation satellite, we also care whether there is a common architecture that can implement the CEM scheme to any phase LUT, and whether there is a flexible implementation architecture that supports online reconstruction of on-board transmission signals. Furthermore, in most of the analysis of a CEM scheme, we are concerned with its multiplexing efficiency, the relative magnitude and relative phase relationship between the useful signal component and the IM term in the composite signal, as well as the nature of the introduced IM signal and its effect on the useful signal components.

Nevertheless, not every CEM design method can obtain an explicit expression of the IM term. The phase LUT obtained according to the PDP design equation only gives the mapping relationship between the phase angle of the synthesized signal and the value of each useful signal component. It is impossible to see the details of the constant envelope synthesis, which is not only inconvenient for the understanding of the constant envelope multiplexing mechanism, but also difficult to see the specific form of the IM term. Moreover, it is also impossible to obtain the analytical PSD expression of the composite signal. The power spectrum can only be analyzed by simulation, which brings inconvenience to the analysis of the interference of the IM term introduced in the CEM on the useful signal components.

Early research [19] has shown that some specific CEM techniques have more than one representation, such as CASM [21] and Interplex [20]. By using the results of [14] and [16], the relationship and mutual transformation of different CEM representation can be generalized. In this section, based on the results in [18], we present several typical representation and implementation forms of CEM signals. We will see that by using the basis vector defined in Sect. 6.5, various CEM signal expressions can be unified. Especially for bipolar signals, every CEM technique has at least three different expressive forms, with their own features in realization or analysis, and they can be converted into each other. Accordingly, three different general-purpose CEM implementation methods are given.

6.9.1 Phase-Mapping-Based Form

Regardless of the design method, every CEM scheme can eventually correspond to a phase LUT. The LUT stores the mapping from every possible value combination of component signals to a phase angle θ_k, in which the phase index k is a function of time. Such phase values are passed to sine and cosine mappers to generate the real and imaginary parts of $A\exp\left(\mathrm{j}\theta_{k(t)}\right)$, and then this complex baseband signal is modulated to radio frequency and transmitted.

Figure 6.5 is a block diagram of the CEM signal generator based on the phase LUT. As shown in Fig. 6.5, a total of N baseband signals $s_1(t), s_2(t), \ldots, s_N(t)$ to be multiplexed are sent to the multiplexed signal generator to complete the mapping between the combination of signal values and the phase angle of the composite signal.

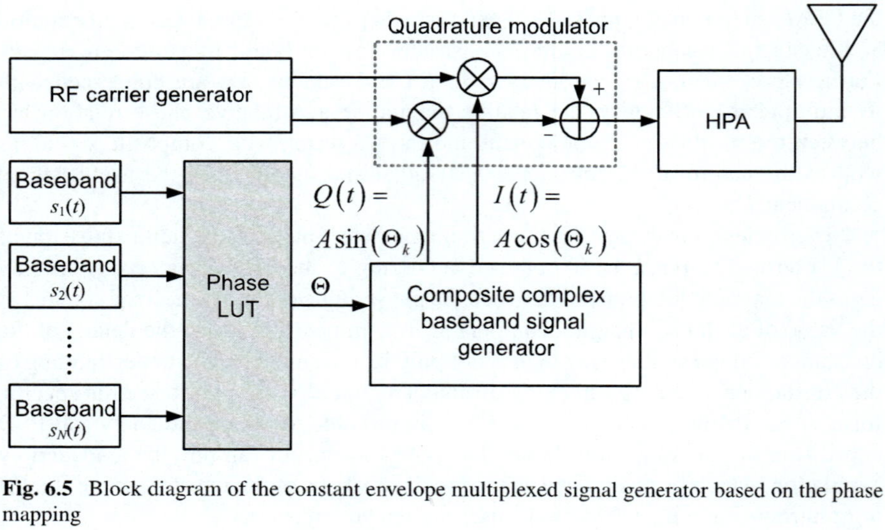

Fig. 6.5 Block diagram of the constant envelope multiplexed signal generator based on the phase mapping

Table 6.2 Phase LUT for CEM of N signal components

s_1	s_2	$\cdots$	s_{N-1}	s_N	$\boldsymbol{\Theta}$
-1	-1	$\cdots$	-1	-1	θ_1
-1	-1	$\cdots$	-1	1	θ_2
-1	-1	$\cdots$	1	-1	θ_3
-1	-1	$\cdots$	1	1	θ_4
$\vdots$	$\vdots$	$\cdots$	$\vdots$	$\vdots$	$\vdots$
1	1	$\cdots$	1	1	θ_{2^N}

The difference between different CEM schemes lies in the difference of the mapping relationship. For N bipolar baseband signals, there can be up to 2^N different combinations of values. In the multiplexed signal generator, there is a LUT in the form of Table 6.2, specifying a phase angle θ_k for each combination of values. According to the value of N signals at the current time, the corresponding phase angle is output. The signal generator produces a composite constant envelope complex baseband signal according to the phase angle, that is

$$\begin{aligned} \tilde{s}(t) = A\mathrm{e}^{\mathrm{j}\theta_k(t)} &= A\left[\cos\left(\theta_k(t)\right) + \mathrm{j}\sin\left(\theta_k(t)\right)\right] \\ &\triangleq I(t) + \mathrm{j}Q(t) \ . \end{aligned} \tag{6.61}$$

The composite constant envelope complex baseband signal is sent to the quadrature modulator and modulated to the radio frequency. The signal at this time can be expressed as

$$\begin{aligned} \tilde{s}_{\mathrm{RF}}(t) &= \mathrm{Re}\left\{A\mathrm{e}^{\mathrm{j}\theta_k(t)} \cdot \mathrm{e}^{\mathrm{j}2\pi f_{\mathrm{RF}} t}\right\} \\ &= I(t)\cos\left(2\pi f_{\mathrm{RF}} t\right) - Q(t)\sin\left(2\pi f_{\mathrm{RF}} t\right) \ . \end{aligned} \tag{6.62}$$

The signal is still a constant envelope signal. It is passed through a high-power amplifier, then band-limited by an analog filter and broadcast through the antenna.

With the above signal generator architecture, if we need to adjust the scheme of the CEM scheme, e.g. changing the power ratio and phase relationship between the signal components, we only need to regenerate a phase LUT and replace the old one in the multiplexed signal generator.

6.9.2 *Phase-Synthesis-Based Form*

Since the phase angle can be completely determined by the combination of the component signals, it can be decomposed into the linear combination of component signals $s_i(t)$ and their IM terms $\xi_i(t)$, i.e.

$$\theta(t) = \sum_{i=1}^{N} x_i s_i(t) + \sum_{i=N+1}^{\digamma} x_i \xi_{i-N}(t) , \tag{6.63}$$

where x_i for $i = 1, 2, \ldots, N$ and x_i for $i = N+1, \ldots, \digamma$ are the weighting coefficients of s_i and ξ_i respectively.

In matrix representation, that is

$$\boldsymbol{\Theta} = \boldsymbol{C}_0 \boldsymbol{x}_1 + \mathring{\boldsymbol{C}} \boldsymbol{x}_2 = \boldsymbol{C}\boldsymbol{x} , \tag{6.64}$$

where $\boldsymbol{C}$ is the augmented signal value matrix defined in (6.35), which is a full rank matrix, and

$$\boldsymbol{x} = \left[x_1, x_2, \ldots, x_{\digamma}\right]^{\mathrm{T}} \tag{6.65}$$

the entries of which correspond to the weighting coefficients and can be obtained by

$$\boldsymbol{x} = \boldsymbol{C}^{-1}\boldsymbol{\Theta} . \tag{6.66}$$

The realization process of the CEM based on the PSB form is that after obtaining the phase map of a given CEM scheme, the linear combination coefficients $x_1, x_2, \ldots, x_{\digamma}$ in the phase domain are calculated according to (6.66). Then, the combined coefficient is used as the gain of the amplifier to realize the linear combination of the basis signals. That combination is then employed to modulate the RF carrier phase to finally achieve constant envelope transmission.

Figure 6.6 shows the implementation of a constant envelope multiplexer based on phase synthesis with three input signals as an example. The IM term generator in the figure obtains the zero-order, second-order, third-order, and other IM signal components by multiplying several useful signals, and the phase synthesizer is responsible for real-time calculation of (6.63). The final transmitted constant envelope RF signal can be expressed as

$$\tilde{s}_{\mathrm{RF}}(t) = A\cos\left(2\pi f_{\mathrm{RF}} t + \theta(t)\right) . \tag{6.67}$$

With the above signal generator architecture, if we need to adjust the scheme of the CEM, e.g. changing the power ratio and phase relationship between the signal components, we can recalculate and replace the combined coefficients x_i of the basis signals, and the rest of the signal generator does not need to be adjusted.

6.9.3 Waveform-Synthesis-Based Form

As discussed in Sects. 6.8.1 and 6.8.2, for any given CEM scheme, the value vector of the complex envelope of the composite signal has two equivalent expressions: (6.48) and (6.55). Integrating these two equations yields

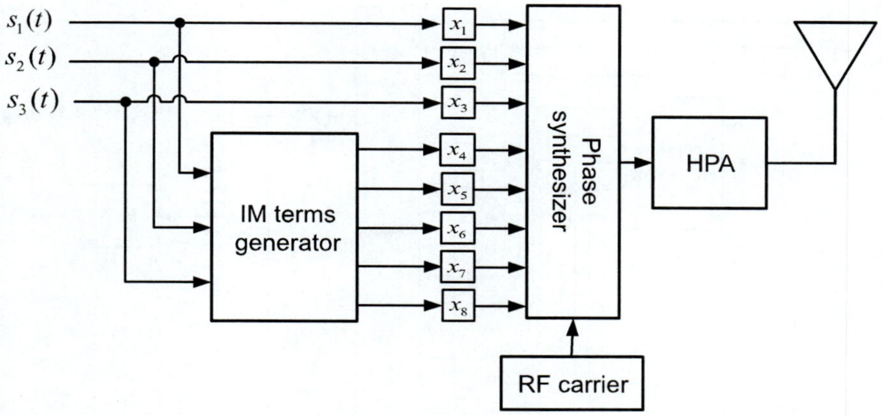

Fig. 6.6 Block diagram of the constant envelope multiplexed signal generator based on phase synthesis

$$\boldsymbol{C}\boldsymbol{w} = A\mathbf{Exp}\left(\boldsymbol{\Theta}\right) . \tag{6.68}$$

Therefore, any CEM scheme expressed in PMB form can be decomposed into

$$s_{\mathrm{CE}}\left(t\right) = \sum_{i=1}^{N} r_i s_i\left(t\right) + \sum_{j=1}^{F-N} w_j \xi_j\left(t\right) \tag{6.69}$$

by calculating

$$\boldsymbol{w} = A\boldsymbol{C}^{-1}\mathbf{Exp}\left(\boldsymbol{\Theta}\right) \tag{6.70}$$

wherethe coefficients r_i and w_j are the first N entries and the last $F - N$ entries of $\boldsymbol{w}$ respectively.

On the other hand, for bipolar signals, using the identities

$$\begin{cases} \cos\left(x_i \cdot \xi_i\left(t\right)\right) = \cos\left(x_i\right) \\ \sin\left(x_i \cdot \xi_i\left(t\right)\right) = \xi_i\left(t\right) \cdot \sin\left(x_i\right) \end{cases} , \tag{6.71}$$

one can also expand the PSB form into the WSB form.

The CEM signal generation method based on the WSB form converts the phase modulation signal into a sum of a series of useful signal components and intermodulation components. The basic method is to determine the combination coefficient $y_i = a_i + \mathrm{j}b_i$ of each basis signal in the waveform domain by (6.70), and put it into the system shown in Fig. 6.7 (taking three signals as an example) as the gain of the amplifier. The synthesizers in the figure add the sub-path signals to obtain an in-phase baseband signal component and an quadra-phase baseband signal component. These two signals are sent to the quadrature modulator and modulated to the radio

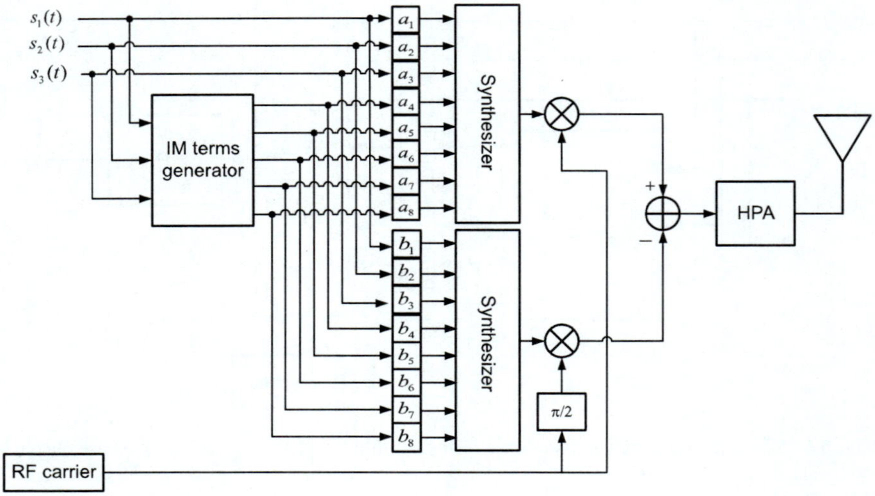

Fig. 6.7 Block diagram of the constant envelope multiplexed signal generator based on the intermodulation construction method

frequency. The signal at this time can be expressed as

$$\tilde{s}_{\mathrm{RF}}(t) = I(t)\cos(2\pi f_{\mathrm{RF}}t) - Q(t)\sin(2\pi f_{\mathrm{RF}}t) \ . \tag{6.72}$$

It is passed through a high-power amplifier then band-limited by an analog filter and broadcast through the antenna.

With the above signal generator architecture, if we need to adjust the scheme of CEM, e.g. changing the power ratio and phase relationship between signal components, we can recalculate and replace the combination coefficients a_i and b_i of each basis signal and the remaining parts of the signal generator does not need to be adjusted.

6.9.4 Relationship Between Different Representations

Figure 6.8 shows the mutual conversion relationship of these three forms [18]. Using this diagram, we can convert a CEM scheme into a form that is easy to handle: Of these three forms, PMB form has the greatest realization flexibility. If it is necessary to adjust the CEM scheme when using this multiplexer architecture, such as changing the power ratio or phase relationship between some component signals, one needs only to regenerate the phase mapping rule, replacing the original LUT. The WSB form has the clearest physical meaning for signal characteristic analysis. From this representation form, one can see the number, composition, and the proportion of IM

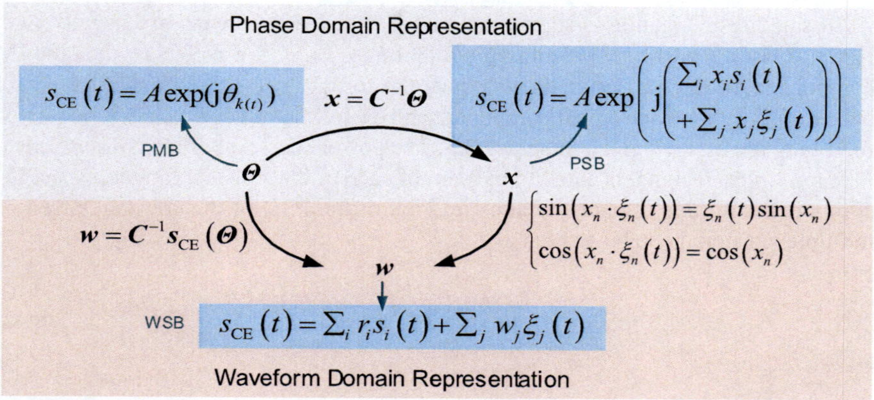

Fig. 6.8 Conversion relationship of three representation forms of a CEM scheme

terms introduced by CEM, and their power and phase relationships with the useful signal components. This allows easier analysis of spectrum characteristics such as the spectrum occupancy of the composite signal and the spectral compatibility between useful signals and the IM terms. Since the frequencies of baseband components are too low to avoid harmonic and intermodulation interference with the desired output during the up-conversion, the PSB form is conceptually useful but presents a series of limitations in a real implementation [28].

6.10 PSD of CEM Signals

Through the WSB form, we transform the constant envelope multiplexed signal, which is originally phase modulated, into a linear superposition of a set of basis signals. For some CEM schemes that directly give phase LUTs, such as the POCET method, because the original expressions cannot give the details of the synthesis, the performance cannot be analyzed in depth, but only be evaluated through simulation. Through the PMB-to-WSB form conversion, the properties of each signal component and each IM term participating in the synthesis can be analyzed separately, and the theoretical PSD and mutual interference between components can be analytically studied as well.

For example, for a specific signal component, all the other signal components and IM terms in the multiplexing process can be regarded as interference signals, so that the inner interference caused by the constant envelope multiplexing process can be analytically studied, and then the SSC provided in Sect. 5.8 can be used to quantitatively analyze the interference caused by all the other signal components and IM terms on a specific signal.

In satellite navigation systems, the useful signal components are usually with BPSK-R, BOC, or other spreading modulations with SCS waveforms, thus the IM components in the basis signals also have the form of the SCS signal. The power spectrum $P_i(f)$ of each signal component and each IM component can be calculated following the method given in Sect. 3.3.5. The power spectrum of the constant envelope multiplexed signal is equal to the sum of each of the signal components and IM components, so that the analytical expression of the PSD of the constant envelope multiplexed signal can be written as

$$P(f) = \sum_{i=1}^{F} w_i^2 P_i(f), \quad (6.73)$$

where the w_i is the ith entry of $\boldsymbol{w}$.

References

1. Pratt AR, Owen J (2001) Signal multiplex techniques in satellite channel availability possible applications to Galileo. In: Proceedings of the 18th international technical meeting of the satellite division of the institute of navigation (ION GNSS 2005), pp 2448–2460
2. ICG (2010) Current and planned global and regional navigation satellite systems and satellite-based augmentations systems, presented at the international committee on global navigation satellite systems providers' forum, New York
3. Fan T, Lin V, Wang G, Dafesh P (2008) Study of signal combining methodologies for future GPS flexible navigation payload (Part II). In: 2008 IEEE/ION position, location and navigation symposium. IEEE, pp 1079–1089
4. Rebeyrol E, Julien O, Macabiau C, Ries L, Delatour A, Lestarquit L (2007) Galileo civil signal modulations. GPS Solut 11(3):159–171
5. Yao Z, Lu M (2013) Constant envelope combination for components on different carrier frequencies with unequal power allocation. In: International technical meeting of the institute of navigation, San Diego, pp 629–637
6. Dafesh P, Cahn C (2011) Application of POCET method to combine GNSS signals at different carrier frequencies. In: Proceedings of ION ITM. San Diego, CA, USA, pp 1201–1206
7. Zhang K, Zhou H, Wang F (2013) Unbalanced AltBOC: a compass B1 candidate with generalized MPOCET technique. GPS Solut 17(2):153–164
8. Yao Z, Lu M (2012) Dual-frequency constant envelope multiplex with non-equal power allocation for GNSS. Electron Lett 48(25):1624–1625
9. Zhang K (2013) Generalised constant-envelope DualQPSK and AltBOC modulations for modern GNSS signals. Electron Lett 49(21):1335–1337
10. Guo F, Yao Z, Lu M (2016) BS-ACEBOC: a generalized low-complexity dual-frequency constant-envelope multiplexing modulation for GNSS. GPS Solu, pp 1–15
11. Yao Z, Zhang J, Lu M (2016) ACE-BOC: dual-frequency constant envelope multiplexing for satellite navigation. IEEE Trans Aerosp Electron Syst 52(1):466–485
12. Lestarquit L, Artaud G, Issler J-L (2008) AltBOC for dummies or everything you always wanted to know about AltBOC, presented at the ION GNSS 2008. Savannah, GA, US
13. Dafesh P, Cahn C (2001) Phase-optimized constant-envelope transmission (POCET) modulation method for GNSS signals. In: Proceedings of the 22nd international technical meeting of the satellite division of the institute of navigation (ION GNSS 2009), pp 2860–2866

14. Zhang X, Zhang X, Yao Z, Lu M (2012) Implementations of constant envelope multiplexing based on extended interplex and inter-modulation construction method In: 25th international technical meeting of the satellite division of the institute of navigation, ION GNSS, Nashville
15. Zhang K, Wang F (2012) Analytical transmission model of POCET technique for Compass B1 and B3 signals. In: Proceedings of ION global navigation satellite system
16. Yao Z, Guo F, Ma J, Lu M (2017) Orthogonality-based generalized multicarrier constant envelope multiplexing for DSSS signals. IEEE Trans Aerosp Electron Syst 53(4)
17. Guo F, Yao Z, Lu M, Zhou H (2016) Generalized transparent constant envelope multiplexing method, presented at the the 2016 international technical meeting of the institute of navigation, Monterey, California, 25–28 Jan 2016
18. Yao Z, Lu M, Signal multiplexing techniques for GNSS: the principle, progress, and challenges within a uniform framework. IEEE Signal Process Mag 34(5):16–26
19. Pratt AR, Owen J (2005) Signal multiplex techniques in satellite channel availability possible applications to Galileo, presented at the ION GNSS 2005. Long Beach, CA, US
20. Butman S, Timor U (1972) Interplex: an efficient multichannel PSK/PM telemetry system. IEEE Trans Commun Technol COM-20(3)
21. Dafesh PA, Nguyen TM, Lazar S (1999) Coherent adaptive subcarrier modulation (CASM) for GPS modernization, presented at the national technical meeting of the institute of navigation, San Diego, CA
22. Abuelmaatti MT (1984) Frequency-dependent nonlinear quadrature model for TWT amplifiers. IEEE Trans Commun 32:982–986
23. Fitzgerald PM, Witherspoon J (1968) Space-ground link subsystem ground station system analysis summary report. Volume I. System design analysis, DTIC Document, technical report
24. Holmes JK (1982) Coherent spread-spectrum systems, vol 1. New York: Wiley-Interscience, p 636
25. Shihabi MM, Nguyen TM, Hinedi A (1994) A comparison of telemetry signals in the presence and absence of a subcarrier. IEEE Trans Electromagn Compat 36(1):60–73
26. Kharisov V, Povalyaev A (2011) Optimal aligning of GNSS navigation signals sum, presented at the the 24th international technical meeting of the satellite division of the institute of navigation, Portland, Oregon
27. James J, Spilker J, Orr RS (1998) Code multiplexing via majority logic for GPS modernization, presented at the ION GNSS 11th International technical meeting of the satellite division, Nashville, TN, US
28. Cangiani GL, Orr RS, Nguyen CQ (2004) Intervote modulator. EU Patent Patent 1334595

Chapter 7
Constant Envelope Multiplexing Techniques for Spread-Spectrum Signals

The best kind of design isn't necessarily an object, a space, or a structure: it's a process—dynamic and adaptable.

— Donald Norman

7.1 Introduction

The development of constant envelope multiplexing (CEM) technology is closely related to the design of the GNSS signal. A critical constraint in GNSS signal design is whether the proposed signals in the same frequency band can be multiplexed together and share a common transmission link. In the previous chapter, the general principles and basic concepts of CEM for spread-spectrum signals were discussed. We found that a CEM scheme can adopt multiple forms of representation and implementation, and different forms can be converted into each other. We also learned that a CEM scheme corresponds to a phase mapping table. However, in the discussion of the CEM design methodologies in the previous chapter, an important problem was not discussed in detail, that is, how to construct this phase map under the given power and phase constraints of each signal component.

In recent years, a large number of novel CEM techniques have emerged as the rising need for satellite navigation. Although the nature of these techniques can be considered to be a mapping relationship between the signal value combinations and the transmission phase, there are differences in the implementation methods. The most typical methods include interplexing methods [1–6] , majority voting methods [7, 8], intermodulation construction methods [9, 10], phase optimization methods [11–13], and cascades of multiple CEM techniques.

Z. Yao and M. Lu, *Next-Generation GNSS Signal Design*, Navigation: Science and Technology 6, https://doi.org/10.1007/978-981-15-5799-6_7

The interplexing methods are developed based on trigonometric identities and represent the composite signal as a linear combination of several useful signals and intermodulation signals. The majority voting method was originally used for the multiplexing of odd-numbered bipolar signals with equal weight [7]. For three signals with non-uniform power, an interlaced transmission method can be adopted to make each received signal satisfy a predetermined power ratio. Phase optimization methods and intermodulation construction methods are two relatively new CEM approaches [9, 11]. These methods directly optimize the phase angle or the intermodulation term of the transmitted signal under the premise of ensuring that the power ratio and relative phase relationship of each signal component satisfy the goal of the system design. In addition to the typical methods described above, several different multiplexing techniques can be cascaded. For example, the Inter-Vote method [14] is a CEM method formed by cascading an interplexing method and a majority voting method. In addition to this cascading method, POCET method, MV method, and time division multiplexing method can also be cascaded with other constant envelope multiplexing methods to form a new multiplexing scheme.

Most of the existing CEM techniques are proposed for the combination of signal components located at the same central frequency. However, for some applications, there is a stronger need for joint multiplexing of signal components at multiple frequencies. For example, some satellite navigation systems may adjust the center frequency of signals during the update process. For the the smooth transition of the system, it is hoped that signals before and after the transition can be simultaneously broadcast by the same transmitter through multiplexing, and some of the signals can be gradually turned off in the future by online reconstruction. In addition, when the center frequencies of two sub-bands of a GNSS are not very far apart, combining signal components on these two sub-bands into a constant envelope composite signal and sharing common amplifier and transmission link can also benefit the design, implementation, and operation of both the transmitter and the receiver. Driven by the above-mentioned demand, dual-frequency and even multi-frequency joint constant-envelope multiplexing technologies have emerged in recent years [15–20].

Although the mathematical nature of all these CEM techniques can be attributed to the mapping between the value combination of signal components and the transmission phase, there is a big difference in the realization approaches of CEM in terms of flexibility, realization complexity on the satellite, and sensitivity to the design error. Therefore, in practical applications, we must evaluate various possible multiplexing methods for the specific constraints of the design project and select the one with the best comprehensive performance.

This chapter offers a detailed discussion of the typical constant envelope multiplexing techniques for satellite navigation spread-spectrum signals. Some of them have been used in satellite navigation systems currently in operation, while others have good application prospects in future system construction. The content of this chapter is not a simple list of the various technical implementations. We will focus on the principles of these constant envelope multiplexing methods and the design concepts behind the development process, in order to understand the implementation details of each technique and display the essential connections between them. This allows us to offer a more systematic understanding of the research field to readers.

7.2 QPSK Multiplexing

The discussion in this chapter begins with one of the simplest multiplexing method. Two bipolar DSSS signals can be multiplexed together by the QPSK technique to form a constant envelope composite signal. In legacy GPS, C/A code and P(Y) code signals are combined and transmitted using this technique, as shown in Example 6.1.

Denote the two bipolar baseband DSSS signals respectively as

$$\begin{cases} s_I(t) = \sqrt{P_I} g_I(t) \\ s_Q(t) = \sqrt{P_Q} g_Q(t) \end{cases} \tag{7.1}$$

where P_I and P_Q are the nominal powers of the two signals respectively, $g_I(t)$ and $g_Q(t)$ are the spreading codes of the two signals respectively. Then, the entire transmitted signal can be expressed as

$$\begin{aligned} s_{\mathrm{RF}}(t) &= s_I(t)\cos(\omega_c t) - s_Q(t)\sin(\omega_c t) \\ &= \sqrt{P_I} g_I(t)\cos(\omega_c t) - \sqrt{P_Q} g_Q(t)\sin(\omega_c t)\,. \end{aligned} \tag{7.2}$$

Using a trigonometric identity, $s_{\mathrm{RF}}(t)$ can also be written as

$$s_{\mathrm{RF}}(t) = \sqrt{P_I + P_Q}\cos(\omega_c t + \theta(t))\,, \tag{7.3}$$

where the additional phase angle is

$$\theta(t) = \operatorname{atan2}\left(\sqrt{P_Q} g_Q(t), \sqrt{P_I} g_I(t)\right). \tag{7.4}$$

Since both $g_I(t)$ and $g_Q(t)$ have only two values, +1 and −1, it is not difficult to obtain a phase map of quadrature phase shift keying (QPSK) multiplexing as shown in Table 7.1. Figure 6.1 in Chap. 6 plots the constellation diagram of QPSK multiplexing when $P_Q : P_I = 2 : 1$. Since the combined signal itself has a constant envelope property regardless of the power ratio of the two signal components, the multiplexing efficiency of the QPSK multiplexing is 100%.

However, if we want to use the same carrier to transmit more than two DSSS signals, we need to use more complex multiplexing techniques. In the construction of the next-generation GNSS, there are several requirements to add new signal components to existing QPSK signals. Superimposing directly on the original QPSK signal is undoubtedly the easiest solution. This method directly adds a new bipolar DSSS

Table 7.1 Phase mapping table for QPSK multiplexing

g_I	g_Q	Θ
-1	-1	$\pi + \arctan\left(\sqrt{P_Q/P_I}\right)$
-1	1	$\pi - \arctan\left(\sqrt{P_Q/P_I}\right)$
1	-1	$2\pi - \arctan\left(\sqrt{P_Q/P_I}\right)$
1	1	$\arctan\left(\sqrt{P_Q/P_I}\right)$

signal to a carrier phase of an existing QPSK signal. We denote the new baseband bipolar signal that is linearly superimposed as $s_N(t)$, where

$$s_N(t) = \sqrt{P_N} g_N(t). \tag{7.5}$$

Without loss of generality, we assume that the signal is superimposed on the in-phase branch of the original QPSK signal (7.2). After the superposition, the in-phase branch and the quadrature-phase branch of the composite signal are

$$\begin{cases} s'_I(t) = \sqrt{P_I} g_I(t) + \sqrt{P_N} g_N(t) \\ s'_Q(t) = \sqrt{P_Q} g_Q(t) \end{cases}. \tag{7.6}$$

The envelope of the signal is

$$A'(t) = \sqrt{P_T + 2\sqrt{P_I P_N} s_I(t) s_N(t)}, \tag{7.7}$$

where $P_T = P_I + P_Q + P_N$ is the total signal power. We find that in addition to a constant value P_T in the above equation, there is a time-varying component $2\sqrt{P_I P_N} s_I(t) s_N(t)$, which means that the envelope of the signal is not a constant value. It is easy to calculate that the envelope of the combined signal has two amplitudes, that is, the constellation points are distributed on two different circumferences on the constellation diagram, as shown in Example 6.2. As described in Chap. 6, this non-constant envelope signal will prevent the transmitter from operating in full saturation mode, otherwise AM/AM and AM/PM distortion will occur [21].

Example 7.1 (*Direct linear combination for three-signal multiplexing*) We use a practical numerical example to see the effects of non-constant envelopes. If the power relationship between the three signals to participate in the multiplexing is

$$\begin{cases} P_{in,1} = -3.0 \text{ dB}, \\ P_{in,2} = 0.0 \text{ dB}, \\ P_{in,3} = 0.5 \text{ dB}, \end{cases} \tag{7.8}$$

then, for non-constant envelope signals, due to the nonlinear nature of the input-to-output of the high-power amplifier, at the output, for a particular amplifier, the ratio of the three may become

$$\begin{cases} P_{out,1} = -6.97 \text{ dB}, \\ P_{out,2} = 0.63 \text{ dB}, \\ P_{out,3} = -0.63 \text{ dB}. \end{cases} \quad (7.9)$$

In order for the direct superposition method to achieve the desired output power ratio, a pre-distortion operation of the input signal is required. For example, Wang et al. [21] finds that for a particular amplifier, when the power ratio of the input components is adjusted to

$$\begin{cases} \tilde{P}_{in,1} = 0.95 \text{ dB}, \\ \tilde{P}_{in,2} = -1.2 \text{ dB}, \\ \tilde{P}_{in,3} = 2.52 \text{ dB}, \end{cases} \quad (7.10)$$

the output can approximate the desired power ratio. The multiplexing efficiency at this time is

$$\eta = \frac{P_{out,1} + P_{out,2} + P_{out,3}}{\tilde{P}_{in,1} + \tilde{P}_{in,2} + \tilde{P}_{in,3}}, \quad (7.11)$$

corresponding to a 1.59 dB multiplexing loss. ■

However, it should be noted that the value of the predistortion in the above example depends entirely on the characteristics of the amplifier. A set of preprocessed values corresponds to only a specific set of amplifiers, so the multiplexing loss is actually a function of the amplifier's characteristics. Therefore, the biggest disadvantage of using the linear superposition multiplexing method is that the in-orbit adjustment signal power ratio requires an extremely complicated pre-aliasing scheme. Moreover, if the characteristics of the amplifier change over time, the pre-distortion scheme must also be adjusted with it, to ensure that the power ratio remains constant. In order to avoid this problem, in the field of satellite navigation, it is necessary to use a CEM technique to multiplex signals.

7.3 Time Division Multiplexing

Time division (TD) multiplexing can realize the composite transmission of multiple bipolar DSSS signals. However, in order to achieve TD multiplexing, the original DSSS signal spreading chip waveform needs to be adjusted.

Using the case of two signals as an example, for simplicity, assume that both signals are BPSK-R(n) signals. Following the DSSS baseband signal definition given in (3.2), these two signals can be written as

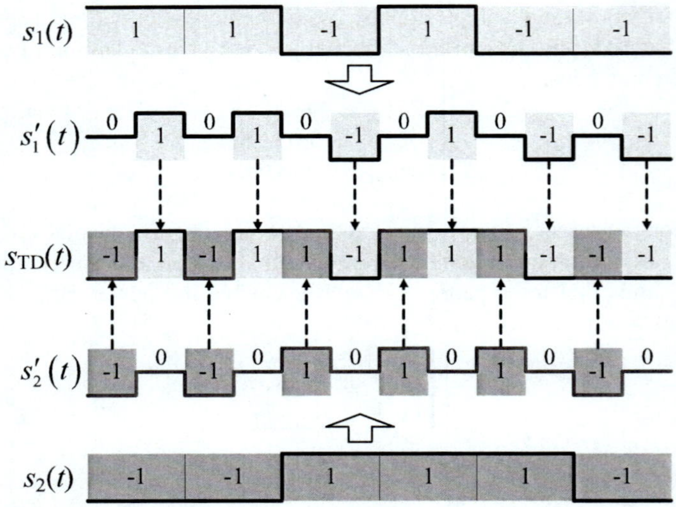

Fig. 7.1 Time division multiplexing of two BPSK signals

$$\begin{cases} s_1(t) = \sum_{n=-\infty}^{+\infty} (-1)^{c_n} p_{\text{BPSK-R}}(t - nT_c) \\ s_2(t) = \sum_{n=-\infty}^{+\infty} (-1)^{e_n} p_{\text{BPSK-R}}(t - nT_c) \end{cases} \tag{7.12}$$

where $p_{\text{BPSK-R}}(t)$ is a rectangular pulse chip defined by (4.1), and $\{c_n\}$ and $\{e_n\}$ are PRN sequences of $s_1(t)$ and $s_2(t)$ respectively, although their periods may be different.

The composite signal after TD multiplexing shortens the chip length of the original two signals and interlaces them chip by chip to form a new bipolar signal

$$\begin{aligned} s_{\text{TD}}(t) &= s_1'(t) + s_2'(t) \\ &= \sum_{n=-\infty}^{+\infty} (-1)^{c_n} p_1(t - nT_c) + \sum_{n=-\infty}^{+\infty} (-1)^{e_n} p_2(t - nT_c), \end{aligned} \tag{7.13}$$

where $p_1(t)$ and $p_2(t)$ are two second-order SCS spreading waveforms (see Sect. 3.3.5), with shape vectors $\boldsymbol{\kappa}_1 = [0, 1]^{\mathrm{T}}$ and $\boldsymbol{\kappa}_2 = [1, 0]^{\mathrm{T}}$ respectively. That is, the TD-multiplexed signal becomes a BPSK-R$(2n)$ signal with half the original code length. The original $s_1(t)$ is broadcast in the odd-numbered chip slots in this signal, while the original $s_2(t)$ is broadcast in the even-numbered ones. Figure 7.1 illustrates this multiplexing process with an example.

Before the emergence of more advanced CEM technologies, TD multiplexing was once considered as a simple and effective multi-signal multiplexing method.

Both the GPS L2C signal [22] and the M-code signal [23] use a TD multiplexing technique to combine the data channel and the pilot channel. However, compared to other constant envelope multiplexing techniques, TD multiplexing has significant drawbacks for navigation signals.

1. Degraded cross-correlation performance

When the time period of the spread-spectrum signal is the same, the use of TD multiplexing significantly shortens the length of the PRN sequence used in one cycle. For example, as shown in Fig. 7.1, if the PRN code rate after TD multiplexing is 10.23 MHz, the spreading code rate of each component in the composite signal is only 5.115 MHz. Then, if the receiver performs the coherent integration with a length of 1 ms, the number of chips participating in the correlation operation is 10,230 for a signal without TD multiplexing, but only 5,115 for a TD-multiplexed signal.

Under the same coherent integration length, the shorter PRN sequence length will undoubtedly deteriorate the cross-correlation performance between the code families and within the code family. This can be seen clearly in the relationship (2.40) between the Welch bound and the length of the PRN sequence (see Sect. 2.9 for detailed analysis). In [24], the above conclusions were verified using PRN sequences that are actually used by various navigation systems. Experiments show that when the sequence length is 10,230, the use of TD multiplexing will cause a 1.5 to 3 dB degradation of the cross-correlation performance. In the design process of the GPS L1C signal and BDS-3 B2 signal, this drawback was recognized, and thus excluded the use of TD multiplexing from the candidate constant envelope multiplexing technologies [24, 25].

2. Poor forward and backward compatibility

Except for TD multiplexing, most of the remaining constant envelope multiplexing techniques are "transparent" to the receiver. That is to say, the kind of multiplexing technique adopted by the transmitter can be independent of the development of receivers. So, when the receiver is processing the signal, it can be considered that the signal components participating the multiplexing are independently broadcast in the original form. For example, if QPSK is used to combine two signal components $s_1(t)$ and $s_2(t)$ given in (7.12), in a receiver, each channel only needs to generate a local BPSK replica to despread the corresponding signal component. If a new signal component is added to the satellite in the future, or if a more advanced multiplexing method is used, the existing receiver can receive and process the original signal component without any modification.

However, if the transmitter uses TD multiplexing, the channel structure of the receiver will be greatly modified. It is not hard to prove that when the system broadcasts a TD multiplexed signal like (7.13), if the receiver still despreads it using $s_1(t)$ and $s_2(t)$, then the power efficiency obtained is only 50%. In order to achieve higher multiplexing efficiency, the signal replicas in the receiver must be changed to $s_1'(t)$ and $s_2'(t)$ as defined in (7.13). That is to say, the correlator of the receiver needs an additional gate switching device so that the $s_1(t)$ processing channel is only integrated in the odd numbered chip time slot, and the $s_2(t)$ processing channel is only integrated in the even numbered chip time slot.

Such a shortcoming of TD multiplexing brings many limitations to its application [26]. TD multiplexing is not compatible with other multiplexing technologies, including advanced ones in the future. Existing non-TD signals should not be multiplexed with a new signal in TD mode, otherwise existing receivers can no longer be used to handle them. Similarly, if several signals are multiplexed by TD multiplexing at the present stage, in order to ensure the normal use of the existing receiver, even if there are other non-TD multiplexing techniques with higher performance in the future, they cannot be used on these signals.

3. The effects of code tracking bias

The non-ideal cross-correlation of the spreading codes among several signal components participating in multiplexing also has a major influence on the receiving of the TD multiplexed signals. Zhu et al. [24] carried out a detailed theoretical derivation and simulation on the principle and influence of this problem. When the TD multiplexed signal received has the form of (7.12), and the local replica is $s_1'(t)$, we can calculate that, for the two signals with the relative delays of T_c and $-T_c$ respectively, the correlation results are respectively

$$R_{\mathrm{TD}}(T_c) = \frac{1}{T}\int_0^T s_{\mathrm{TD}}(t)\, s_1'(t+T_c)\,\mathrm{d}t = \frac{2}{N}\sum_{n=0}^{N/2-1}(-1)^{c_n+e_n}, \tag{7.14}$$

and

$$R_{\mathrm{TD}}(-T_c) = \frac{1}{T}\int_0^T s_{\mathrm{TD}}(t)\, s_1'(t-T_c)\,\mathrm{d}t = \frac{2}{N}\sum_{n=0}^{N/2-1}(-1)^{c_n+e_{n+1}}, \tag{7.15}$$

where N is the number of chips included in the integration time. If the cross-correlation values of the PRN sequences $\{c_n\}$ and $\{e_n\}$ at the relative offsets of $+1$ and -1 chip are unequal, the shape of the signal correlation peak will be asymmetrical, resulting in an inherent deviation of the output of the code tracking loop.

Calculations and simulations show that for a signal with a code rate of 10.23 MHz, even if the difference between $R_{\mathrm{TD}}(T_c)$ and $R_{\mathrm{TD}}(-T_c)$ is only -35 dB of the autocorrelation main peak, it will bring a 0.25 m inherent deviation to the output of the tracking loop [24]. Liu et al. [27] and Yan et al. [28] also observed this phenomenon of TD multiplexing under actual testing and simulation conditions. This inherent bias is unacceptable for navigation signals where accurate ranging is of the primary concern.

The main solution to this problem is to add additional constraints when designing the PRN sequence, requiring that the cross-correlation between the PRN sequences used by the two signal components in TD multiplexing is equal when offsetting $+1$ and -1 chip. However, this additional constraint will undoubtedly deteriorate the PRN code's cross-correlation performance. Therefore, the TD multiplexing technique should be avoided as much as possible when other multiplexing means are available.

In summary, even though TD multiplexing technique is relatively simple in terms of the transmitter's implementation, there are some obvious defects for receivers. The more advanced constant-envelope multiplexing technologies discussed in following sections guarantee the transparency to the user and avoids the aforementioned drawbacks of TD multiplexing.

7.4 POCET Technique

The design of the phase-optimized constant-envelope transmission (POCET) technique [11] can be treated as solving an optimization problem under a set of power and phase constraints. As we pointed out in Chap. 6, any CEM scheme can be represented by a phase mapping table. The optimization goal of phase-optimized constant-envelope transmission (POCET) is to find a mapping relationship between a set of signal components and the phase angle of the transmission

$$\boldsymbol{\Theta} = [\theta_1, \theta_2, \ldots, \theta_{2^N}]^{\mathrm{T}} \tag{7.16}$$

letting the N bipolar signal components participating in the multiplexing satisfy a given relationship of power ratio and phase constraint at the output of receiver correlators.

It is assumed that the N baseband signals participating in the multiplexing are completely random, and the occurrence probability of $+1$ and -1 are both 50%, while the occurrence probability of the 2^N value combination is $1/2^N$. Thus, the statistic average of coherent integrated output values of the ith signal can be written as

$$\tilde{R}_i = \frac{A}{2^N} \sum_{k=1}^{2^N} b_i(k) \exp(\mathrm{j}\theta_k), \tag{7.17}$$

where $b_i(k)$ is the value of the ith signal component in the kth combination of values. In fact, it is not difficult to find that (7.17) is the decomposition form of (6.53).

The expected power value constraint P_i of the ith signal component can be written as

$$P_i(\boldsymbol{\Theta}) = \left|\tilde{R}_i\right|^2. \tag{7.18}$$

In addition, because the phase angle $\Delta\phi_{ij}$ between the ith signal component and the jth signal component is equivalent to the relative phase angle of the correlator output value $\tilde{R}_i$ and $\tilde{R}_j$, we have

$$\tilde{R}_i(\boldsymbol{\Theta})\tilde{R}_j^*(\boldsymbol{\Theta}) = \left|\tilde{R}_i(\boldsymbol{\Theta})\tilde{R}_j(\boldsymbol{\Theta})\right|\exp\left(\mathrm{j}\Delta\phi_{ij}\right), \tag{7.19}$$

i.e.

$$\left|\tilde{R}_i(\boldsymbol{\Theta})\tilde{R}_j(\boldsymbol{\Theta})\right| = \tilde{R}_i(\boldsymbol{\Theta})\tilde{R}_j^*(\boldsymbol{\Theta})\exp\left(-\mathrm{j}\Delta\phi_{ij}\right). \tag{7.20}$$

Thus, the phase angle relationship $\Delta\phi_{ij}$ between the ith signal component and the jth signal component can be expressed by constraining the real part of the right end of (7.20) to be greater than zero and the imaginary part to be equal to zero, i.e.

$$\begin{cases} \mathrm{Im}\left\{\tilde{R}_i(\boldsymbol{\Theta})\tilde{R}_j^*(\boldsymbol{\Theta})\exp\left(-\mathrm{j}\Delta\phi_{ij}\right)\right\} = 0, \\ \mathrm{Re}\left\{\tilde{R}_i(\boldsymbol{\Theta})\tilde{R}_j^*(\boldsymbol{\Theta})\exp\left(-\mathrm{j}\Delta\phi_{ij}\right)\right\} > 0. \end{cases} \tag{7.21}$$

The optimization goal is to minimize the total signal power A^2, that is, to maximize the multiplexing efficiency under the power constraint P_i of each useful signal power and the phase angle relationship $\Delta\phi_{ij}$ between components. That is

$$\begin{cases} \min\ A^2 \\ \text{s.t.}\ \left|\tilde{R}_i(\boldsymbol{\Theta})\right|^2 = P_i \\ \qquad \mathrm{Im}\left\{\tilde{R}_j(\boldsymbol{\Theta})\tilde{R}_k^*(\boldsymbol{\Theta})\exp\left(-\mathrm{j}\Delta\phi_{jk}\right)\right\} = 0, \ \text{for } i \in \mathcal{V} \text{ and } j, k \in \mathcal{U} \end{cases} \tag{7.22}$$

where $\mathcal{V}$ and $\mathcal{U}$ are the sets of component indices which have power and phase constraints, respectively.

We can use the penalty function method to convert the above constrained problem into the following unconstrained problem:

$$\begin{aligned} \min F(\boldsymbol{\Theta}) = A^2 &+ \mu_a \sum_{i\in\mathcal{V}} \left(\left|\tilde{R}_i(\boldsymbol{\Theta})\right|^2 - P_i\right) \\ &+ \mu_b \sum_{i,j\in\mathcal{U}} \mathrm{Im}\left\{\tilde{R}_i(\boldsymbol{\Theta})\tilde{R}_j^*(\boldsymbol{\Theta})\exp\left(-\mathrm{j}\Delta\phi_{ij}\right)\right\}. \end{aligned} \tag{7.23}$$

The penalty factors μ_a and μ_b are positive numbers. As the penalty factor becomes larger, the result approaches the optimal.

In the phase optimizing process of the POCET method, the penalty function method is influenced by the penalty factors. It is difficult to control the optimization process when optimizing amplitude and phase parameters simultaneously. Since the constraints and the objective function are both nonlinear functions and the forms are complex, the POCET method generally needs to use a numerical search algorithm to find a quasi-optimal solution. When the number N of signal components participating in the composite is large, the search space will increase rapidly, and the number of iterations is also intensified.

7.5 Quadrature Product Subcarrier Modulation

Through the analysis in Chap. 6, we found that in view of PDP, the design of the CEM scheme is the process of finding the phase mapping table with the highest possible multiplexing efficiency under the given power and phase constraints. Different design methods might give the same result, but because of the various means adopted, they show different characteristics.

The POCET technique introduced in the previous section completely uses the numerical optimization method to construct the phase mapping table based on a set of optimization equations. However, the process of finding the optimal phase when N is large is very complicated, and the optimal solution may not be found. In this section, we turn to a class of constant envelope multiplexing methods based on trigonometric function expansion. In the process of system upgrade, sometimes it is necessary to superimpose new signal components on the existing multiplexed scheme. Moreover, with the continuous development of satellite navigation spreading modulation technology, many new modulations adopt non-bipolar spreading chip waveforms such as CBCS and CBOC. For these cases, the CEM method based on trigonometric function expansion may show some unique advantages.

The quadrature product subcarrier modulation (QPSM) technique [1, 2] can transmit a modulated carrier including multiple subcarrier signals through a constant envelope waveform. The general form of QPSM can be regarded as superimposing a new subcarrier signal on the phase angle of an existing QPSK multiplexed signal while keeping the envelope constant. Therefore, one of the biggest advantages of QPSM is that it can add new signal components to the multiplexed signals of existing systems while ensuring that they have good spectral separation characteristics from the original signals. The power distribution of each signal can be achieved by selecting appropriate modulation index.

Under some specific signal component quantities, there are variants of the specific implementation of QPSM technique, such as coherent adaptive subcarrier modulation (CASM) and Interplex. The main difference is whether the composite signal is implemented using the PDP generation method given in Sect. 6.8.2, or the WDP method given in Sect. 6.8.1. The analysis in Sects. 6.8 and 6.9 offers a full elaboration of the essential difference and relationship between these two design patterns. In this section, we will consider the general QPSM technique, a large class of CEM methods, as a whole, within a unified framework.

Consider a composite signal constructed by QPSK multiplexing, of which the baseband complex envelope can be written as

$$s(t) = \sqrt{P_I} s_I(t) + \mathrm{j}\sqrt{P_Q} s_Q(t), \tag{7.24}$$

where $s_I(t)$ and $s_Q(t)$ are both bipolar signals with amplitude of one. The envelope $A = \sqrt{P_I + P_Q}$ can be guaranteed to be constant. Unlike the linear superposition method, QPSM technology does not directly superimpose the newly added signal on the existing signal, but adopts the following method

$$s_{\mathrm{QPSM}}(t) = s(t) \cdot \exp[\phi_s(t)], \tag{7.25}$$

where

$$\phi_s(t) = \sum_{n=1}^{N} m_n s_n(t), \tag{7.26}$$

in which N is the total number of new signals to be added, $s_n(t)$ is the nth new bipolar signal to be modulated, and m_n is the corresponding modulation index, which determines the power allocation of each signal. Obviously, the envelope of this composite signal of the form (7.25) is still constant.

7.5.1 QPSM Multiplexing of Three-Signal Case

For the sake of simplicity, we first consider the case of only one new signal $s_N(t)$ added. At this time, (7.26) is simplified to

$$\phi_s(t) = m s_N(t). \tag{7.27}$$

Equation (7.25) can be reorganized as

$$s_{\mathrm{QPSM}}(t) = I(t) + \mathrm{j}Q(t), \tag{7.28}$$

where

$$\begin{cases} I(t) = \sqrt{P_I} s_I(t) \cos[\phi_s(t)] - \sqrt{P_Q} s_Q(t) \sin[\phi_s(t)], \\ Q(t) = \sqrt{P_I} s_I(t) \sin[\phi_s(t)] + \sqrt{P_Q} s_Q(t) \cos[\phi_s(t)]. \end{cases} \tag{7.29}$$

Letting $m = \theta_1 s_I(t)$, and utilizing the characteristic that both $s_I(t)$ and $s_N(t)$ are bipolar signals satisfying the following identities

$$\begin{cases} \cos[\phi_s(t)] = \cos(\theta_1) \\ \sin[\phi_s(t)] = \sin(\theta_1) s_I(t) s_N(t) \end{cases}, \tag{7.30}$$

we can rewrite (7.28) as

$$s_{\mathrm{QPSM}}(t) = \left[\sqrt{P_I} s_I(t) \cos(\theta_1) - \sqrt{P_Q} s_I(t) s_Q(t) s_N(t) \sin(\theta_1)\right] + \mathrm{j}\left[\sqrt{P_I} s_N(t) \sin(\theta_1) + \sqrt{P_Q} s_Q(t) \cos(\theta_1)\right]. \tag{7.31}$$

The above identity transformation process is essentially a special case of PSB-to-WSB form conversion, as discussed in Sect. 6.9.4. If we assume that the spreading codes of the three signals are completely orthogonal with each other, then the power of the newly added signal is all in the Q branch, and the three signals can be separately despread and demodulated.

We note from (7.31) that after QPSM multiplexing, the amplitudes of both of the original two mutually orthogonal signals $s_I(t)$ and $s_Q(t)$ are reduced by a factor of $\cos(\theta_1)$, but the power ratio and phase relationship between them remain the same. The transmitted signal has two additional components, one of which is the newly added signal $s_N(t)$ and the other is the product of the three signals $s_{\mathrm{IM}} = s_I(t) s_Q(t) s_N(t)$, that is, the inter-modulation (IM) component. In general, the IM component does not transmit any useful information. However, if the three codes are designed to be an appropriate structure, this IM component can be used to transmit an additional information [4].

We can calculate from (7.31) that, in the QPSM multiplexed signal, the power of the original in-phase branch component $s_I(t)$, the original orthogonal branch component $s_Q(t)$, the newly added signal $s_N(t)$, and the IM term are $P_I \cos^2(\theta_1)$, $P_Q \cos^2(\theta_1)$, $P_I \sin^2(\theta_1)$, and $P_Q \sin^2(\theta_1)$ respectively. If the power ratio of $s_N(t)$ to $s_Q(t)$ is given, the corresponding coefficient θ_1 can be easily obtained. Once θ_1 is given, the multiplexing efficiency of QPSM

$$\eta = \frac{P_I + P_Q \cos^2(\theta_1)}{P_I + P_Q} \tag{7.32}$$

is also determined.

7.5.2 QPSM Multiplexing of Arbitrary Number of Signals

The analysis of QPSM multiplexing when only one signal is added can be generalized. In cases where a total of N signals are to be multiplexed, the synthesized baseband signal can be expressed as

$$s(t) = \sqrt{P} \exp(\mathrm{j}\phi_s(t)), \tag{7.33}$$

where

$$\phi_s(t) = \theta_1 s_1(t) + \sum_{n=2}^{N} \theta_n s_1(t) s_n(t) \tag{7.34}$$

in which each θ_i can be calculated by [5]

$$\begin{cases} \theta_1 = \dfrac{\pi}{2} \\ \theta_k = \arctan\sqrt{\dfrac{P_k}{P_1}} \end{cases} \tag{7.35}$$

and the multiplexing efficiency is

$$\eta = \frac{\sum_{k=1}^{N} (P_k/P_1)}{\prod_{k=2}^{N} (1 + P_k/P_1)}. \tag{7.36}$$

From the above equation, we find that the multiplexing efficiency of QPSM decreases rapidly at an exponential rate as the number of N signals participating in the multiplexing increases. Therefore, QPSM multiplexing is often used to process three signals at most in practical applications.

7.5.3 QPSM Multiplexing for CBCS Signals

The CBCS signal (see Sect. 4.5) is a spread-spectrum signal with a four-level waveform. The CBOC signal can be seen as a special case of CBCS. If we want to use QPSM technique to combine some CBCS signals with several other bipolar signals into a constant envelope composite signal, we need to constrain the form of the signal involved in multiplexing.

Because standard QPSM multiplexing is only suitable for the multiplexing of bipolar signals, general non-bipolar signals do not satisfy the identities (7.30). Therefore, even if the multilevel signals can be multiplexed in the combined form (7.34) using QPSM technique, for the multiplexed signal, it is not necessarily guaranteed that the signal components can be despread separately in the receiver.

In the signal design of the Galileo system E1 band, one challenge faced by CEM is how to synthesize two CBCS signals $s_{\mathrm{CBCS1}}(t)$ and $s_{\mathrm{CBCS2}}(t)$ with another bipolar signal $s_3(t)$ into a constant envelope signal. To achieve this, the design limits the power of the two CBCS signals participating in the multiplex to be equal, and to be located on the same phase of the carrier. One uses in-phase CBCS modulation and the other uses anti-phase CBCS modulation, i.e.

$$s_{\mathrm{CBCS1}}(t) = c(t)\left[\sqrt{1-\gamma}\, s_{\mathrm{BCS}(\kappa_1,n)}(t) + \sqrt{\gamma}\, s_{\mathrm{BCS}(\kappa_2,n)}(t)\right], \tag{7.37}$$

$$s_{\mathrm{CBCS2}}(t) = c'(t)\left[\sqrt{1-\gamma}\, s_{\mathrm{BCS}(\kappa_1,n)}(t) - \sqrt{\gamma}\, s_{\mathrm{BCS}(\kappa_2,n)}(t)\right], \tag{7.38}$$

where $c(t)$ and $c'(t)$ are the spreading codes of the two CBCS signals respectively, and their values are ± 1, $s_{\mathrm{BCS}(\kappa_1,n)}(t)$ and $s_{\mathrm{BCS}(\kappa_2,n)}(t)$ are the subcarriers of $\mathrm{BCS}(\kappa_1, n)$ and $\mathrm{BCS}(\kappa_2, n)$ respectively.

Under the above constraints, the superposition of these two signals on the baseband can be expressed as

$$\begin{aligned} s'(t) &= \sqrt{P_1}\left[s_{\mathrm{CBCS1}}(t) + s_{\mathrm{CBCS2}}(t)\right] \\ &= \sqrt{(1-\gamma)P_1}\left(c(t) + c'(t)\right) s_{\mathrm{BCS}(\kappa_1,n)}(t) \\ &+ \sqrt{\gamma P_1}\left(c(t) - c'(t)\right) s_{\mathrm{BCS}(\kappa_2,n)}(t). \end{aligned} \tag{7.39}$$

Note that in the above equation, either $c(t) + c'(t)$ or $c(t) - c'(t)$ is zero, so in the actually transmitted signal, the $\mathrm{BCS}(\kappa_1, n)$ chip and the $\mathrm{BCS}(\kappa_2, n)$ chip do not appear at the same time. Thus, if another bipolar signal $s_3(t)$ is placed on a carrier phase quadrated to the phase of these two CBCS signals, by (7.34), the phase angle of the synthesized baseband QPSM signal can be written as

$$\begin{aligned} \phi_s(t) &= -\frac{\pi}{2}s_3(t) + \theta_1\frac{c(t) + c'(t)}{2}s_{\mathrm{BCS}(\kappa_1,n)}(t)\, s_3(t) \\ &+ \theta_2\frac{c(t) - c'(t)}{2}s_{\mathrm{BCS}(\kappa_2,n)}s_3(t), \end{aligned} \tag{7.40}$$

where $s_{\mathrm{BCS}(\kappa_1,n)}(t)$, $s_{\mathrm{BCS}(\kappa_2,n)}$ and $s_3(t)$ are bipolar signals with amplitudes of 1. It is not difficult to verify that if a number a has only three values of ± 1 and 0, and $s(t)$ is a bipolar signal, then the following identities hold:

$$\begin{cases} \cos\left[a\sqrt{\gamma}s(t)\right] = \begin{cases} \cos\left(\sqrt{\gamma}\right), & a = \pm 1 \\ 1 & a = 0 \end{cases} \\ \sin\left[a\sqrt{\gamma}s(t)\right] = as(t)\sin\left(\sqrt{\gamma}\right) \end{cases}. \tag{7.41}$$

Using (7.41), the baseband QPSM composite signal can be expanded into

$$\begin{aligned} s(t) &= \sqrt{P}\left[\cos\left(\phi_s(t)\right) + \mathrm{j}\sin\left(\phi_s(t)\right)\right] \\ &= \sqrt{P}c(t)\left[\frac{\sin\theta_1}{2}s_{\mathrm{BCS}(\kappa_1,n)}(t) + \frac{\sin\theta_2}{2}s_{\mathrm{BCS}(\kappa_2,n)}(t)\right] \\ &+ \sqrt{P}c'(t)\left[\frac{\sin\theta_1}{2}s_{\mathrm{BCS}(\kappa_1,n)}(t) - \frac{\sin\theta_2}{2}s_{\mathrm{BCS}(\kappa_2,n)}(t)\right] \\ &- \mathrm{j}\sqrt{P}\left[\frac{\cos\theta_1 + \cos\theta_2}{2}s_3(t) - s_{\mathrm{IM}}(t)\right], \end{aligned} \tag{7.42}$$

where

$$s_{\mathrm{IM}}(t) = \frac{\cos\theta_1 - \cos\theta_2}{2}c(t)\,c'(t)\,s_3(t) \tag{7.43}$$

is the IM term introduced to maintain the constant envelope of the composite signal.

By selecting the appropriate θ_1 and θ_2 values, one can change the power ratio of $s_{\mathrm{CBCS1}}(t)$, $s_{\mathrm{CBCS2}}(t)$, $s_3(t)$ and the IM term in the total signal, and also change the proportion of the two BCS signal components in CBCS signals. Therefore, for a given power ratio of each signal component, the corresponding composite signal expression can be obtained by inversely solving the values of θ_1 and θ_2.

7.6 Multiplexing Based on Majority Voting Logic

7.6.1 Majority Voting Logic

The basic principle of majority voting (MV) is to judge the polarity of all the bipolar spread-spectrum signals participating in the multiplexing at each moment and take the polarity of the majority as the output of the majority voting (MV) logic. For example, for N bipolar spread-spectrum signals $c_1, c_2, \ldots, c_N$, at time t, if the number of components taking $+1$ is greater than the number of components taking -1, the MV logic outputs $c_{\mathrm{MV}} = +1$, otherwise the output is $c_{\mathrm{MV}} = -1$. In fact, this rule is equivalent to calculating the algebraic sign of the values of these bipolar chips, i.e.

$$c_{\mathrm{MV}} = \operatorname{sgn}\left(\sum_{i=1}^{N} c_i\right). \tag{7.44}$$

In order to avoid an equal number of signals taking the value of $+1$ and -1 at a certain time, it is necessary to limit the total number N of signals participating in MV multiplexing to an odd number $2M+1$.

It is easy to see that c_{MV} is also a bipolar signal. MV multiplexing technique uses the c_{MV} signal instead of $2M+1$ original signal components for transmission. After receiving the c_{MV}, the receiver can recover the information in c_i by correlating c_{MV} with the local replica of c_i.

In order to understand the principle of this multiplexing technique, it is necessary to analyze the correlation value between the c_{MV} signal and each specific signal component c_i, and the attenuation of the receiving power when replacing the $2M+1$ signals with c_{MV}.

7.6.2 Uniform Weighting MV Multiplexing

Assuming that $2M+1$ signal components $c_1, c_2, \ldots, c_{2M+1}$ are statistically balanced, and the chip values of which can be modeled as independent and identically distributed bipolar random variables, the correlation result of c_{MV} and c_i can be replaced by the ensemble average of $c_{\text{MV}} \cdot c_i$ if the receiver's coherent integration time is long enough to traverse each value combination of the signal component.

Let $\chi_i = c_{\text{MV}} \cdot c_i$, assuming that the chips in the correlation operation are strictly aligned. Then, at some moment, the majority voting chip c_{MV} has the same polarity as the local reference chip c_i, i.e. $\chi_i = 1$ if and only if at least M chips of the remaining $2M$ chips have the same polarity as c_i. Thus, for a certain signal component c_i, the mathematical expectation of χ_i is

$$\begin{aligned}\mathbb{E}[\chi_i] &= \Pr(c_i(+), c_{\text{MV}}(+)) + \Pr(c_i(-), c_{\text{MV}}(-)) \\ &\quad - \Pr(c_i(+), c_{\text{MV}}(-)) - \Pr(c_i(-), c_{\text{MV}}(+)) \\ &= \Pr(c_i(+)) \Pr(c_{\text{MV}}(+) | c_i(+)) + \Pr(c_i(-)) \Pr(c_{\text{MV}}(-) | c_i(-)) \\ &\quad - \Pr(c_i(+)) \Pr(c_{\text{MV}}(-) | c_i(+)) - \Pr(c_i(-)) \Pr(c_{\text{MV}}(+) | c_i(-)),\end{aligned} \tag{7.45}$$

where we use $c_i(+)$ and $c_i(+)$ to represent $c_i = +1$ and $c_i = -1$ respectively. In (7.45), according to the assumptions, we have

$$\Pr(c_i(+)) = \Pr(c_i(-)) = \frac{1}{2} \tag{7.46}$$

and

$$\begin{cases} \Pr(c_{\text{MV}}(+) | c_i(+)) = \Pr(c_{\text{MV}}(-) | c_i(-)) = \sum\limits_{i=M+1}^{2M} p_i^{2M}(+) \\ \Pr(c_{\text{MV}}(+) | c_i(-)) = \Pr(c_{\text{MV}}(-) | c_i(+)) = \sum\limits_{i=M+1}^{2M} p_i^{2M}(-) \end{cases} \tag{7.47}$$

where $p_i^{2M}(+)$ and $p_i^{2M}(-)$ are the probabilities that exactly i chips in $2M$ chips take $+1$ and -1 respectively. Obviously, we have

$$p_i^{2M}(+) = p_i^{2M}(-) = \frac{1}{2^{2M}} \binom{2M}{i}. \tag{7.48}$$

Substituting (7.46), (7.47), and (7.48) into (7.45), and using Stirling's formula [29], there is

$$\mathbb{E}[\chi_i] = p_M^{2M}(+) = \frac{1}{2^{2M}} \binom{2M}{M} \approx \frac{1}{\sqrt{\pi M}} \mathrm{e}^{-\frac{1}{8M}}. \tag{7.49}$$

It can be seen that in the uniform weighting MV multiplexing, when one bipolar MV multiplexed signal c_{MV} is used for transmission instead of the $2M + 1$ original spread spectral signals, the power of each signal component in the receiver at the correlator output is

$$P_i = |\mathbb{E}[\chi_i]|^2. \tag{7.50}$$

Table 7.2 Uniform weighting MV multiplexing of three signals

	c_1	c_2	c_3	c_{MV}	χ_1	χ_2	χ_3
Value	+	+	+	+	+	+	+
	+	+	−	+	+	+	−
	+	−	+	+	+	−	+
	+	−	−	−	−	+	+
	−	+	+	+	−	+	+
	−	+	−	−	+	−	+
	−	−	+	−	+	+	−
	−	−	−	−	+	+	+
Expectation $\mathbb{E}[\chi]$					3/4	3/4	3/4

Then, we can calculate the multiplexing efficiency of the equal weight MV, which is

$$\eta_{MV} = (2M+1)\left|\mathbb{E}[\chi_i]\right|^2 = \frac{2M+1}{\pi M}e^{-\frac{1}{4M}}. \tag{7.51}$$

Example 7.2 (*Uniform weighting MV multiplexing of three signals*) Consider the case where MV multiplexing is used to simultaneously transmit three signals c_1, c_2, and c_3. Table 7.2 lists the combination of the three signals and the value of the corresponding MV multiplexed signal c_{MV}, and also provides the correlation between each signal and the majority vote signal. We can see that although the result of MV multiplexing does not appear to be the same as any of the three signals, in its correlation results with the three signals, 3/4 of the total chips can be matched and only 1/4 of the chips cannot be matched. Therefore, $\mathbb{E}[\chi] = 1/2$ and the multiplexing efficiency is 75%, corresponding to a 1.25 dB power loss. This numerical simulation result is consistent with the calculation result of (7.51). ■

An interesting question is how the power loss of uniform weighting MV multiplexing varies with the number of signals participating in the multiplexing. Let $M \to \infty$ in (7.51), we can obtain by approximation expression that the multiplexing efficiency will gradually approach $2/\pi$. That is, the power loss of the uniform weighting MV multiplexing will increase with the signals participating in the synthesis but will not exceed 1.96 dB.

Figure 7.2 shows the power loss with respect to the number of signals participating in the MV multiplexing. The results obtained using the exact formula and the approximation formula are depicted together. It can be seen that even when $2M+1$ is small, the difference between the approximation formula and the exact formula is not significant.

The power loss of MV multiplexing can be understood as being caused by the spreading waveform mismatch. As we can see in Example 7.2, when the receiver uses a local signal to correlate with the received MV multiplexed signal, some chips cannot be matched, meaning that the correlation result cannot reach the maximum.

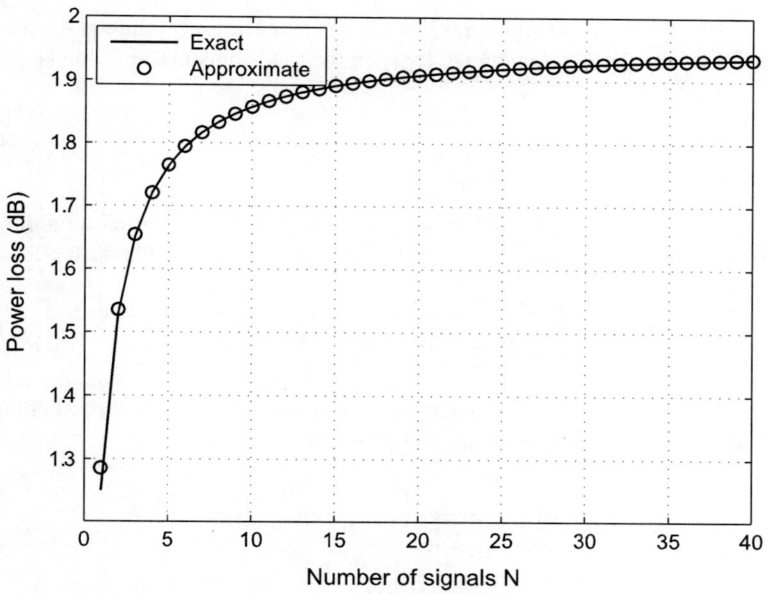

Fig. 7.2 Power loss of uniform weighting MV multiplexing

Such power loss is acceptable as the cost of CEM. The majority vote is simple to implement, which is an advantage. The shortcoming is mainly reflected in the fact that the relative power of the signal components participating in the multiplexing cannot be adjusted, and the phase of these signal components can only be identical or opposite. Besides, this multiplexing technique does not allow sufficient spectral separation between signal components.

7.6.3 Non-uniform Weighting MV Multiplexing

If we want to use uniform weighting MV to achieve signal multiplexing with any power ratio, one of the simplest methods is to cascade it with a time division multiplexing device, which is called interlaced MV multiplexing [7]. For the multiplexing of $2M+1$ signals, the transmitted signal can have $2M+2$ choices at each decision time: transmitting the MV signal c_{MV}, or one of the $2M+1$ codes participating in the multiplexing. By controlling the number of times each signal appears in the final transmitted signal, the power allocation ratio between the multiplexed signals can be adjusted.

Assume that the desired power ratio of these $2M+1$ signals is $P_1 : P_2 : \cdots : P_{2M+1}$. For the following discussion, the index of the specified signal is arranged in non-decreasing order of power, i.e. $P_{i+1} \geq P_i$, while $P_1 = 1$. The proportion of the time slot occupied by the signal c_{MV} in the total transmission time is t_{MV}, and the proportion of the time slot occupied by c_i in the total transmission time is t_i, so we have

$$\sum_{i=1}^{2M+1} t_i + t_{\mathrm{MV}} = 1\,. \tag{7.52}$$

We set that the weakest signal has no chance of being transmitted alone, but can only be transmitted with c_{MV}, i.e. $t_1 = 0$. Then, we can easily obtain the following equations

$$\begin{cases} \sqrt{P_1} = A\sqrt{2M+1}\bar{\chi} \cdot t_{\mathrm{MV}} = 1 \\ \sqrt{P_i} = \sqrt{P_1} + A\sqrt{2M+1} \cdot t_i \end{cases}, \tag{7.53}$$

where $\bar{\chi} = \mathbb{E}[\chi]$. Next, by the constraint condition (7.52), the proportion of the total transmission time of each signal can be solved as

$$\begin{cases} t_{\mathrm{MV}} = \dfrac{1}{\bar{\chi}\sum_{i=2}^{2M+1}\sqrt{P_i} - (2M\bar{\chi} - 1)}, \\ t_i = \dfrac{\sqrt{P_i} - 1}{\sum_{i=2}^{2M+1}\sqrt{P_i} - \left(2M - \bar{\chi}^{-1}\right)}. \end{cases} \tag{7.54}$$

When the receiver correlates the local replica of signal c_i with the received interlaced MV multiplexed signal, the average correlation value is

$$\bar{\chi}_i' = \bar{\chi} t_{\mathrm{MV}} + 1 \cdot t_i = \frac{\sqrt{P_i}}{\sum_{i=2}^{2M+1}\sqrt{P_i} - \left(2M - \bar{\chi}^{-1}\right)}. \tag{7.55}$$

Therefore, the multiplexing efficiency is

$$\eta = \sum_{i=1}^{2M+1} \left(\bar{\chi}_i'\right)^2 = \frac{1 + \sum_{i=2}^{2M+1} P_i}{\left[\sum_{i=2}^{2M+1}\sqrt{P_i} - \left(2M - \bar{\chi}^{-1}\right)\right]^2}. \tag{7.56}$$

Example 7.3 (*Interlaced MV multiplexing of three signals*) In this example, we consider three signals for non-uniform weighting MV multiplexing, where $M = 1$. According to (7.54), when the expected transmission powers of these three signals are P_1, P_2, and P_3 respectively, the time ratio of transmitting the MV signal, the second signal, and the third signal is

$$\begin{cases} t_{\mathrm{MV}} = \dfrac{2}{\sqrt{P_2} + \sqrt{P_3}}, \\ t_2 = \dfrac{\sqrt{P_2} - 1}{\sqrt{P_2} + \sqrt{P_3}}, \\ t_3 = \dfrac{\sqrt{P_3} - 1}{\sqrt{P_2} + \sqrt{P_3}}. \end{cases} \tag{7.57}$$

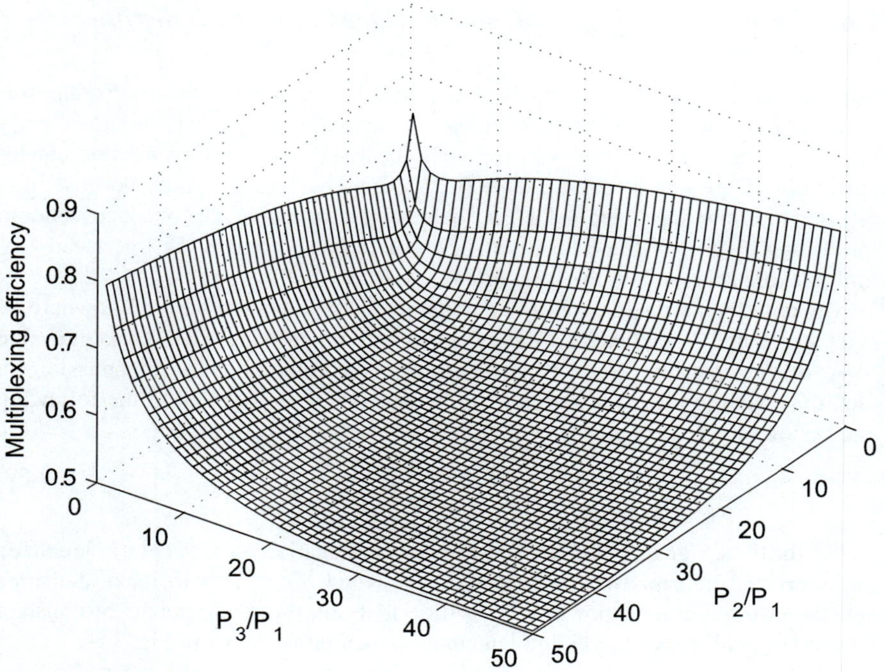

Fig. 7.3 Efficiency for three-signal interlaced MV multiplexing

The multiplexing efficiency in this example is

$$\eta = \frac{1 + P_2 + P_3}{\left(\sqrt{P_2} + \sqrt{P_3}\right)^2}. \tag{7.58}$$

Figure 7.3 shows the curve of the multiplexing efficiency changing with power distribution. When the power of one signal is much larger than that of the other two signals, the multiplexing efficiency can reach 100%. This conclusion can also be obtained in (7.58) that when $P_2 \to \infty$, we have $\eta \to 1$. However, this extreme case is equivalent to the case where only one signal is transmitted. When the power of two signals is similar and both are much larger than the power of the other signal, the multiplexing efficiency of the interlaced majority vote is only 50%. When the powers of the three signals are close to each other, the multiplexing efficiency is 75%. ■

Although the above example is for three signal multiplexing, a similar phenomenon occurs for multiple signal MV multiplexing. The interlaced MV technique can solve the problem of the non-uniform power ratio. However, when a small number of the signals participating in the multiplexing occupy most of the power, the multiplexing efficiency drops significantly.

7.6.4 Phase Mapping Table of Interlaced MV Multiplexing

For the class of MV-based methods, when the $2M + 1$ signals have uniform power allocation, the MV logic has an equivalent explicit expression (7.44) to directly obtain the phase LUT. However, the interlaced MV multiplexing method cannot easily obtain the phase look-up table (LUT) directly. This is mainly because time division multiplexing is applied. Nevertheless, by introducing the concept of random interlacing control of the transmit switch, we can also convert the interlaced MV multiplexing scheme into the form of a phase LUT.

The random interlacing controlled transmit switch is shown in Fig. 7.4, where W is a control signal and takes a value of $+1$ or -1. When W is $+1$, S takes the value of A for transmission. When W is -1, S takes the value of B for transmission. In fact, the switching logic of Fig. 7.4 can be equivalently converted to the following expression:

$$S = \left(\frac{1+W}{2}\right) \cdot A + \left(\frac{1-W}{2}\right) \cdot B\,. \tag{7.59}$$

For the three signals with non-uniform power allocation, Wang et al. [21] realized the interlaced transmission by means of the switch in Fig. 7.4 to make the three signals have the desired power ratio while satisfying the requirements of constant enveloping multiplexing. The implementation method is shown in Fig. 7.5

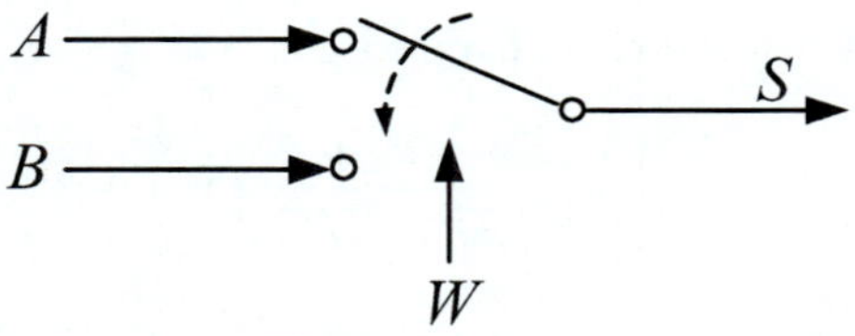

Fig. 7.4 Random interlacing controlled transmit switch

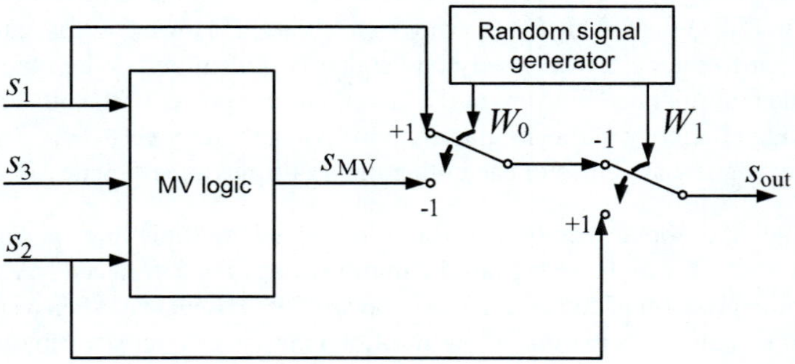

Fig. 7.5 Majority voting method for three signals with non-uniform power

Table 7.3 Distribution of the value of the switch variable

W_0	W_1	Probability
1	1	$\mathcal{P}_3$
1	−1	$\mathcal{P}_2$
−1	1	$\mathcal{P}_1$
−1	−1	$\mathcal{P}_0$

Using (7.59), the random signal generator that implements interlaced transmission can be equivalently replaced. The final multiplexed signal can be written as an explicit expression as follows

$$s_{out} = \left(\frac{1+W_1}{2}\right)s_2 + \left(\frac{1-W_1}{2}\right)\left[\left(\frac{1+W_0}{2}\right)s_1 + \left(\frac{1-W_0}{2}\right)s_{\mathrm{MV}}\right] \tag{7.60}$$

where

$$s_{\mathrm{MV}} = \frac{1}{2}\left(s_1 + s_2 + s_3 - s_1 s_2 s_3\right). \tag{7.61}$$

Here, the mapping from the useful signal to the final composite signal depends not only on the input value of the useful signal, but also on the value of the switch control signals. The probability distribution of the control signal W_0 and W_1 of the two switches is shown in Table 7.3

The values of $\mathcal{P}_0$, $\mathcal{P}_1$, $\mathcal{P}_2$, and $\mathcal{P}_3$ in Table 7.3 are in fact consistent with the component transmission time slot ratios in the previous section. If we want the power ratio of the three signals to be −3 dB: 0 dB: 0.5 dB, we can set

$$\begin{cases} \mathcal{P}_0 = 0.6876\,, \\ \mathcal{P}_1 = 0.1706\,, \\ \mathcal{P}_2 = 0.1418\,, \\ \mathcal{P}_3 = 0\,. \end{cases} \tag{7.62}$$

If we add the values of switch control signals W_0 and W_1 to the phase map as two random signals, we can obtain the final phase map, as shown in Table 7.4. According to the mapping table, the phase of the composite signal appears at only two points of 0° or 180°. Therefore, the final transmitted signal is a BPSK signal. Using the phase map, the multiplexing efficiency of this multiplexing scheme can be calculated, which is 61.86%.

Table 7.4 Corresponding phase mapping table of Fig. 7.5

s_1	s_3	s_2	W_0	W_1	Transmission phase	State occurrence probability
−1	−1	−1	−1	−1	π	$\frac{1}{8}\mathcal{P}_0$
−1	−1	−1	−1	1	π	$\frac{1}{8}\mathcal{P}_1$
−1	−1	−1	1	−1	π	$\frac{1}{8}\mathcal{P}_2$
−1	−1	−1	1	1	π	$\frac{1}{8}\mathcal{P}_3$
−1	−1	1	−1	−1	π	$\frac{1}{8}\mathcal{P}_0$
−1	−1	1	−1	1	0	$\frac{1}{8}\mathcal{P}_1$
−1	−1	1	1	−1	π	$\frac{1}{8}\mathcal{P}_2$
−1	−1	1	1	1	0	$\frac{1}{8}\mathcal{P}_3$
−1	1	−1	−1	−1	π	$\frac{1}{8}\mathcal{P}_0$
−1	1	−1	−1	1	π	$\frac{1}{8}\mathcal{P}_1$
−1	1	−1	1	−1	π	$\frac{1}{8}\mathcal{P}_2$
−1	1	−1	1	1	π	$\frac{1}{8}\mathcal{P}_3$
−1	1	1	−1	−1	0	$\frac{1}{8}\mathcal{P}_0$
−1	1	1	−1	1	0	$\frac{1}{8}\mathcal{P}_1$
−1	1	1	1	−1	π	$\frac{1}{8}\mathcal{P}_2$
−1	1	1	1	1	0	$\frac{1}{8}\mathcal{P}_3$
1	−1	−1	−1	−1	π	$\frac{1}{8}\mathcal{P}_0$
1	−1	−1	−1	1	π	$\frac{1}{8}\mathcal{P}_1$
1	−1	−1	1	−1	0	$\frac{1}{8}\mathcal{P}_2$
1	−1	−1	1	1	π	$\frac{1}{8}\mathcal{P}_3$
1	−1	1	−1	−1	0	$\frac{1}{8}\mathcal{P}_0$
1	−1	1	−1	1	0	$\frac{1}{8}\mathcal{P}_1$
1	−1	1	1	−1	0	$\frac{1}{8}\mathcal{P}_2$
1	−1	1	1	1	0	$\frac{1}{8}\mathcal{P}_3$
1	1	−1	−1	−1	0	$\frac{1}{8}\mathcal{P}_0$
1	1	−1	−1	1	π	$\frac{1}{8}\mathcal{P}_1$
1	1	−1	1	−1	0	$\frac{1}{8}\mathcal{P}_2$
1	1	−1	1	1	π	$\frac{1}{8}\mathcal{P}_3$
1	1	1	−1	−1	0	$\frac{1}{8}\mathcal{P}_0$
1	1	1	−1	1	0	$\frac{1}{8}\mathcal{P}_1$
1	1	1	1	−1	0	$\frac{1}{8}\mathcal{P}_2$
1	1	1	1	1	0	$\frac{1}{8}\mathcal{P}_3$

7.7 Constant Envelope Multiplexing via Intermodulation Construction (CEMIC)

In all the CEMs discussed so far, most of the typical techniques are dedicated to signals on the same central frequency. And most of them can only apply to bipolar signals. With other signal elements becoming more diverse and complex in future GNSSs, the demand for the flexibility of a CEM becomes greater. The high-flexibility multi-carrier CEM is gaining more and more attention in recent years as a new hot topic in the signal multiplexing field. Several CEM design and implementation methods with a higher flexibility for different types of signal components have emerged.

In order to loosen constraints of involved signals as much as possible, a high efficiency generalized multicarrier joint CEM technique for multilevel DSSS signals, termed constant envelope multiplexing via intermodulation construction (CEMIC), is proposed in [9]. Compared with existing CEM techniques, CEMIC has a much higher design flexibility in the number of sub-bands, the number of signal components, power ratio, and phase relationship among components, and the spreading chip waveforms. It can combine any number of bipolar or multilevel DSSS signals with arbitrary power allocation at one or more carriers into a constant envelope integrated signal. Moreover, this technique has the highest multiplexing efficiency of all existing CEM techniques with the same signal constraints.

Complying with the definitions and notations given in Sects. 6.4 and 6.5, consider combining N spreading spectrum signal components located at several sub-bands, $s_i(t)$, for $i = 1 \sim N$, into a composite signal with constant envelope. The power and initial phase of $s_i(t)$ are P_i and θ_i respectively. The core of constant envelope multiplexing via intermodulation construction (CEMIC) is to find an optimal mapping rule, which constructs an IM term that can guarantee the optimal power efficiency, minimal impact on the correlation characteristics of useful components, and the envelope constancy of the composite signal.

Generally, using CEMIC to construct a CEM composite signal has the following four main steps:

(1) According to the chip length T_c, the frequency offset of the ith component signal from the carrier frequency f_i, sub-chip number M and the shapes of SCS signal waveform p_i, for $i = 1 \sim N$, list all F possible combinations of values of $\{\tilde{c}_1, \tilde{c}_2, \ldots, \tilde{c}_N\}$, and construct the component weight vectors $\boldsymbol{c}_i = \left[\tilde{c}_i^{(1)}, \tilde{c}_i^{(2)}, \ldots, \tilde{c}_i^{(F)}\right]$, for $i = 1 \sim N$, where $\tilde{c}_i^{(\ell)}$ is the value of $\tilde{c}_\ell$ in the ℓth combination.

(2) Based on component weight vectors $\boldsymbol{c}_1, \boldsymbol{c}_2, \ldots, \boldsymbol{c}_N$, using the G-S orthogonalizing method or other methods, construct a set of orthogonal vectors $\left\{\mathring{\boldsymbol{c}}_1, \mathring{\boldsymbol{c}}_2, \ldots, \mathring{\boldsymbol{c}}_{F-N}\right\}$ to make span $\left\{\mathring{\boldsymbol{c}}_1, \mathring{\boldsymbol{c}}_2, \ldots, \mathring{\boldsymbol{c}}_{F-N}\right\}$ be the orthogonal complement space of span $\{\boldsymbol{c}_1, \boldsymbol{c}_2, \ldots, \boldsymbol{c}_N\}$. A construction algorithm is provided in Algorithm 7.1. More general construction algorithm for this step can be found in [9].

Algorithm 7.1: Basis vectors construction algorithm for $\mathcal{W}^{\perp}$

Input: $\boldsymbol{c}_1, \boldsymbol{c}_2, \ldots, \boldsymbol{c}_N$
Output: $\mathring{\boldsymbol{c}}_1, \mathring{\boldsymbol{c}}_2, \ldots, \mathring{\boldsymbol{c}}_{F-N}$

Initialisation:
$\boldsymbol{I}_{F\times F} \leftarrow \text{diag}\,(1, 1, \ldots, 1) \triangleq \left(\boldsymbol{e}_1, \boldsymbol{e}_2, \ldots, \boldsymbol{e}_F\right)^{\mathrm{T}}$;
$n \leftarrow 1$;
$k \leftarrow 1$;
$\boldsymbol{u}_0 \leftarrow \boldsymbol{0}_{F\times 1}$;

Loop Process:
while $n \leqslant F - N$ *and* $k \leqslant F$ **do**
  $\boldsymbol{u}_k \leftarrow \boldsymbol{e}_k - \sum_{i=1}^{N} \boldsymbol{c}_i^{\mathrm{H}} \boldsymbol{e}_k \boldsymbol{c}_i - \sum_{i=0}^{k-1} \boldsymbol{u}_i^{\mathrm{H}} \boldsymbol{e}_k \boldsymbol{u}_i$;
  if $\|\boldsymbol{u}_k\|_2 \neq 0$ **then**
    $\mathring{\boldsymbol{c}}_n \leftarrow \boldsymbol{u}_k$;
    $n \leftarrow n + 1$;
  end
  $k \leftarrow k + 1$;
end

(3) Define $\boldsymbol{s}_{\mathrm{CE}} = \boldsymbol{C}_0 \boldsymbol{w}_{\mathrm{s}} + \mathring{\boldsymbol{C}} \boldsymbol{w}_{\mathrm{m}}$, where $\boldsymbol{C}_0 = [\boldsymbol{c}_1, \ldots, \boldsymbol{c}_N]$, $\mathring{\boldsymbol{C}} = \left[\mathring{\boldsymbol{c}}_1, \ldots, \mathring{\boldsymbol{c}}_{F-N}\right]$, and $\boldsymbol{w}_{\mathrm{s}} = \left[\sqrt{P_1}\mathrm{e}^{\mathrm{j}\theta_1}, \sqrt{P_2}\mathrm{e}^{\mathrm{j}\theta_2}, \ldots, \sqrt{P_N}\mathrm{e}^{\mathrm{j}\theta_N}\right]^{\mathrm{T}}$, and solve the following constraint minimization problem

$$\begin{cases} \min\limits_{\boldsymbol{w}_{\mathrm{m}} \in \mathbb{C}^{F-N}} & \|\boldsymbol{w}_{\mathrm{m}}\|^2, \\ \text{s.t.} & \left|s_{\mathrm{CE}}^{(1)}\right| = \left|s_{\mathrm{CE}}^{(2)}\right| = \cdots = \left|s_{\mathrm{CE}}^{(F)}\right| \end{cases} \tag{7.63}$$

to obtain the optimal coefficient vector $\boldsymbol{w}_{\mathrm{opt}}$, where $s_{\mathrm{CE}}^{(\ell)}$ is the ℓth entry of $\boldsymbol{s}_{\mathrm{CE}}$.

(4) Let $\boldsymbol{\lambda} = \mathring{\boldsymbol{C}} \boldsymbol{w}_{\mathrm{opt}} = \left[\lambda^{(1)}, \lambda^{(2)}, \ldots, \lambda^{(F)}\right]^{\mathrm{T}}$. Then we obtain the optimal mapping rule from the value combination of N signal components to the IM term $I_{\mathrm{IM}}(t)$. In every moment, if the values of $\{\tilde{c}_1(t), \tilde{c}_2(t), \ldots, \tilde{c}_N(t)\}$ correspond to the ℓth value combination, $I_{\mathrm{IM}}(t)$ takes the value $\lambda^{(\ell)}$, and

$$s_{\mathrm{CE}}(t) = \sum_{i=1}^{N} s_i(t) + I_{\mathrm{IM}}(t). \tag{7.64}$$

A case study of CEMIC method for single-frequency CEM of bipolar signals has been given in Example 6.3 to demonstrate the high efficiency and high flexibility of this technique in which we can see that for single frequency bipolar components CEM, CEMIC has the same efficiency as POCET, which is the currently acknowledged most efficient single-frequency CEM method.

In contrast with POCET, which can only output the optimized phase LUT, a noteworthy feature of CEMIC is that the composition of the composite signal and the explicit expression of the IM term can be directly obtained, allowing for easier analysis of spectrum characteristics such as the spectrum occupancy of the composite signal and the spectral compatibility between useful signals and the IM term. Nevertheless, compared with POCET, CEMIC in the single frequency CEM case also has some drawbacks. CEMIC requires that all of the power and phase relationships have been previously determined, whereas in POCET, some of the power and phase constraints do not need to be specified as input. They can be treated as parameters to be optimized in the maximum multiplexing efficiency sense. However, CEMIC needs to traverse all possible values for those unconstrained power and phase relationships to find the same final result.

More performance analysis and typical case studies of CEMIC for multi-frequency CEM of multilevel signals can be found in [9].

7.8 Multi-frequency Constant Envelope Multiplexing

In the CEM techniques discussed so far, except for CEMIC, most of the typical CEM techniques are proposed to meet the combined emission requirements of signal components at the same central frequency. Nevertheless, in some situations, there is more demand for multi-frequency CEM, which means combining signals located on two or more different central frequencies into a constant envelope composite signal. In multi-carrier CEM, the object is transmitting a constant envelope signal to achieve an effect at correlator outputs similar to that of transmitting $\sum_{i=1}^{N} \sqrt{P_i} \mathrm{e}^{\mathrm{j}\phi_i} s_i(t) \mathrm{e}^{\mathrm{j}2\pi f_i t}$, where f_i is the frequency offset of the ith component signal from the carrier frequency. An example of this multicarrier joint multiplexing for just two central frequencies is termed dual-frequency CEM, or DCEM.

The first demand for DCEM in satellite navigation arose from the Galileo E5 signal design [30]. The interval between the main lobes of the signals located on two sub-bands of E5 is only 10.23 MHz. In order to solve problems in filter design, AltBOC [31] is proposed and employed. Afterward, the BDS signal design also made a similar request for DCEM. To achieve interoperability improvement during the system updating, the central frequencies of the B1 and B2 bands were adjusted. Considering the need for a smooth system transition, it is necessary to allow signals of two phases to coexist over a period of time to ensure on-line reconstruction in the satellite payload transmitter. To achieve this, it is desired that those signals be combined in digital baseband and share the amplifier chain.

Additionally, if the central frequencies of two sub-bands are not far from each other, such as the BDS B2a and B2b that are 30.69 MHz apart, then combining signals on these two frequencies into a constant envelope signal can not only save the number of HPAs, but also ensure high coherence of link characteristics of the signals on those two adjacent frequencies. For receivers, signals located at those two frequencies can be received and processed separately. The DCEM-integrated signal also offers the possibility of wideband composite receiving [32, 33].

7.8.1 Sideband Modulation by Using Complex Subcarriers

Considering the satellite transmitter broadcasts different signal components at two RF frequencies. The two RF signals can be written as $s_{\mathrm{RF,L}}(t) = \mathrm{Re}\left\{s_L(t)\,\mathrm{e}^{\mathrm{j}2\pi f_L t}\right\}$ and $s_{\mathrm{RF,U}}(t) = \mathrm{Re}\left\{s_U(t)\,\mathrm{e}^{\mathrm{j}2\pi f_U t}\right\}$ respectively, where $s_L(t)$ and $s_U(t)$ are the complex envelopes of the two signals, and f_L and f_U are their central frequencies, with $f_L < f_U$. It is assumed that the spreading chip waveforms of the signals at the two frequencies participating in the multiplexing are all BCS waveforms.

Combining $s_{\mathrm{RF,L}}(t)$ and $s_{\mathrm{RF,U}}(t)$ into an integrated signal and transmitting it on a common carrier is equivalent to constructing a new complex envelope $s_{\mathrm{MUX}}(t)$ and modulating it onto carrier frequency f_0, while minimizing the difference to the original dual-frequency independent transmission from the receiving perspective. This equivalence relation is presented as

$$\begin{aligned} &s_{\mathrm{RF,MUX}}(t) \\ &= \mathrm{Re}\left\{s_{\mathrm{MUX}}(t)\,\mathrm{e}^{\mathrm{j}2\pi f_0 t}\right\} \approx s_{\mathrm{RF,L}}(t) + s_{\mathrm{RF,U}}(t) \\ &= \mathrm{Re}\left\{\left[s_L(t)\,\mathrm{e}^{\mathrm{j}2\pi(f_L - f_0)t} + s_U(t)\,\mathrm{e}^{\mathrm{j}2\pi(f_U - f_0)t}\right]\mathrm{e}^{\mathrm{j}2\pi f_0 t}\right\}. \end{aligned} \tag{7.65}$$

Choosing a different f_0 in (7.65) results in a different complex envelope $s_{\mathrm{MUX}}(t)$, while a typical approach is to place the carrier frequency at $f_0 = (f_L + f_U)/2$ and let $s_U(t)$ multiply with a complex subcarrier

$$\gamma_{sc}(t) = \mathrm{e}^{\mathrm{j}2\pi f_s t} = \cos(2\pi f_s t) + \mathrm{j}\sin(2\pi f_s t) \tag{7.66}$$

in which $f_s = (f_U - f_L)/2$, meanwhile let $s_L(t)$ multiply with the complex conjugate of (7.66), so as to shift the spectrum of these two signals to the upper sideband and the lower sideband respectively. Then, the composite baseband complex envelope is

$$s_{\mathrm{MUX}}(t) = s_L(t)\,\gamma_{sc}^*(t) + s_U(t)\,\gamma_{sc}(t). \tag{7.67}$$

Another typical implementation is to place the carrier frequency at the center frequency of one of original signals. Without loss of generality, let $f_0 = f_L$, then by using the complex subcarrier

$$\gamma_{sc}'(t) = \mathrm{e}^{\mathrm{j}4\pi f_s t} = \cos(4\pi f_s t) + \mathrm{j}\sin(4\pi f_s t), \tag{7.68}$$

the synthesized baseband complex envelope is

$$s_{\mathrm{MUX}}(t) = s_L(t) + s_U(t)\,\gamma_{sc}'(t). \tag{7.69}$$

7.8.2 DCEM Based on Square Wave Complex Subcarriers

The composite signal of (7.67) does not have a constant envelope. To make it simple to generate on digital circuits and achieve dual-frequency multiplexing using existing CEM techniques, we can use a binary subcarrier

$$\tilde{\gamma}_{sc}(t) = \frac{\sqrt{2}}{2}\operatorname{sgn}[\cos(2\pi f_s t)] + \mathrm{j}\frac{\sqrt{2}}{2}\operatorname{sgn}[\sin(2\pi f_s t)] \tag{7.70}$$

to replace $\gamma_{sc}(t)$. Then, since each component in the composite signal is bipolar, existing single-frequency CEM techniques can be used to achieve constant envelope reconstruction of $s_{\mathrm{MUX}}(t)$.

For example, when both $s_L(t)$ and $s_U(t)$ are real bipolar components, there are four real bipolar components participating in the multiplexing. At this time, QPSM or multicomponent signals with phase modulation (MSPM) technique [26] can be used to achieve constant envelope reconstruction of $s_{\mathrm{MUX}}(t)$. When both $s_L(t)$ and $s_U(t)$ are QPSK signals, i.e.

$$s_i(t) = s_{iI}(t) + \mathrm{j}s_{iQ}(t), i \in \{L, U\}, \tag{7.71}$$

$s_{\mathrm{MUX}}(t)$ contains eight real bipolar components. At this point, the POCET and CEMIC techniques can be used to perform constant envelope reconstruction of $s_{\mathrm{MUX}}(t)$.

Such DCEM methods based on the combination of square wave subcarriers and existing single-frequency CEM are simple and straightforward, but in some cases their effects are not satisfactory. The efficiency of QPSM multiplexing to process bipolar signals of more than three components is not high. The MSPM technique has a strict limitation of the power ratio of the signal components, where the signal components participating in the multiplexing must have equal power. Although the POCET technique is flexible, when processing dual-frequency combination of two QPSK signals, the search space of the phase optimization of eight bipolar components is huge. While the search process is difficult to converge, the final constant envelope multiplexed signal is a 256-PSK phase-modulated signal, which is too complex to implement.

In fact, when the phase and power relationships of the signal components participating in the multiplexing are relatively regular, some DCEM design methods based on signal phase rotating or signal waveform reconstruction are more convenient. Moreover, the multiplexing efficiency of the constant envelope multiplexing scheme obtained based on these advanced methods can also often be optimized. In the next two sections, two typical DCEM techniques, rotating POCET technique and asymmetric constant envelope BOC (ACE-BOC) multiplexing technique, are introduced in detail.

7.9 Rotating POCET Technique

In the viewpoint of PDP, the phase relation between signals located on different sub-frequencies is continually changing at a rate equal to the frequency difference. Considering two components located on two sub-frequencies f_0 apart, the phase difference between these two components will increase by $\Delta\theta = 2\pi f_0 \Delta t$ every Δt.

Based on this characteristic, Dafesh et al. [34] showed that continuous phase rotation can be quantized into finite phase steps. Further, at each phase step, a LUT for the fixed phase relation between the signals can be obtained through a PDP method, thus extending the single LUT into multiple LUTs, or a time-variable LUT. The transmitter can generate a dual-frequency or multicarrier CEM signal by cyclically employing different LUTs at different phase steps.

Consider an example of combining four bipolar signals, s_{UI}, s_{UQ}, s_{LI}, and s_{LQ}, located on two sub-frequencies, f_U and f_L, into a constant envelope composite signal, where there are two quadrature signals in each carrier, as shown in Fig. 7.6. Assume an implementation with sampling rate $4\,(f_U - f_L)$ so that the f_U carrier phase rotates at $\pi/2$ per phase step relative to the f_L carrier phase. Therefore, s_{UI} and s_{UQ} rotate forward with respect to s_{LI} and s_{LQ} by $\pi/2$ per phase step, and each phase relationship corresponds to a LUT, as shown in Fig. 7.7.

If the phase angles of s_{LI} and s_{LQ} are fixed, four LUTs, or a more compact time-variable LUT, are needed, and the central frequency of the composite signal is f_L. However, if in each phase step, the phases of s_{LI} and s_{LQ} rotate backwards with respect to the zero phase by $\pi/4$ while the phases of s_{UI} and s_{UQ} rotate frontwards with respect to the phase of zero by $\pi/4$, the central frequency of the composite signal is $(f_U + f_L)\,/2$, and eight LUTs are needed.

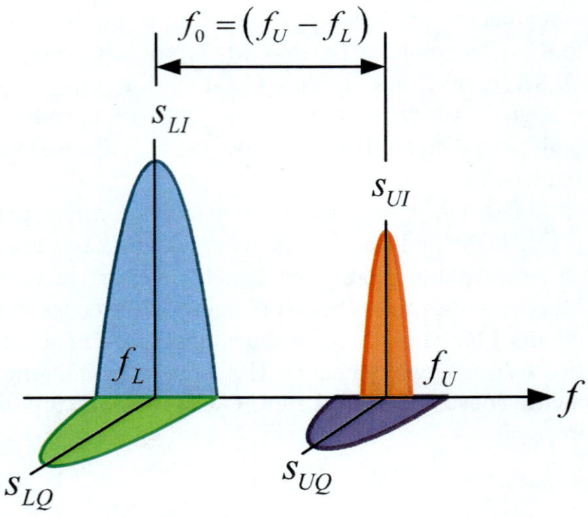

Fig. 7.6 Spectra sketches of four component signals in DCEM and their central frequencies and phase relationships

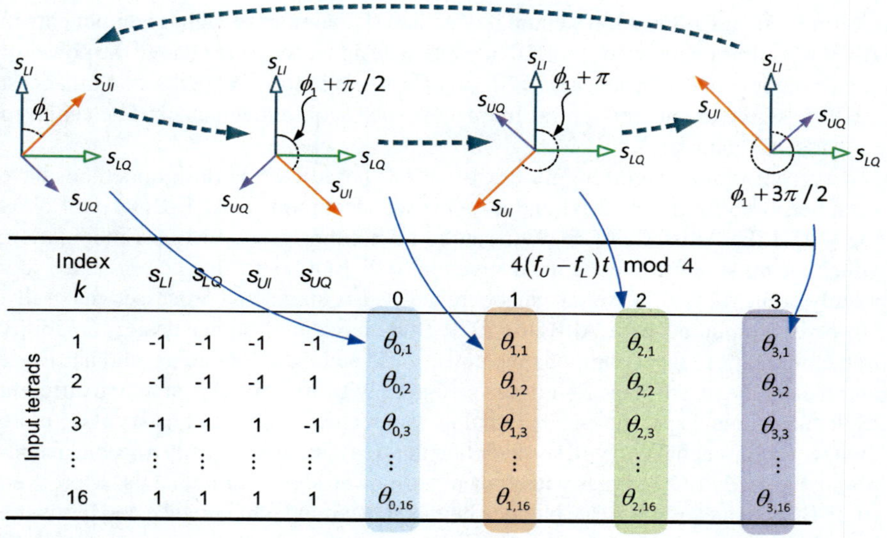

Fig. 7.7 Relative phase relationships between components under four-step sampling and the corresponding time-variable phase LUT

As pointed out in [34], in the above rotating POCET approach, an overlong phase step may cause some additional power loss. To reduce such loss, a higher sampling rate can be considered, which means finer step size, higher driving clock rate, and more sub-tables stored in the transmitter. Moreover, when the sampling rate becomes higher, the complexity of the optimization algorithm is significantly increased, and with some relative phase relationships, the convergence of the solution cannot be guaranteed.

7.10 ACE-BOC Modulation/Multiplexing Technique

The asymmetric constant envelope BOC (ACE-BOC) modulation/multiplexing technique can be regarded as a generalized AltBOC technique. In Sect. 4.9.1, we explained AltBOC as a form of spreading modulation. In fact, AltBOC technique can also be viewed as a DCEM technique.

The two-component AltBOC technique can combine two equal-power BPSK-R signal components located at two central frequencies into one QPSK composite signal, while the four-component AltBOC technique can combine four equal-power BPSK-R signal components located in pairs at two central frequencies into an 8-PSK composite signal. As a DCEM technique dedicated to the specific needs of the Galileo E5 band, AltBOC is good enough. Nevertheless, as a general DCEM technique, it still has flexibility limitations. Components involved in an AltBOC multiplexed signal must have equal power, and the number of signal components in AltBOC is limited to two or four. If the system designer wants to change the power of some components, or broadcast QPSK signals at one central frequency and broadcast a BPSK signal at another central frequency, such requirements cannot be achieved by AltBOC technique.

In recent years, several highly flexible WDP-based DCEM design methods have been proposed for the BDS B1 and B2. One such technique is ACE-BOC multiplexing [32]. Like AltBOC, ACE-BOC can be seen either as a CEM or a modulation, which means signal components involved in ACE-BOC can either be processed separately or the ACE-BOC signal can be treated as one signal and processed integrally. However, compared with AltBOC, ACE-BOC has much higher design flexibility in the number of signal components, power ratio among components, and hardware complexity. It can combine four or fewer signals with arbitrary power at two different frequencies into a spectrum-split signal in which each sideband consists of no more than two components. We will see later that the AltBOC signal is only a special example of the ACE-BOC signal with four equal-power signal components. Moreover, ACE-BOC achieves a flexible balance between receiving performance and hardware complexity. As will be shown, such high-level design flexibility allows the signal designer to provide more diversified signal schemes for different applications.

7.10.1 Direct Form of the ACE-BOC Signal

Without loss of generality, let us consider the constant envelope combination of four dual-frequency components. Following the signal representation of the Sect. 7.8, the center frequencies of the two bands are f_L and f_U respectively. Both $s_L(t)$ and $s_U(t)$ are QPSK signals in the form of (7.71), and each bipolar component can be further written as

$$s_\ell(t) = \sqrt{P_\ell} \sum_{n=-\infty}^{+\infty} c_n^{(\ell)} p_\ell \left(t - nT_c^{(\ell)}\right) d^{(\ell)}(t), \tag{7.72}$$

where $\ell \in \{LI, LQ, UI, UQ\}$, P_ℓ is the nominal power of the component $s_\ell(t)$, $d^{(\ell)}(t)$ is the navigation data or the secondary code modulated on $s_\ell(t)$. For a pilot channel without a modulated message, $d^{(\ell)}(t) \equiv 1$. In the above equation, $c_n^{(\ell)} \in \{\pm 1\}$ is the PRN code with the rate of $f_c^{(\ell)} = 1/T_c^{(i)}$. We set the chip waveform $p_i(t)$ to have an amplitude of 1 and a duration of $t \in \left[0, T_c^{(i)}\right)$, and the type is limited to the BCS waveform. The spreading rate of the four components is not required to be exactly equal, and the least common multiple rate of the spreading code rate defining the four components is $f_u = 1/T_u$. It should be noted that there is no limit to the value of P_ℓ in the above signal model. If we set one or several $P_\ell = 0$, it is equivalent to the fact that the corresponding signal component does not exist in the multiplexed signal. Therefore, the above model is valid for DCEM cases of two, three, and four bipolar components.

The baseband complex envelope of the ACE-BOC signal can be written as

$$s_{\mathrm{ACE}}(t) = \frac{\sqrt{2}}{2}\alpha_I(t)\gamma_I(t) + \mathrm{j}\frac{\sqrt{2}}{2}\alpha_Q(t)\gamma_Q(t), \tag{7.73}$$

where $\gamma_I(t) \triangleq \mathrm{sgn}\left[\sin\left(2\pi f_s t + \varphi_I\right)\right]$ and $\gamma_Q(t) \triangleq \mathrm{sgn}\left[\sin\left(2\pi f_s t + \varphi_Q\right)\right]$ are the waveforms of its real and imaginary parts, which are in the form of a square wave in one period. The amplitude and phase of the square wave are respectively

$$\begin{cases} \alpha_I = -\sqrt{(s_{UI} + s_{LI})^2 + \left(s_{UQ} - s_{LQ}\right)^2} \\ \alpha_Q = \sqrt{(s_{UI} - s_{LI})^2 + \left(s_{UQ} + s_{LQ}\right)^2} \end{cases} \tag{7.74}$$

and

$$\begin{cases} \varphi_I = -\operatorname{atan2}\left(s_{UI} + s_{LI}, s_{UQ} - s_{LQ}\right) \\ \varphi_Q = \operatorname{atan2}\left(s_{UQ} + s_{LQ}, s_{UI} - s_{LI}\right) \end{cases}, \tag{7.75}$$

both of which are functions of the four component values participating in the multiplexing, where $\operatorname{atan2}(\cdot,\cdot)$ is a four-quadrant arctangent function.

The constant envelope properties of the ACE-BOC signal are easily verified using (7.73). The amplitude of an ACE-BOC signal

$$A_{\mathrm{ACE}} = \left|s_{\mathrm{ACE}}(t)\right| = \sqrt{\sum\nolimits_{\ell} P_{\ell}} \tag{7.76}$$

is a constant that is only related to the sum of the nominal powers of the components. Since the power ratio of each component is not limited, it can be optimized flexibly in the ACE-BOC signal. In addition, it can be easily found that if we set some $P_i = 0$, the constant envelope characteristic of the ACE-BOC signal is not changed. It means that when several signal components are combined and transmitted using the ACE-BOC technique, any of the signal components can be temporarily turned off without affecting the constant envelope characteristics of the total signal. In the next section we will offer some typical ACE-BOC solutions with different component amount.

The ACE-BOC signal has a variety of equivalent implementations. The method of generating the ACE-BOC baseband complex envelope directly using (7.73) is called direct form implementation. When f_u is an integer multiple of f_s, the values of s_{UI}, s_{LI}, s_{UQ}, and s_{LQ} remain unchanged during a complete period of $\gamma_I(t)$ and $\gamma_Q(t)$. That is to say, in the time period $t \in [mT_u, (m+1)T_u)$, the real part and the imaginary part of the ACE-BOC complex envelope waveform are square waves of fixed amplitude and initial phase. However, in different time periods, since the combination of the four component values may change, the amplitudes and initial phases of $\gamma_I(t)$ and $\gamma_Q(t)$ might change accordingly. In [17], the features of ACE-BOC direct form of time domain waveform are analyzed in detail, and the relationship between amplitude and phase and the power ratio of each component is given.

When the power of each component participating in the multiplexing is given, the complex envelope waveform of ACE-BOC in $t \in [mT_u, (m+1)\,T_u)$ is completely determined by the sign of each component value. We define

$$\boldsymbol{v} = \left[\mathrm{sgn}\,(s_{UI})\,,\, \mathrm{sgn}\,(s_{LI})\,,\, \mathrm{sgn}\left(s_{UQ}\right),\, \mathrm{sgn}\left(s_{LQ}\right)\right]^{\mathrm{T}}\,. \tag{7.77}$$

Assuming in each $s_\ell\,(t)$, $+1$ and -1 are equally likely, $\boldsymbol{v}$ has a total of 16 equally probable states, denoted as $\boldsymbol{v}^{(k)}$, $k = 1, 2, \ldots, 16$, and then α_I, α_Q, φ_I, and φ_Q can all be uniquely determined by k. Thus, at any time, the ACE-BOC complex envelope waveform can be uniquely determined by state number k and time t. Therefore, the transmitter can pre-calculate a LUT which is mapped from $\boldsymbol{v}^{(k)}$ to $\varphi_i\,(k)$ and $\alpha_i\,(k)$ based on (7.74) and (7.75), where $i \in \{I, Q\}$. When $t \in [mT_u, (m+1)\,T_u)$, if the value combination of s_{UI}, s_{LI}, s_{UQ}, and s_{LQ} corresponds to the k_0th state $\boldsymbol{v}^{(k_0)}$, the corresponding waveform of the ACE-BOC complex envelope is

$$\begin{aligned} s_{\mathrm{ACE}}\,(t) = \frac{\sqrt{2}}{2}\, & \{\alpha_I\,(k_0)\,\mathrm{sgn}\,[\sin\,(2\pi f_s t + \varphi_I\,(k_0))] \\ & + \mathrm{j}\alpha_Q\,(k_0)\,\mathrm{sgn}\left[\sin\left(2\pi f_s t + \varphi_Q\,(k_0)\right)\right]\}\,. \end{aligned} \tag{7.78}$$

7.10.2 Phase Rotation Form of the ACE-BOC Signal

The principle of direct form implementation is simple, and easy to implement in software. However, when a digital logic circuit is used to implement a transmitter of an ACE-BOC signal, the implementation using the phase rotation form is simpler and more feasible.

We can write (7.73) as a representation of the amplitude and phase angle $s_{\mathrm{ACE}}\,(t) = A_{\mathrm{ACE}}\mathrm{e}^{\mathrm{j}\Theta(t)}$, where A_{ACE} is a constant, as given by (7.76), and the phase angle

$$\Theta\,(t) = \mathrm{atan2}\left(\alpha_Q\,(t)\,\gamma_Q\,(t)\,,\, \alpha_I\,(t)\,\gamma_I\,(t)\right) \tag{7.79}$$

can be uniquely determined by the state index k and the time t. In $t \in [mT_u, (m+1)\,T_u)$, the value combination of $\boldsymbol{v}^{(k)}$ remains unchanged, while $\Theta\,(t)$ will switch between up to four phases during this time period. The values of the four phases and the switching time points are all determined by the state index k and the power ratio of the components. Thus, if an LUT mapping $\boldsymbol{v}^{(k)}$ to $\Theta\,(t, k)$ is pre-stored in the transmitter, then at any time, after determining the value of $\Theta\,(t, k)$ according to k and t, we can construct an ACE-BOC complex envelope by

$$s_{\mathrm{ACE}}\,(t) = A_{\mathrm{ACE}}\,\{\cos\,(\Theta\,(t, k)) + \mathrm{j}\sin\,(\Theta\,(t, k))\}\,. \tag{7.80}$$

The biggest advantage of using the phase rotation form to generate an ACE-BOC signal is simplicity and flexibility, and it facilitates online reconstruction of the signal. In the previous chapters, we offered a detailed discussion of the phase map implementation structure of a general constant envelope multiplexed signal. Under this signal generator architecture, if we need to adjust the scheme of the CEM, for example, changing the power ratio and phase relationship between the signal components, we can simply regenerate an LUT and upload it to the transmitter, replacing the old table.

The high flexibility of ACE-BOC on power ratio and component numbers makes this technique useful for addressing many of the requirements in signal design. A general ACE-BOC signal can be represented by ACE-BOC(f_s, f_c, $\boldsymbol{p}$), where $\boldsymbol{p} = \left[P_{UI}, P_{LI}, P_{UQ}, P_{LQ}\right]$ indicates the power ratio of this ACE-BOC signal.

Example 7.4 (*Waveform and phase LUT for ACE-BOC* ([1, 1, 3, 3])) The high flexibility of the ACE-BOC allows for a certain difference in signal waveforms and phase LUTs at different power ratios. In order to provide a more intuitive understanding, in this example, we use a specific ACE-BOC signal with the power relationship $P_{UQ} = P_{LQ} = 3P_{UI} = 3P_{LI}$. We may as well set the nominal values of these four component powers as 3/8, 3/8, 1/8, and 1/8 respectively.

The polarities of the four components are changing as the modulated spreading code changes. If we have exactly $\boldsymbol{v}^{(1)} = [1, 1, 1, 1]^{\mathrm{T}}$ in a certain period of time, we can derive from (7.73) that the complex envelope waveform of the ACE-BOC signal during that interval is

$$s_{\mathrm{ACE}}\left(t; \boldsymbol{v}^{(1)}\right) = \left(\frac{1}{2} + \mathrm{j}\frac{\sqrt{3}}{2}\right) \mathrm{sgn}\left[\cos\left(2\pi f_s t\right)\right]. \tag{7.81}$$

In another time period, if $\boldsymbol{v}^{(2)} = [1, 1, 1, -1]^{\mathrm{T}}$, then the complex envelope waveform of the ACE-BOC signal becomes

$$s_{\mathrm{ACE}}\left(t; \boldsymbol{v}^{(2)}\right) = \mathrm{sgn}\left[\sin\left(2\pi f_s t - \frac{\pi}{6}\right)\right]. \tag{7.82}$$

Similarly, it is easy to enumerate ACE-BOC complex envelope waveform expressions for all 16 $\boldsymbol{v}^{(k)}$ values. If we draw these waveforms together, we can see that each of them is a square wave with a period of $1/f_s$. For $s_{\mathrm{ACE}}\left(t; \boldsymbol{v}^{(1)}\right)$, within a subcarrier period, the value of the signal oscillates between two phase points $\pi/3$ and $4\pi/3$ on the unit circle. For $s_{\mathrm{ACE}}\left(t; \boldsymbol{v}^{(2)}\right)$, the signal value oscillates between the phase 0 and π on the unit circle. If we record the constellation point and the switching time point for each $\boldsymbol{v}^{(k)}$, we can obtain the LUT mapping $\boldsymbol{v}^{(k)}$ to $\Theta\left(t, k\right)$. For this example, the LUT can be written in the form shown in Table 7.5. ■

Using the above analytical construction method, the results of this particular example can be easily extended to ACE-BOC signals with other power allocations.

Table 7.5 ACE-BOC phase LUT for the power allocation in Example 7.4

sgn (s_{UI})	1	1	1	1	1	1	1	1	−1	−1	−1	−1	−1	−1	−1	−1
sgn (s_{LI})	1	1	1	1	−1	−1	−1	−1	1	1	1	1	−1	−1	−1	−1
sgn (s_{UQ})	1	1	−1	−1	1	1	−1	−1	1	1	−1	−1	1	1	−1	−1
sgn (s_{LQ})	1	−1	1	−1	1	−1	1	−1	1	−1	1	−1	1	−1	1	−1
$\frac{t \bmod T_s}{T_s}$	Index x, corresponding to $s_{\mathrm{ACE}}\left(t; \boldsymbol{v}^{(k)}\right) = \exp\left(\mathrm{j}x\frac{\pi}{6}\right)$															
$\left[0, \frac{1}{12}\right)$	2	12	12	10	3	5	1	9	3	7	11	9	4	6	6	8
$\left[\frac{1}{12}, \frac{2}{12}\right)$	2	6	12	10	3	5	1	9	3	7	11	9	4	6	12	8
$\left[\frac{2}{12}, \frac{3}{12}\right)$	2	6	12	10	3	5	1	3	9	7	11	9	4	6	12	8
$\left[\frac{3}{12}, \frac{4}{12}\right)$	8	6	12	4	3	5	1	3	9	7	11	9	10	6	12	2
$\left[\frac{4}{12}, \frac{5}{12}\right)$	8	6	12	4	9	5	1	3	9	7	11	3	10	6	12	2
$\left[\frac{5}{12}, \frac{6}{12}\right)$	8	6	6	4	9	5	1	3	9	7	11	3	10	12	12	2
$\left[\frac{6}{12}, \frac{7}{12}\right)$	8	6	6	4	9	11	7	3	9	1	5	3	10	12	12	2
$\left[\frac{7}{12}, \frac{8}{12}\right)$	8	12	6	4	9	11	7	3	9	1	5	3	10	12	6	2
$\left[\frac{8}{12}, \frac{9}{12}\right)$	8	12	6	4	9	11	7	9	3	1	5	3	10	12	6	2
$\left[\frac{9}{12}, \frac{10}{12}\right)$	2	12	6	10	9	11	7	9	3	1	5	3	4	12	6	8
$\left[\frac{10}{12}, \frac{11}{12}\right)$	2	12	6	10	3	11	7	9	3	1	5	9	4	12	6	8
$\left[\frac{11}{12}, 1\right)$	2	12	12	10	3	11	7	9	3	1	5	9	4	6	6	8

7.10.3 ACE-BOC Solutions with Typical Power Allocations

The flexibility of ACE-BOC, especially in the number of involved components and their power allocations, makes it a promising solution for many special scenarios in the next-generation GNSS signal design. In this section, we provide some typical ACE-BOC solutions.

7.10.3.1 Symmetric Dual-Frequency Unbalanced QPSK

For some practical applications, it is desired that the in-phase component and quadrature-phase component in each sideband have different power levels, while the total power of upper sideband and lower sideband are the same, i.e.,

$$P_{UQ} = P_{LQ} = \beta^2 P_{UI} = \beta^2 P_{LI} \tag{7.83}$$

where β is a constant to control the power ratio of the I and Q components of each sideband I and Q. A schematic diagram of the spectrum of such signal is shown in Fig. 7.8a.

As discussed in the previous sections, the Galileo system and BDS will each broadcast a set of QPSK signals in the two adjacent bands of the center frequencies 1207.14 MHz (E5a/B2a) and 1176.45 MHz (E5b/B2b). For each side band, the I

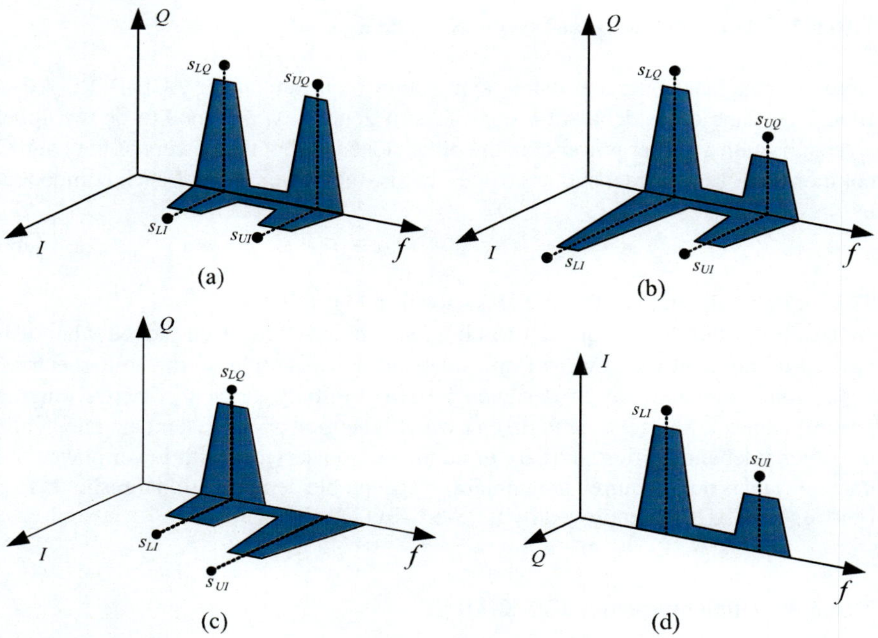

Fig. 7.8 Spectra sketches of four typical ACE-BOC power allocation schemes

component serves as the data channel, and the Q component serves as the pilot channel. The ACE-BOC signal scheme given above is an ideal solution to meet this need. With the B2 frequency signal design as the application background, the ACE-BOC signal scheme can be used to allocate more power to the pilot components of each sideband, thereby improving the signal acquisition and tracking sensitivity as well as the code pseudorange and carrier pseudorange measurement accuracy.

In Sect. 2.10.2, we offered a detailed discussion of the rationality and necessity of allocating more power to the pilot channel. For the B2 band, the signal service is mainly to cooperate with the B1 signal to achieve dual-frequency or tri-frequency high-precision applications. Since navigation data messages can be provided simultaneously by B1, B2a, and B2b signal components, the data transmission burden of B2a/b signals is relatively light, while the optimization of the measurement accuracy is particularly important. The assessment in [35] shows that if the power ratio of the pilot channel to the data channel is 3 : 1 on each sideband of B2, when the pilot component is used for acquisition tracking, both sensitivity and accuracy can be improved by 1.8 dB compared with equal power allocation.

7.10.3.2 Asymmetric Dual-Frequency QPSK

Here we consider another possible application requirement for ACE-BOC. ACE-BOC can make one sideband (without loss of generality, assumed to be the upper sideband) with a higher power than the other sideband, but the powers of the I and Q components of each sideband are equal, that is, the power ratio of each component is

$$P_{UQ} = P_{UI} = \beta^2 P_{LQ} = \beta^2 P_{LI} \,. \tag{7.84}$$

The spectrum diagram of the signal is shown in Fig. 7.8b.

This proposed dual-frequency combination scheme can be employed when it is desired to augment the power of one sideband. For example, in the split-spectrum signal, one sideband can be designed to carry primary service of better ranging performance and shorter time to first fix with higher power and faster data rate, while the other sideband carries auxiliary or augmentation services with lower power and data rate and is only required to maintain an acceptable demodulation threshold. This type of ACE-BOC is employed by BDS-3 MEO B2 signal.

7.10.3.3 Dual-Frequency BPSK+QPSK

It is not always necessary to provide four different services within a frequency band. In some cases, a three-signal CEM is a preferable option. For some requirements, e.g., a smooth transition issue in the update process of an operating GNSS, one might want to leave the legacy BPSK component in one sideband unchanged while placing a new QPSK component, consisting of data channel and pilot channel, on the other sideband. The advantage of three components lies in the fact that transmitting two pilot channels on different frequencies is beneficial to providing two sets of carrier phase measurements. However, there is no evident enhancement by transmitting two data channels for demodulation. Therefore, the data channel on one sideband can be cancelled, the power can be allocated to more useful components, and a dual-frequency BPSK+QPSK composite signal can be constructed using ACE-BOC technique. The spectrum sketch of this signal is shown in Fig. 7.8c.

The signal schemes that can be achieved by ACE-BOC technique go far beyond the three given above. If we make $P_{UQ} = P_{LQ} = 0$ and $P_{UI} = \beta^2 P_{LI}$, we can simplify the ACE-BOC signal to an asymmetric dual-frequency BPSK signal (see Fig. 7.8d). In a more extreme case, if $P_{UQ} = P_{UI} = 0$, we can construct a one-sided QPSK signal. The specific implementations are not enumerated here. Interested readers can refer to [17, 32], where the signal waveforms, constellation structure, and LUT of the above typical schemes are analyzed in detail.

Example 7.5 (*Four-component AltBOC signal*) If we make $P_{UQ} = P_{UI} = P_{LQ} = P_{LI}$, we can see from (7.73) that when $\boldsymbol{\nu}^{(1)} = [1, 1, 1, 1]^{\mathrm{T}}$, the complex envelope waveform of the signal is

Table 7.6 Phase look-up table for the four-component AltBOC signal

sgn (s_{UI})	1	1	1	1	1	1	1	1	−1	−1	−1	−1	−1	−1	−1	−1
sgn (s_{LI})	1	1	1	1	−1	−1	−1	−1	1	1	1	1	−1	−1	−1	−1
sgn (s_{UQ})	1	1	−1	−1	1	1	−1	−1	1	1	−1	−1	1	1	−1	−1
sgn (s_{LQ})	1	−1	1	−1	1	−1	1	−1	1	−1	1	−1	1	−1	1	−1
$\frac{t \bmod T_s}{T_s}$	Index x correspongding to $s_{\mathrm{ACE}}\left(t;\boldsymbol{v}^{(k)}\right)=\exp\left(\mathrm{j}x\frac{\pi}{4}\right)$															
$\left[0,\frac{1}{8}\right)$	5	4	4	3	6	3	1	2	6	5	7	2	7	8	8	1
$\left[\frac{1}{8},\frac{2}{8}\right)$	5	4	8	3	2	3	1	2	6	5	7	6	7	4	8	1
$\left[\frac{2}{8},\frac{3}{8}\right)$	1	4	8	7	2	3	1	2	6	5	7	6	3	4	8	5
$\left[\frac{3}{8},\frac{4}{8}\right)$	1	8	8	7	2	3	1	6	2	5	7	6	3	4	4	5
$\left[\frac{4}{8},\frac{5}{8}\right)$	1	8	8	7	2	7	5	6	2	1	3	6	3	4	4	5
$\left[\frac{5}{8},\frac{6}{8}\right)$	1	8	4	7	6	7	5	6	2	1	3	2	3	8	4	5
$\left[\frac{6}{8},\frac{7}{8}\right)$	5	8	4	3	6	7	5	6	2	1	3	2	7	8	4	1
$\left[\frac{7}{8},1\right)$	5	4	4	3	6	7	5	2	6	1	3	2	7	8	8	1

$$s\left(t;\boldsymbol{v}^{(1)}\right)=\mathrm{e}^{\mathrm{j}\frac{\pi}{4}}\,\mathrm{sgn}\left[\cos\left(2\pi f_s t\right)\right]. \tag{7.85}$$

When $\boldsymbol{v}^{(2)}=[1,1,1,-1]^{\mathrm{T}}$, the complex envelope waveform of the signal is

$$s\left(t;\boldsymbol{v}^{(2)}\right)=\mathrm{sgn}\left[\sin\left(2\pi f_s t-\frac{\pi}{4}\right)\right]. \tag{7.86}$$

Similarly, it is easy to write ACE-BOC complex envelope waveform expressions of the other 14 values of $\boldsymbol{v}^{(k)}$. If we record the constellation points and switching time points under each $\boldsymbol{v}^{(k)}$, we can obtain the LUT in the form of Table 7.6. This is exactly the four-component AltBOC signal used by the Galileo E5 signal. ■

As can be seen from the above discussion, the ACE-BOC signal multiplexing technique has a high degree of design flexibility, and either the four-component AltBOC or the two-component AltBOC can be regarded as a special case of ACE-BOC. Moreover, the ACE-BOC signal can be realized by using a phase LUT, which creates conditions for online reconstruction of the on-board signal. The power ratio of each component of the ACE-BOC signal can be adjusted online by modifying the content of the LUT, thereby enhancing a certain component or temporarily turning off a component, without affecting the constant envelope characteristic of the entire signal. For a more in-depth analysis of the principles and performance of the ACE-BOC technology, see [32].

On the basis of ACE-BOC, a series of low-complexity DCEM techniques have emerged, such as equal spacing ACE-BOC (ES-ACEBOC) [20, 32] and binary subcarrier ACE-BOC (BS-ACEBOC) [36]. These techniques further reduce the implementation complexity of the transmitter while maintaining the flexibility and transparency of the original ACE-BOC.

7.11 CEM for Multilevel Signals

Most of the existing CEM techniques can only apply to bipolar signals. However, some research has shown that multilevel spreading chip waveforms, which have more degrees of freedom on design compared with bipolar waveforms, can bring better radio frequency compatibility (RFC) [37] and higher potential ranging accuracy [38], and provide the possibility of higher transmission rate per single signal component. The CBOC signal is a typical multilevel signal, the chip waveform of which is the superposition of two square waves of different frequencies. The multilevel amplitude characteristic of this signal results in the unsuitability of earlier CEM techniques. As we discussed in Sect. 7.5.3, in the Galileo E1 signal design, in order to combine two CBOC signals and another bipolar signal into a constant envelope signal, designers must limit these two CBOC signals to be of equal power and both in phase. Their high frequency square wave components have to have opposite polarities to make their superposition take on a pseudo-random time multiplexing form so that a modified QPSM technique (See Sect. 7.5.3) can be applied.

However, this case cannot be treated as a general multilevel signals' multiplexing problem. Later studies extend the solutions to more general cases based on both PDP and WDP design patterns. For CEM of multilevel signals, the largest difference from the bipolar case is the increase of the possible values of component signals. Without loss of generality, consider the multiplexing of N M-level signals. The number of value combinations will increase from 2^N to M^N.

Following the design philosophy of PDP, we can extend the number of rows in the phase LUT to M^N, and look for the optimal phase angle for each value combination via numerical optimization [15]. Design can also start with WDP. Although the complete orthogonal basis of $\mathcal{W}^{\perp}$ can no longer be constructed by IM terms of component signals for multilevel signals, a more general orthonormal basis expansion can be used to obtain bases [9].

Nevertheless, whether based on PDP or WDP, it is clear that the effect of the increase in amplitude level amount of component signals on the vector combination amount is very significant, which increases the difficulty of optimization. For example, multiplexing eight four-level signals increases the number of value combinations to $4^8 = 65,536$, corresponding to the number of angles that need to be optimized in PDP and the dimensions of $\boldsymbol{C}$ in WDP.

Zhang et al. [15] pointed out that in most practical cases, due to the regularity of the chip waveform that signals used, there is a strong correlation between the value combinations of different time slots, and some combinations do not even appear. Making full use of such regular patterns can effectively reduce the number of value combinations. Nevertheless, such reduction is realized case by case, which requires sufficient analysis of signal characteristics. In summary, there is still a lot of work to be done to find a more effective general CEM design method for multilevel signals.

7.12 Cascading of Constant Envelope Multiplexing Techniques

In the previous sections, we have discussed several typical constant envelope multiplexing techniques. In this section we will see that some of these techniques can be combined with each other to construct a new constant envelope multiplexing scheme.

Note that multiplexing techniques based on MV logic and TD multiplexing can multiplex multiple bipolar signals into one bipolar signal. Therefore, it is possible to combine several signal components into one bipolar signal by MV multiplexing or TD multiplexing, and then use other multiplexing techniques to perform secondary combining with other bipolar signal components.

7.12.1 InterVote

The InterVote technique [14] is essentially a cascade of the QPSM technique and MV multiplexing. As mentioned in Sect. 7.5.2, the QPSM technique can achieve CEM with any power ratio in any number of bipolar signals. However, the multiplexing efficiency decreases rapidly at an exponential rate as the number of signals participating in the multiplexing increases. Therefore, it is most common to use QPSM multiplexing to combine three signals.

On the other hand, when the signal components participating the MV multiplexing have equal power, the decrease in the multiplexing efficiency is not very fast as the number of signals increases, and there is a upper bound of 1.96 dB (see Sect. 7.6.2). However, the relative phase relationship selection of signal components in MV multiplexing is very limited. Moreover, if the power difference between the signal components is large, the efficiency of the interlaced MV multiplexing is significantly deteriorated.

Therefore, for some specific multiplexing constraints, cascading QPSM technique with MV multiplexing may result in higher multiplexing efficiency. For example, if five signals c_1, c_2, c_3, c_4, and c_5 participate in multiplexing, we can first merge c_1, c_2, and c_3 into a new bipolar signal c' by using interlaced MV multiplexing, then use the QPSM technique to multiplex c', c_4, and c_5 into a PSK signal. The key to using the InterVote technique is to determine which signals will be MV merged and the combined weight calculations. These issues are discussed in detail in [14].

7.12.2 POCET-Vote

The POCET-Vote technique is essentially a cascade of POCET and MV multiplexing. The POCET technique uses numerical optimization to map 2^N different combinations of a total of N signal components to 2^N different phase angles of the PSK signal. Since the constraint condition and the objective function are both nonlinear functions and complicated, when the number of signals participating in multiplexing is large, the search space expands rapidly, and a convergent solution may not be found. Similar to the InterVote, the POCET technique and MV multiplexing can also be integrated to achieve better results for some given multiplexing constraints. First using interlaced MV multiplexing to combine some signal components and then using the POCET technique to multiplex the interlaced MV multiplexed signal with other signal components can greatly reduce the optimization complexity of POCET and simplify the final structure of the constant envelope composite signal.

Similarly, other cascading methods based on MV multiplexing include integrating MV multiplexing into ACE-BOC, integrating MV multiplexing into MSPM, and even combining MV with rotating POCET. There are many such possible combinations, which will not be listed here.

7.12.3 TD-AltBOC

Similar to MV multiplexing, TD multiplexing can also combine multiple bipolar signals into one bipolar signal. When multiple signals need to be combined, some of the signals may be time-division multiplexed into a new signal component, and then subjected to secondary CEM with the remaining signal components using other multiplexing techniques.

As we discussed in the MV-multiplexing-based CEM cascade, TD multiplexing can also be combined with QPSK, QPSM, POCET, ACE-BOC, and other means of multiplexing. TD multiplexing can even be combined with MV multiplexing, and the resulting bipolar composite signal is further cascaded with other multiplexing techniques. Our discussion does not exhaust every cascading approach. In this section, we use TD-AltBOC as an example to analyze a possible way to combine TD multiplexing with other multiplexing techniques.

As its name implies, TD-AltBOC is a cascade of TD multiplexing and AltBOC multiplexing. From the previous explanation, we know that AltBOC is a special case of ACE-BOC when each signal component has equal power. Two AltBOC techniques, two-component AltBOC, and a four-component AltBOC are presented in [31], which correspond to the power ratios of ACE-BOC$\left(f_s, f_c, \left[\frac{1}{2}, \frac{1}{2}, 0, 0\right]\right)$ and ACE-BOC$\left(f_s, f_c, \left[\frac{1}{4}, \frac{1}{4}, \frac{1}{4}, \frac{1}{4}\right]\right)$ respectively. Among them, the two-component AltBOC can carry two bipolar components, denoted as $s_U(t)$ and $s_L(t)$ respectively, in the upper sideband and the lower sideband of the composite signal, and the expression of the whole signal can be written as

$$s_{\text{Alt}-2}(t) = s_L(t)\,\gamma_{sc}^{*}(t) + s_U(t)\,\gamma_{sc}(t)\,, \tag{7.87}$$

where $\gamma_{sc}(t)$ has the form of (7.70).

In order to use the two-component AltBOC technique to carry four bipolar signal components, Tang et al. [39] propose to cascade TD multiplexing to AltBOC multiplexing. TD-AltBOC first inserts two signals on each sideband by chip interlacing to form a bipolar signal. The interlaced signals in upper sideband and lower sideband are then combined using two-component AltBOC to further construct a constant envelope composite signal. A major benefit of this method is that the complexity of the transmitter is lower than that of the four-component AltBOC. The drive clock for generating the four-component AltBOC signal should be at least eight times the subcarrier frequency. However, the drive clock rate for generating the two-component AltBOC signal can be half that of the four-component case.

While reducing the implementation complexity of the transmitter, the inherent defects of TD multiplexing pose some potential problems for this cascaded multiplexing technique for receiving. As detailed in Sect. 7.3, first, with the same time period of the spread-spectrum signal, TD-AltBOC multiplexing significantly shortens the length of the PRN sequence used, while deteriorating the correlation performance of the spreading code [24, 25]. Second, TD-AltBOC will amplify the influence of the non-ideal correlation on the inherent deviation of the code tracking error [24, 27, 28]. Moreover, TD-AltBOC does not satisfy the transparency constraint of multiplexing (See Sect. 6.5). It reduces the forward and backward compatibility of the signal [26].

7.13 Remaining Challenges in CEM

Research on CEM for GNSS is still at an early stage and extensive work is required to improve its applicability and theoretical understanding. This section presents possible future directions to promote further research.

1. Existence of the CEM solutions

The necessary and sufficient conditions for the existence of non-trivial solutions of the problem (6.49) or (6.58) are still lacking in-depth theoretical study. Dafesh et al. [34] found that, although for some specific power and phase constraints, (6.58) has no solution, introducing a so-called "phantom" signal without power and phase constraints into the original CEM problem can often render it solvable. Zhang et al. [40] and Guo et al. [10] discussed the relaxation of the composite signal phase angle symmetry that, in some cases, can not only make an unsolvable CEM equation become solvable, but can also produce solutions with higher power efficiency. However, as of now, discussions on the existence of CEM solutions are all based on experience rather than rigorous theoretical analysis. Further investigation in this area may lead to a more general and rigorously based multiplexing theory which could, therefore, provide more flexibility and applicability for CEM designs.

2. Joint optimization of signature waveforms and multiplexing

Most of the existing CEM design approaches are based on the assumption that the spreading sequences of different component signals are ideally orthogonal. However, in practice, the non-ideal orthogonality of sequences may introduce a slight distortion on the correlation function after multiplexing and can, in some cases, result in an inherent bias in code tracking. A promising future research direction is the joint design of spreading sequences, spreading chip waveforms, and CEM. A preliminary attempt was made in [41], where the symmetry of the code sequence is added as a constraint to the PDP optimization equation, significantly reducing the inherent code tracking bias.

3. Diversified receiving process strategies of multiplexed signals

Multiplexing techniques have long been working behind the scene and are invisible to most receivers due to transparency requirements. However, emerging studies point to the benefit of utilizing prior knowledge of the CEM composite signal structure where not only the relative power ratio and the phase relationship between each component signal are strictly deterministic, but the errors introduced by thermal noise, multipath, as well as the dynamic stress also have high coherence. Comparing the separate processing modes, joint acquisition and tracking of the multiplexed composite signal with the structural feature of the CEM can make more effective use of the transmitter power and improve the performance [42]. Multicarrier signals such as Galileo E5 and BDS B2 signals provide even more favorable conditions for the diversification of receiving strategies [33]. We believe that more algorithms will emerge in this research direction.

4. CEM with more general modulations

Note that all of the CEMs used in GNSS we have discussed so far are for memoryless DSSS signals with multilevel chip waveform, in which instantaneous waveform value transition is assumed. This means that the components to be multiplexed retain sufficient side lobes outside of its main spectrum lobe. Since the multiplexing operation is done in the digital baseband part via digital logic, and the envelope constancy feature only needs to be maintained until the high power amplifier (HPA) input, as long as the bandwidth of the up-converter before HPA is wide enough, the instantaneous transition assumption is approximately tenable. Therefore, the envelope distortions with the chip transition have not been considered as an issue for a long time. However, due to increasingly sparse spectrum resources, more and more complex chip waveforms with high spectral efficiency, such as the PSWF waveform discussed in Sect. 4.9.5, are proposed to be used in future satellite navigation signals, especially for C band [43]. Some advanced modulations, such as continuous phase-modulation (CPM) [44], are also suggested as potential solutions instead of DSSS. For such signals whose waveform has continuous transition and whose modulation is with memory, existing CEM techniques cannot be employed directly. Developing CEM techniques which can apply to signals with a more general spreading waveform or even non-DSSS signals is worth investigating. Some preliminary research results are emerging. For example, in [44], two components were combined into a CPM signal.

7.14 Summary

Signal multiplexing is a promising area that is not widely treated in most GNSS ICDs but is the basis for a variety of position, navigation, and timing (PNT) services for today and the future. In the previous chapter and this chapter, we provide readers with a global overview of CEM techniques highlighting similarities and differences of the various methods. Throughout the development process in the past 15 years, great efforts have been made to enhance the adaptability and flexibility of GNSS multiplexing techniques to enable more diversified signal broadcasting requirements. A more general multiplexing technique can provide system designers more choices of the number of sub-bands, the number of signals, the spreading modulation and power allocation. With signal elements in future GNSSs becoming more diverse and complex, multiplexing techniques will face new challenges. We believe that a uniform CEM representation framework and a more general and rigorously based multiplexing design theory can offer a good starting point to meet these challenges. In this research direction, we still have a lot to do.

References

1. Dafesh PA (1999) Quadrature product subcarrier modulation (QPSM). IEEE Aerosp Conf Proc 5:175–182
2. Dafesh PA, Nguyen TM (2006) Quadrature product subcarrier modulation system. US Patent 7,120,198
3. Dafesh PA, Nguyen TM, Lazar S (1999) Coherent adaptive subcarrier modulation (CASM) for GPS modernization. In: Proceedings of the 1999 national technical meeting of the institute of navigation, January 25–27, pp 649–660
4. Dafesh PA (2002) Coherent adaptive subcarrier modulation method. US Patent 6,430,213
5. Butman S, Timor U (1972) Interplex: an efficient multichannel PSK/PM telemetry system. IEEE Trans Commun Technol COM-20(3):415–419
6. Cangiani GL (2005) Methods and apparatus for multi-beam, multi-signal transmission for active phased array antenna. US Patent 6,856,284
7. James J, Spilker J, Orr RS (1998) Code multiplexing via majority logic for GPS modernization, pp 265–273
8. Orr RS, Veytsman B (2002) Methods and apparatus for multiplexing signal codes via weighted majority logic, US Patent WO/2002/069516
9. Yao Z, Guo F, Ma J, Lu M (2017) Orthogonality-based generalized multicarrier constant envelope multiplexing for DSSS signals. IEEE Trans Aerosp Electron Syst 53(4):1685–1698
10. Guo F, Yao Z, Lu M, Zhou H (2016) Generalized transparent constant envelope multiplexing method. In: Presented at the the 2016 international technical meeting of the institute of navigation, Monterey, California, January 25–28
11. Dafesh PA, Cahn CR (2009) Phase-optimized constant-envelope transmission (POCET) modulation method for GNSS signals, pp 2860–2866
12. Cahn CR, Dafesh PA (2014) Phase-optimized constant envelope transmission (POCET) method, apparatus and system. US Patent
13. Dafesh PA, Cahn CR (2011) Application of POCET method to combine GNSS signals at different carrier frequencies. In: Proceedings of the 2011 international technical meeting of the institute of navigation, San Diego, CA, January 2011, pp 1201–1206

14. Cangiani GL, Orr RS, Nguyen CQ (2001) Methods and apparatus for generating a constant-envelope composite transmission signal. US Patent 09/963669
15. Zhang K, Zhou H, Wang F (2013) Unbalanced AltBOC: a Compass B1 candidate with generalized MPOCET technique. GPS Solut 17(2):153–164
16. Zhu L, Yin H, Yao Z, Lu M (2011) Non-Symmetrical ALTBOC multiplexing for compass B1 signal design. In: Fifth European workshop on GNSS signals and signal processing
17. Yao Z, Lu M (2013) Constant envelope multiplexing technique for components on different carrier frequencies with unequal power allocation. In: Proceedings of the institute of navigation international technical meeting ITM-2013
18. Yao Z, Lu M (2012) Dual-frequency constant envelope multiplex with non-equal power allocation for GNSS. Electron Lett 48(25):1624–1625
19. Yao Z, Lu M (2013) Design, implementation, and performance analysis of ACE-BOC modulation. In: Proceedings of the 26th international technical meeting of the satellite division of the institute of navigation (ION GNSS+ 2013), Nashville, TN, September 2013, pp 361–368
20. Zhang J, Yao Z, Lu M (2014) Applications and low-complexity implementation of ACE-BOC multiplexing. In: Proceedings of the 2014 international technical meeting of the institute of navigation, San Diego, California, January 2014, pp 781–791
21. Wang GH, Lin VS, Fan T, Maine KP, Dafesh PA, Myers B (2004) Study of signal combining methodologies for GPS III's flexible navigation payload. In: Proceedings of the 17th international technical meeting of the satellite division of the institute of navigation (ION GNSS 2004), Long Beach, CA, September 2004, pp 2207–2218
22. Fontana RD, Cheung W, Novak PM, Stansell TA (2001) The new L2 civil signal. In: Proceedings of the 14th international technical meeting of the satellite division of the institute of navigation (ION GPS 2001), Salt Lake City, UT, September 2001, pp 617–631
23. Barker BC, Betz JW, Clark JE, Correia JT, Gillis JT, Steven L, Rehborn KA, Straton JR (2000) Overview of the GPS M code signal. In: Proceedings of the 2000 national technical meeting of the institute of navigation, Anaheim, CA, January 2000, pp 542–549
24. Zhu Y, Yao Z, Cui X, Lu M (2014) Comparative analysis of dual-frequency constant envelop multiplexing techniques for ARNS band. In: Proceedings of the 2014 international technical meeting of the institute of navigation, San Diego, California, January 2014, pp 792–801
25. Betz J, Cahn CR, Dafesh PA et al (2006) L1C signal design options. In: National technical meeting of the institute of navigation. Monterey, CA, pp 685–697
26. Kharisov V, Povalyaev A (2011) Optimal aligning of GNSS navigation signals sum. In: Proceedings of the 24th international technical meeting of the satellite division of the institute of navigation (ION GNSS 2011), Portland, OR, September 2011, pp 3141–3155
27. Liu Y, Tang X, Ge R, Wang F (2013) Analysis for cross correlation in multiplexing. In: China satellite navigation conference (CSNC) 2013 proceedings, pp 81–90
28. Yan T, Wei J, Tang Z, Zhou Z (2013) Performance analysis on single sideband of TD-AltBOC modulation signal. In: China satellite navigation conference (CSNC) 2013 proceedings, pp 91–100
29. Paris RB, Kaminsky D (2001) Asymptotics and the Mellin–Barnes integrals. Cambridge University Press, Cambridge
30. Rebeyrol E, Julien O, Macabiau C et al (2007) Galileo civil signal modulations. GPS Solut 11(3):159–171
31. Lestarquit L, Artaud G, Issler J-L (2008) AltBOC for dummies or everything you always wanted to know about AltBOC, pp 961–970
32. Yao Z, Zhang J, Lu M (2016) ACE-BOC: dual-frequency constant envelope multiplexing for satellite navigation. IEEE Trans Aerosp Electron Syst 52(1):466–485
33. Zhang J, Yao Z, Lu M (2016) Generalized theory and decoupled evaluation criteria for unmatched despreading of modernized GNSS signals. Sensors 16(6):1128
34. Dafesh PA, Cahn CR (2011) Application of POCET method to combine GNSS signals at different carrier frequencies, t in ION ITM 2011. San Diego, CA, USA
35. Yao Z, Lu M (2013) Multiplexing ACED, application its, on BeiDou B2 band. In: Proceedings China satellite navigation conference (CSNC). Lecture notes in electrical engineering 244(2013):25–34

36. Guo F, Yao Z, Lu M (2016) BS-ACEBOC: a generalized low-complexity dual-frequency constant-envelope multiplexing modulation for GNSS. GPS Solut 21:1–15
37. Pratt AR, Owen JI (2003) BOC modulation waveforms. In: Proceedings of the 16th international technical meeting of the satellite division of the institute of navigation, ION-GPS/GNSS-2003, pp 1044–1057
38. Zhang X, Yao Z, Zhang X, Lu M (2011) A method to optimize the spreading code chip waveform in sense of gabor bandwidth. In: Proceedings of the 24th international technical meeting of the satellite division of the institute of navigation (ION GNSS 2011), pp 1299–1304
39. Tang Z, Zhou H, Wei J et al (2011) TD-AltBOC: a new COMPASS B2 modulation. Sci China Phys Mech Astron 54(6): 1014–1021
40. Zhang K, Li Y, Zhou H, Wang F (2012) Analytical transmission model of POCET technique for compass B1 and B3 signals. In: Proceedings of the 25th international technical meeting of the satellite division of the institute of navigation (ION GNSS 2012), Nashville, TN, September 2012, pp 277–285
41. Zhang J, Yao Z, Lu M (2016) On the inherent tracking error caused by cem and imperfect spreading code of GNSS signal. In: Presented at the ION GNSS+ 2016, Portland, Oregon, September 25–29
42. Zhou H, Yao Z, Lu M (2016) A quasi-coherent receiving algorithm and its performance verification for composite signals of next generation GNSS. In: Presented at the the 2016 international technical meeting of the institute of navigation, Monterey, California, January 25–28
43. Avila-Rodriguez J-A, Wallner S, Hein GW, Eissfeller B, Irsigler M, Issler J-L (2007) A vision on new frequencies, signals and concepts for future GNSS systems. In: Proceedings of the 20th international technical meeting of the satellite division of the institute of navigation. Fort Worth TX, pp 517–534
44. Emmanuele A, Zanier F, Boccolini G, Luise M (2012) Spread-spectrum continuous-phase-modulated signals for satellite navigation. IEEE Trans Aerosp Electron Syst 48(4):3234–3249

Chapter 8
Multicarrier Constant Envelope Composite Signal

The secret of breaking rules in a way that works is understanding what the rules are in the first place.

— Rick Wakeman

8.1 Introduction

As a significant infrastructure, GNSS has a long development cycle. It may take several years from the time of initial signal design to the full operation of the system. This fact forces signal designers to have sufficient foresight, using existing technology, to enable unknown service requirements over future decades of operation. Therefore, although most of the signals in current GNSSs have been defined, the evolution of GNSS signal design will not stop there. The full operation of the current GNSS is the beginning of the design work of the next generation GNSS.

From the reality of GNSS design one can find that expanded applications and refined services prompt the new generation systems to broadcast more signals with more complicated structure, which on the one hand, makes more efficient use of limited spectrum resources, already crowded with GNSS signals, and on the other hand makes the spectrum crowding situation even worse. Moreover, such a complicated signal structure may make the realization of signal multiplexing more difficult for satellite payloads. Additionally, the limitation of receiving complexity, the requirement for backward compatibility, as well as the demand of interoperability among systems, add many constraints into the GNSS signal design optimization problem. The conflicts between application expansions and resource scarcity cause the signal design of the next generation GNSS to face a series of challenges. At present, it is necessary to take a hard look at the technical challenges in future GNSS signal design and look for possible solutions in advance.

In order to get out of the cycle of conflicts in future GNSS signal design, the concept of using multicarrier modulation and multiplexing joint design is introduced in this chapter. In the proposed multicarrier constant-envelope composite (MCC)

Z. Yao and M. Lu, *Next-Generation GNSS Signal Design*, Navigation: Science and Technology 6, https://doi.org/10.1007/978-981-15-5799-6_8

signal, several types of signal components located at multiple different central frequencies are treated as a whole, and their modulation and multiplexing are designed and optimized jointly. The multiplexing is transparent for receivers, so that it enables multiple receiving strategies for receivers. Performance analysis with some typical case studies of this technique demonstrates that the MCC signal can not only achieve outstanding ranging accuracy without significantly increasing RF interference to the existing signals in the same band, but also provide users with diversified and targeted service without significantly deteriorating the onboard multiplexing efficiency. Therefore, the MCC signal provides a promising solution for the next generation GNSS signal design.

8.2 Challenges in Future GNSS Signal Design

Compared with wireless communication signal design, the major feature of navigation signal design is the pursuit of high accuracy ranging capability. If we compare the wireless communication signal, of which the most concerned targets are the capacity and reliability of data transmission, to a paper envelope, then the satellite navigation signal can be compared to a ruler, since its main concern is the accuracy and robustness of the ranging measurement.

In the design process of this "ruler", the selection of the carrier frequency, the optimization of the spreading modulation, and the multiplexing of signal components are the top three critical and challenging parts.

8.2.1 Carrier Frequency Selection

As detailed in Sect. 2.4, the carrier frequency, as is the material of a ruler, determines many attributes of a navigation signal, including the propagation characteristics, the cost of transmitting and receiving hardware, signal Doppler shift, and possible interference with other radio systems. Among all available spectrum resources, the L-band has many advantages for satellite navigation applications, such as good propagation characteristics, moderate antenna size, and relatively small atmosphere influence, and therefore becomes the preferred frequency band for satellite navigation signals.

From Fig. 2.4 we have seen that the vast majority of GNSS signals are gathered in the upper L band (1559 ∼1610 MHz), and now it is hard to find any unoccupied contiguous segment in this frequency band. There are only a few scattered frequency fragments available between main lobes of existing signals. Interference arises among different signals. Although it is possible to reduce the spectral overlap of the signal located at the same central frequency to a certain extent by using different subcarriers or adjusting the spreading chip waveforms, it is still becoming increasingly difficult to find a suitable central frequency for a newly added navigation signal in L-band.

8.2.2 Spreading Modulation Design

Spreading modulation is to a satellite navigation what scale is to a ruler. From the discussion in Chap. 4 we can see that in the new generation GNSS, there are two new trends emerging in spread modulation design. Firstly, signal power distribution is changing from concentrating near the carrier frequency to a splitting spectrum form, which is represented by BOC modulation. Research indicates that splitting spectrum characteristics result in a spectrum separation from legacy signals located at the same central frequency and a wide RMS bandwidth which results in the potential advantage of improved ranging accuracy and inherent multipath resisting ability. Secondly, in addition to bipolar waveforms, more and more multi-level spread modulation waveforms are emerging, such as those in CBOC and AltBOC modulations. Relaxing the constraint of waveform levels can provide greater freedom for spread modulation waveform optimization, thus providing more possibilities for improvement of the signal performance [1, 2].

However, these two trends bring increased complexity to both the transmitter and the receiver. The larger RMS bandwidth of the splitting spectrum signal requires a higher subcarrier frequency, which means a wider receiver front-end bandwidth and the increasing of complexity of the receiver. Although the sophisticated receiving strategy is acceptable for high-end applications such as surveying and mapping, the cost and complexity is hard to justify for low-end consumer electronics devices. A direct way to address this problem is by transmitting multiple signals from the satellite, providing high-end users with a wideband signal that uses a complex chip waveform, while allowing the low-end users to have a simple receiving strategy for a narrowband signal with a simple chip waveform. Unfortunately, the increment in the number of signals at the same frequency increases the MAI between the system and the inter-system signals, which deteriorates the receiving performance. Furthermore, the increase in the number of signals and the complexity of the signal waveform pose a challenge to constant envelope multiplexing, which is another critical part of satellite navigation signal design.

8.2.3 Multiplexing

As discussed in Chaps. 6 and 7, in the CEM on the navigation satellites, as the number of signals increases, more IM terms should be added to keep the envelope of the multiplexed signal constant. Since the information carried on IM terms is redundant for a receiver, a higher proportion of IM terms means less useful power output, which is expressed as a lower multiplexing efficiency, thus reducing the received CNR. Though increasing the transmitting power can compensate for the multiplexing loss, it will further deteriorate MAI between the system and the inter-system signals.

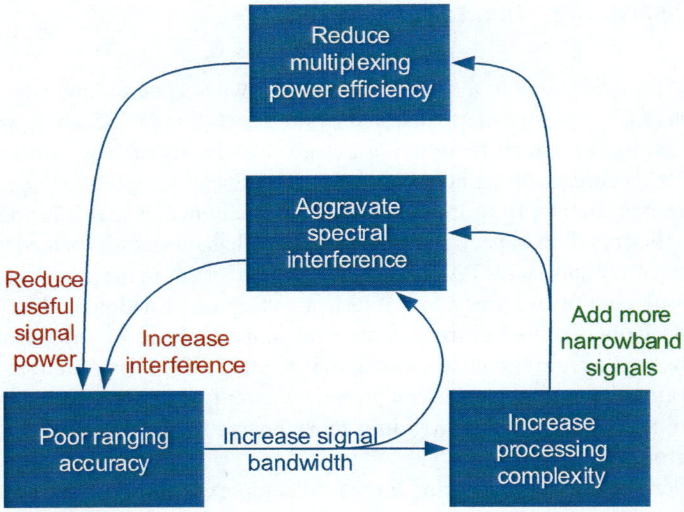

Fig. 8.1 A potential vicious cycle in future GNSS signal design

8.2.4 A Gordian Knot

Under the conventional idea of separately optimizing carrier frequency, spreading modulation, and multiplexing, a Gordian knot is emerging in future GNSS signal design. As shown in Fig. 8.1, in the independent design of these three key elements, it is difficult to reconcile the contradictions among service diversity, ranging accuracy, receiving complexity, RF compatibility and multiplexing efficiency.

Users always want signal ranging performance to be as high as possible while the receiving complexity is as low as possible. However, one cannot have both at the same time. The most straightforward way to cope with the high-performance demand is to increase the signal bandwidth and move the main spectrum component of signal away from the carrier frequency. However, since there are almost no contiguous segments of unoccupied spectrum remaining in the upper L-band, increasing the signal bandwidth will aggravate spectral interference between the new signal and existing signals.

Furthermore, a wider signal bandwidth and a complex subcarrier structure also results in a higher processing burden on receivers, which is unacceptable to low-end users. The direct way to further support low-end users is to add more narrowband signals with simple structures. Nevertheless, increasing signal numbers not only further increases spectrum interference, but also further reduces the power efficiency of multiplexing. This means that the useful signal power is reduced and the interference is increased. As a result, although the original intention is to improve the overall performance, the actual effect is to degrade the performance of each signal component.

In order to get out of the cycle of contradictions among measurement accuracy, services variety, RF compatibility, as well as multiplexing efficiency in satellite navigation signal design, it is necessary to break the routine, optimizing the carrier frequency, spreading modulation, and multiplexing jointly. The MCC signal, proposed in [3], is a promising solution to cut the above mentioned Gordian knot.

8.3 Multicarrier Constant-Envelope Composite Signal

The concept of multi-carrier signals originates from the field of wireless communications. Typical multicarrier communication signals include multi-tone signals, orthogonal frequency division multiplexing (OFDM) signals, and multicarrier code division multiple access (MC-CDMA) signals. However, the satellite navigation signal has two major differences compared with the communication signal: First, the core mission of the navigation signal is the high accuracy TOA estimation but not the data transmission. The TOA estimation processes based on multi-carrier communication signals, such as OFDM signal, are complex [4] and the inherent high RMS bandwidth performance advantage of multi-carrier signals is difficult to be adequately brought into play. Second, the vast majority of existing multi-carrier signals have a high peak-to-average ratio (PAR), which hinders their application for satellite transmission [5, 6].

Unlike the above-mentioned multi-carrier communication signals, as shown in Fig. 8.2, the MCC signal is like a "spectral transparent adhesive". It "glues" a plurality of narrowband signal components located in multiple spectral gaps together to form a wideband constant envelope signal, sharing a common up-converter, amplifier chain and antenna aperture. The core features of the MCC signal are:

- Sparsity in the frequency domain;
- Envelope constancy in the time domain;
- High flexibility on design elements such as the number of sub-bands, sub-band frequency spacing, the number of signal components in each sub-band, shape of the spreading waveform, the power ratio and relative phase of signal components;
- Transparency of the compositing to the receiver.

Compared with existing solutions, the MCC signal has the following unique advantages:

On the one hand, multicarrier is one of the most effective ways to utilize spectrum gap resources. As previously mentioned, in the increasingly crowded satellite navigation band, the absolute bandwidth of the newly added signal is severely limited. There are only some scattered frequency fragments available between main lobes of existing signals, as illustrated in Fig. 8.2a. However, the frequency difference between signal components in different sub-bands of the multi-carrier signal can transform this unfavorable factor into a favorable one. As illustrated in Fig. 8.2b, c, placing multiple sets of narrowband signals components in the fragment band

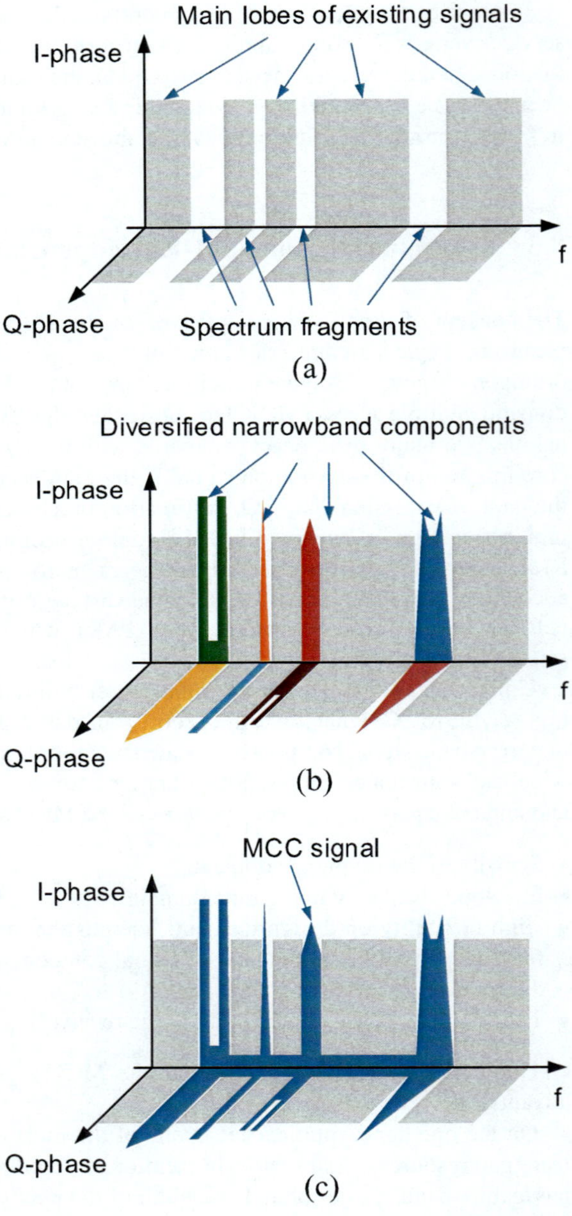

Fig. 8.2 The frequency domain diagrams of multicarrier constant-envelope composite signal: **a** The spectrum occupancy of existing signals in the band; **b** Inserting multiple diversified narrowband components into the spectrum gaps; **c** Multiplexing these narrowband components into a wideband constant-envelope composite signal

gaps and combining them into a wideband constant envelope signal can construct an MCC signal. The spectrum sparsity of such a signal can not only ensure adequate spectral separation with existing signals in the same band, but also provide a large RMS bandwidth for better ranging performance, resolving the contradiction between spectral efficiency and ranging performance.

On the other hand, the different narrowband components in the MCC signal can be optimized for targeted PNT services, with different spreading sequences, different spreading waveforms, different power allocations, and different data message structures and contents, to meet future diversified PNT requirements. At the same time, in MCC signals, those components are combined into a whole signal by CEM that is "transparent" enough for receivers, not only allowing narrowband receivers to process each component separately with low-complexity, but also allowing wideband receivers to process the total or partial components of this signal with a wide RMS bandwidth. That is, the MCC signal has inherent features of diversified receiving and processing strategies, which addresses the contradiction between power efficiency and service diversity.

In addition, the integrated structure of the MCC signal ensures a strong coherence of the transmission channel effects on each sub-band component, which creates conditions for joint processing of components in multi sub-bands, such as joint acquisition, joint tracking, and joint pseudorange extraction.

Given the above, an MCC signal can not only achieve outstanding ranging accuracy without significantly increasing the RF interference to the existing signals in the same band, but also provide users with diversified and targeted service without noticeably deteriorating the multiplexing efficiency. It provides a promising technique solution for the next generation GNSS signal design.

The key to the MCC is determining how to combine several flexible signals located at multiple different central frequencies with arbitrary power, chip rate, and spreading waveform into an integral signal with a constant envelope. CEMIC [7], as described in Sect. 7.7, presents the possibility for the realization of an MCC signal. Compared with other CEM techniques, CEMIC has a much higher design flexibility in the number of sub-bands, the number of signal components, power ratio and phase relationship among components, and the shape of spreading chip waveforms. CEMIC can be applied to any number of bipolar or multilevel spreading spectrum signals with arbitrary power distribution at one or more subcarrier frequencies. Such a high degree of design flexibility provides system designers great room in signal scheme optimization for varied navigation applications in the future.

8.4 Case Study of Adding a MCC Signal in L1 Band

As a sample application, consider adding a new MCC signal in the L1 band. Although current GNSSs do not have such a plan yet, through this specific example, one can clearly see the design process of an MCC signal, and the characteristics and advantages of this signal in both the transmitter and receiver.

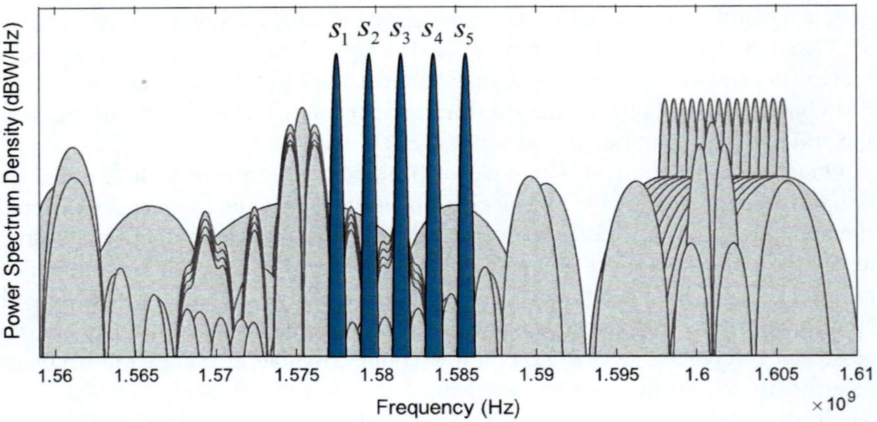

Fig. 8.3 Spectra of a newly added MCC signal in upper L band

8.4.1 Signal Description

As mentioned, the upper L-band has been overcrowded. All GNSSs broadcast their open and authorized service signals in this band. If a new wideband signal is added to this band, it will be hard to avoid significant spectrum overlapping with the existing signals. However, it is noted that most of the existing signals in this band have spectral nulls at 1575.42 ± 2.046, ±4.092, ±8.184, and ±10.23 MHz, etc. Thus, under the premise of ensuring good RF compatibility, a possible new signal solution is to place multiple narrowband components in these spectral gaps.

For simplicity, consider the case of multiplexing five narrowband components with BPSK-R(0.5) spreading modulation in this example. As shown in Fig. 8.3, the center frequencies of these five components are set to 1577.466 MHz, 1579.512 MHz, 1581.558 MHz, 1583.604 MHz, and 1585.65 MHz, respectively. In the transmitter, the carrier frequency of the composite MCC signal can be 1581.558 MHz. Thus, the subcarrier frequencies of components are $f_1 = -4.092$ MHz, $f_2 = -2.046$ MHz, $f_3 = 0$, $f_4 = 2.046$ MHz, and $f_5 = 4.092$ MHz, respectively.

Under the equal power assumption, the signal coefficient vector $\boldsymbol{r}$ in CEMIC method can be set to $[1, 1, 1, 1, 1]^{\mathrm{T}}$. In this design case, considering the constraint of implementation complexity, the number of SCS segment M should not be too large. Here we take $M = 8$, the shortest segment length $T_s = (32 \times 1.023\text{e}6)^{-1}$ s.

The PSD of these five narrowband components before the constant envelope reconstruction is shown in Fig. 8.4a. Following four steps of the CEMIC method presented in Sect. 7.7, the MCC signal is constructed, of which the theoretical and simulated PSDs are shown in Fig. 8.4b, c, respectively.

By comparing Fig. 8.4a, b, it is observed that after constant envelope reconstruction, the PSD of the MCC signal is different from that of the direct superposition of these five components. The difference is mainly reflected in the appearance of the IM

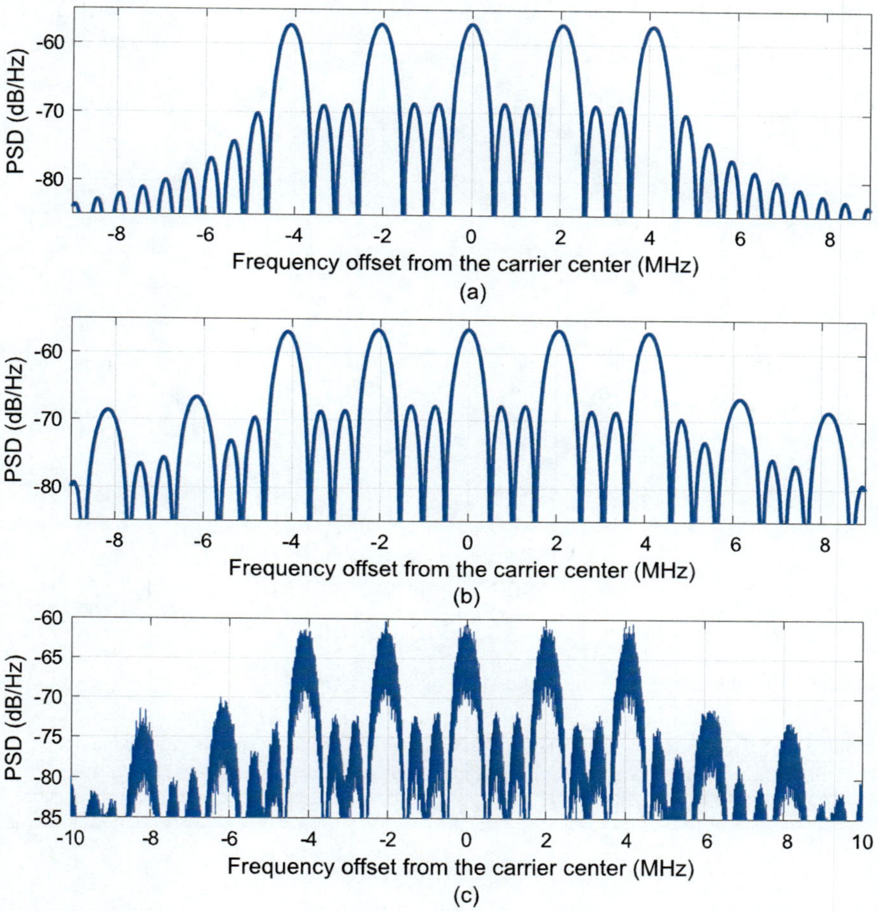

Fig. 8.4 **a** Theoretical PSD before constant envelope reconstruction; **b** Theoretical PSD after constant envelope reconstruction; **c** Simulated PSD after constant envelope reconstruction

term in MCC signal. It can be seen that the power of the IM term is much lower than that of the useful signal components, and the difference is at least 10 dB. In fact, the multiplexing efficiency of this example is 80.41%. That is, the newly added IM term accounts for only 19.6% of the total signal power, and its spectrum is distributed far away from the carrier center frequency.

Figure 8.5 shows the modulation constellation of the MCC signal. As illustrated in the figure, all the phase points are distributed on a circle. This feature enables the payload HPA to operate in its full-saturation mode to maximize power conversion efficiency.

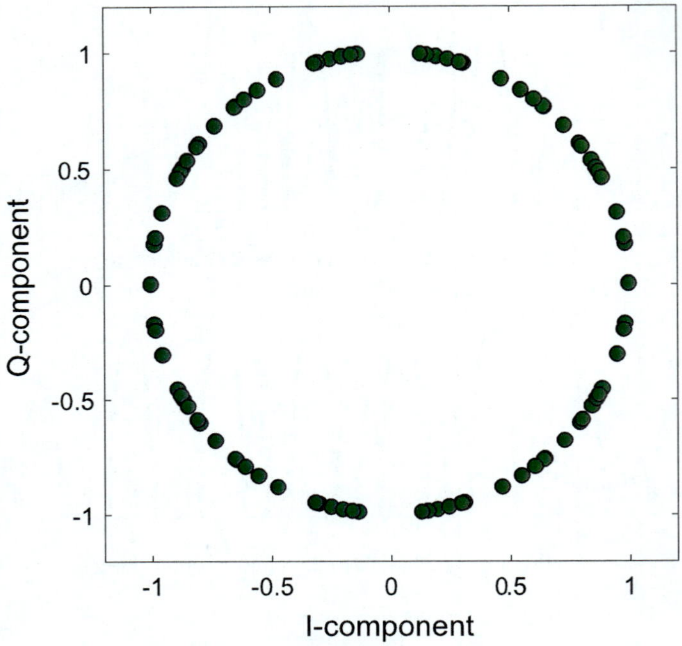

Fig. 8.5 Modulation constellation of the MCC signal

Table 8.1 SSC between the newly added MCC signal and the other signals in L1 band

SSC(dB)	BPSK-R(1)	MBOC(6,1,4/33)	BPSK-R(10)	BOC(10,5)
MCC	−77.8131	−79.5120	−75.1385	−75.5252

8.4.2 RF Compatibility Analysis

To evaluate the RF compatibility between the newly added MCC signal and the existing BPSK-R(1), MBOC(6,1,1/11), BPSK-R(10), as well as BOC(10,5) signals in the same frequency band, we calculate their SSC. The receiver front-end filter is assumed to be centered at 1575.42 MHz, with a single side bandwidth of 12 MHz, which is sufficient to receive the highest frequency component of the MCC signal. Table 8.1 shows the SSC of the existing signals with MCC signal.

It can be seen that the MCC signal maintains good RF compatibility with the existing signals by effectively utilizing the fragment band gaps between the main lobes of existing signal spectrum. It can be verified that if more sub-bands are employed, or moving f_5 from 1575.42 + 10.23 MHz to 1575.42 + 12.276 MHz, the RF compatibility between MCC signal and the BOC(10,5) signal can be further improved.

8.4.3 Diversified Processing Strategies

As previously mentioned, in addition to the effective utilization of the spectrum resource, another key advantage of the MCC signal is that it inherently has multiple receive modes, providing a variety of processing strategies for receivers with different performance and complexity constraints.

Since the MCC signal is composited in the digital baseband, the subcarrier phase of each component is completely coherent, and components within the MCC signal pass through the same transmission channel, the errors introduced by thermal noise, multipath, as well as the dynamic stress also have strong coherence. The receiver can either process these signal components separately, or jointly process multiple components or even the entire composite signal as a whole.

The simplest processing mode is treating the narrowband components in the MCC signal as different signals. Such a processing mode requires minimal processing complexity. If narrowband components employ BPSK-R spreading modulation, as discussed in this example, their acquisition and tracking methods can be directly inherited from the traditional cases, where the rectangular pulse spreading chip can be employed in the local replica. The CCF between the received MCC signal and the local replica of each signal component in this processing mode is shown in Fig. 8.6a.

If the receiver jointly processes three signal components, which are $s_2(t)$, $s_3(t)$, and $s_4(t)$, without loss of generality, the local replica can be

$$s_{\text{local}}(t) = s_2(t)\,\mathrm{e}^{\mathrm{j}2\pi f_2 t} + s_3(t) + s_4(t)\,\mathrm{e}^{\mathrm{j}2\pi f_4 t}\,. \tag{8.1}$$

The CCF between the received MCC signal and this local replica is shown in Fig. 8.6b. Further, the whole MCC signal can even be used as the local replica to realize the matching receiving, of which the CCF is shown as Fig. 8.6c.

In order to quantitatively compare the performance of above three processing modes, equivalent RMS bandwidth, correlation loss, and average multipath error envelope are used as evaluation indicators to measure the code tracking accuracy, the performance of acquisition and demodulation, and the multipath resisting performance, respectively.

Figures 8.7 and 8.8 show the equivalent RMS bandwidth and the correlation loss of these three processing strategies with respect to the front-end double-sided bandwidth, respectively. Figure 8.9 shows the average multipath error envelopes of these three processing strategies, with front-end double-sided bandwidth of 10 MHz, and MDR of −5 dB.

One can see from Figs. 8.7, 8.8 and 8.9 that in different processing strategies, the receiving performance presents an obvious graded characteristic. With a narrow bandwidth, the single-component processing mode has the minimum processing complexity, but the largest correlation loss and the lowest ranging accuracy. However, as an increasing number of components are processed jointly, for wideband receivers, not only is the correlation power loss decreased, but also a higher ranging accuracy as well as a better multipath resisting ability can be obtained. That means

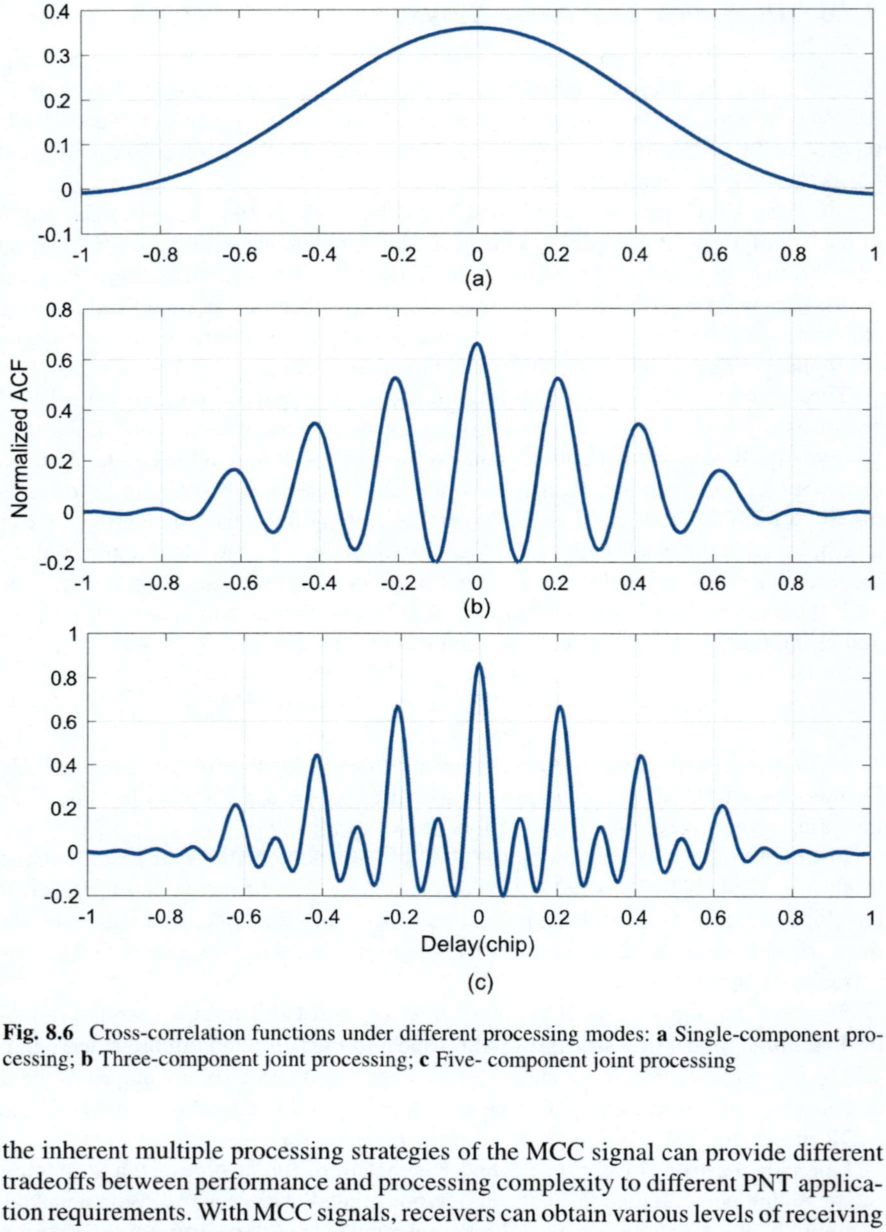

Fig. 8.6 Cross-correlation functions under different processing modes: **a** Single-component processing; **b** Three-component joint processing; **c** Five- component joint processing

the inherent multiple processing strategies of the MCC signal can provide different tradeoffs between performance and processing complexity to different PNT application requirements. With MCC signals, receivers can obtain various levels of receiving performance by jointly processing different subsets of signal components. This is one of the major advantages of the MCC signal.

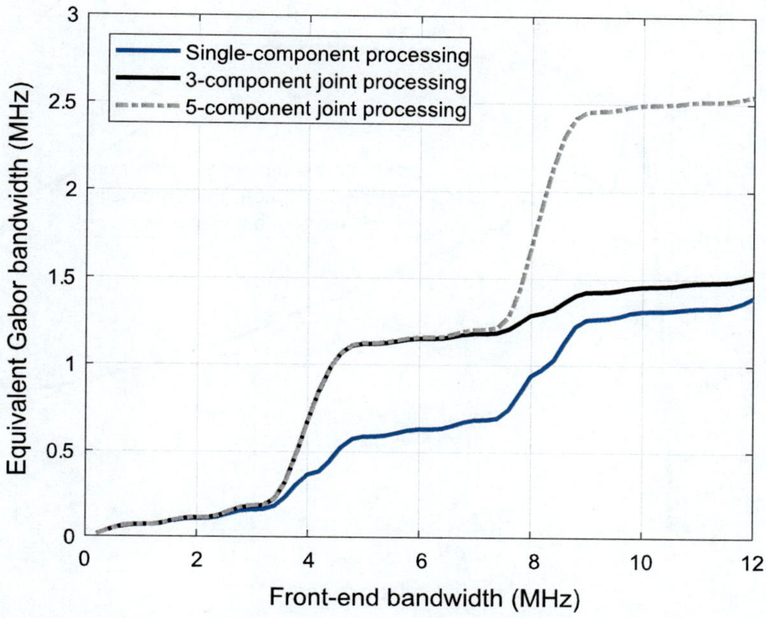

Fig. 8.7 Equivalent RMS bandwidth of different processing strategies

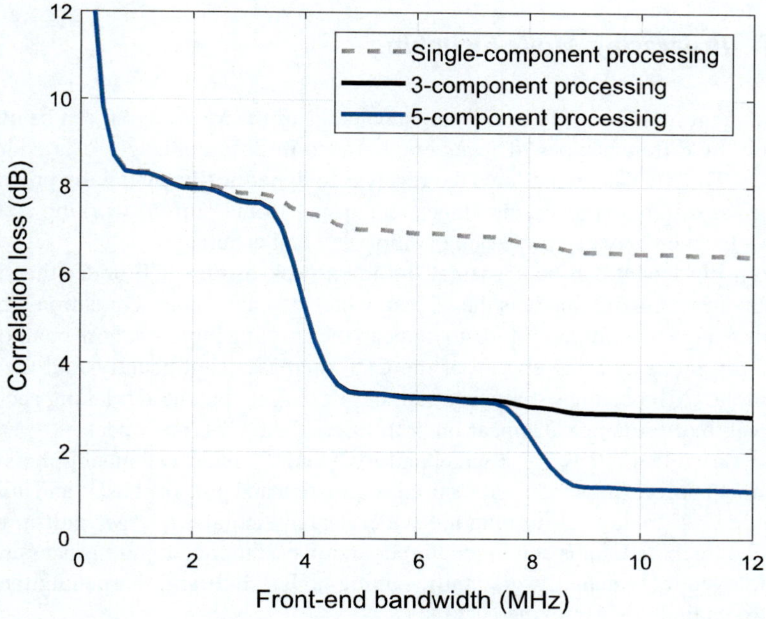

Fig. 8.8 Correlation loss of different processing strategies

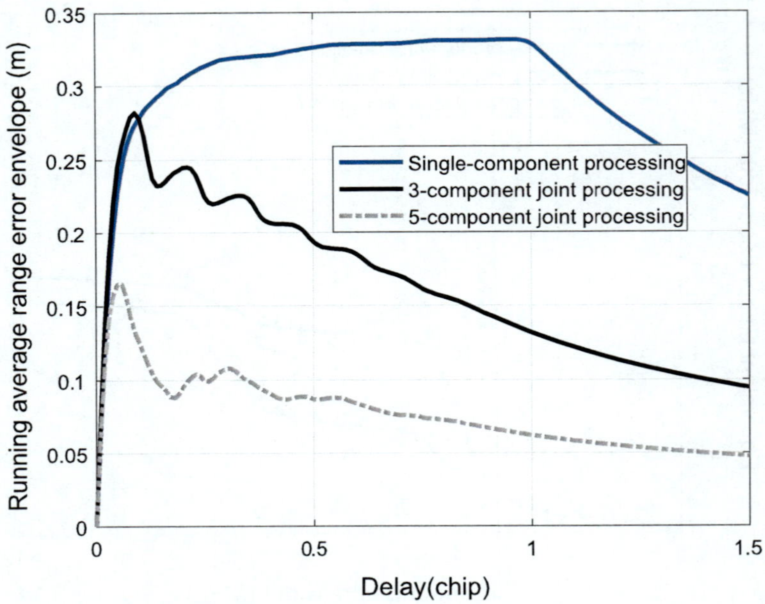

Fig. 8.9 Running average multipath error of different processing strategies

8.4.4 *Processing Mode Switching*

The inherent multi-strategy processing advantage of the MCC signal can be utilized not only by different types of receivers, but also in different stages of a wideband receiver. The MCC signal allows the receiver to dynamically switch the processing strategy at different processing stages, according to the current working status to achieve balance between processing complexity and accuracy.

From Fig. 8.6, it can be seen that the main peak of the CCF under the single-component processing mode is the widest, which can widen the acquisition bins and provide a large unambiguous pull-in range to the tracking loop. As more components are jointly processed, the energy of the CCF increases significantly, and the main peak of the CCF becomes sharper, which implies higher potential tracking accuracy. However, more side peaks appear on both sides of the CCF main peak.

One possible strategy for a wideband MCC signal receiver is using the single-component processing mode in the initial acquisition and pull-in phases, and utilizing the wide CCF main peak to obtain a wider search step and a larger pull-in range. After the tracking loop is stabilized, three- and five-component joint processing can be employed incrementally, gradually gaining higher SNR and sharpening the CCF peak to obtain higher tracking accuracy.

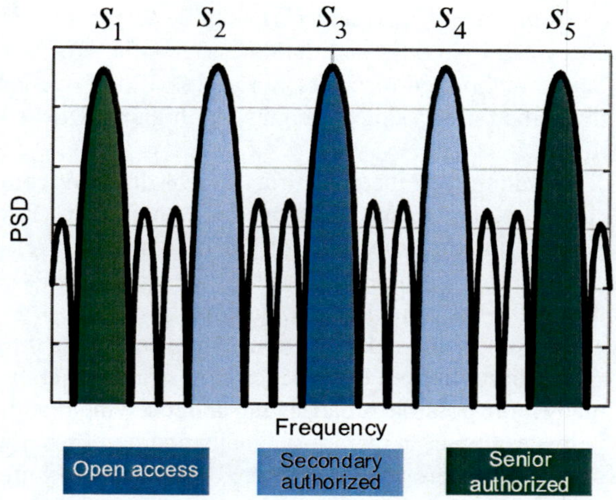

Fig. 8.10 A possible access permission configuration of MCC signal

The switching strategies of the processing mode of MCC signals are not limited to this simple mode. In fact, the multi-component multi-subcarrier structure of the MCC signal provides the possibility for the future receivers to explore the diverse switching strategies.

8.4.5 Selective Availability

Since the multiplexing used in MCC signal construction is sufficiently flexible, different signal components in the MCC signal can be configured with different PRN codes and different spreading modulation waveforms and modulated with different data messages. Therefore, the service provider can control the access permissions and providing selective performance to different user levels. The receiver selects the corresponding processing mode according to its own privilege level and thus obtains the available acquisition, tracking, and demodulation performances.

For example, as shown in Fig. 8.10, in the five-component design case provided in this section, $s_3(t)$, which is located on the carrier frequency, can be assigned to be the open access signal, of which the PRN code generation method and the data message structure are fully open. All receivers can access this component with the single-component processing mode, thus obtaining a relatively low SNR, and a basic ranging accuracy level.

Components $s_2(t)$ and $s_4(t)$, which are with relatively low subcarrier frequencies, can be assigned as the secondary authorized signals. Their PRN code generating information and data message structures are only provided to the authorized secondary users. These users can access three components, so that joint processing and dynamic mode switching strategies can be used to obtain the improved performance.

Components $s_1(t)$ and $s_5(t)$ can be assigned as senior authorized signals, with encrypted PN codes and data message structures, serving authorized senior users. Senior authorized receivers can access all the five components, to obtain the most diversified processing strategies, the highest SNR, and the highest ranging performance.

In addition, if the data structures of different components are well-designed to carry complementary messages, for authorized users who can access multiple components, the time to first fix (TTFF) can be effectively reduced. In theory, the TTFF of a receiver that jointly processes five components can be shortened by 80% over that of the basic single-component receiver.

The case study in this section demonstrates that the MCC signal has a high degree of flexibility in both the broadcasting strategy and the receiving strategy. There are many more possible broadcasting and receiving modes of MCC than those discussed in this example. In fact, this signal structure offers a wide design space for both the system providers and the receiver developers in future.

8.5 Conclusions

As a significant infrastructure, GNSS has a long development cycle. This characteristic means that we can only employ existing techniques to meet the demands over the next few decades. Although it is impossible to envision GNSS products and services further out in time, we can enable future development by implementing excellent signal designs with higher adaptability and flexibility.

The contradiction between the need for performance improvement and the fact that power and spectrum resources are limited will be more serious in the next generation of GNSS signal designs. In order to resolve this contradiction, this chapter introduces the concept of multi-carrier constant envelope signal and studies its feasibility as the next generation satellite navigation signal. A corresponding design method based on the CEMIC technique is given, and an example is presented to demonstrate the RF compatibility, typical receiving strategies, corresponding performances and selection availability of MCC signals. The analyses show that the MCC signal can make full use of the existing spectrum resources, providing both various broadcast strategies and multiple receiving strategies with a variety of performance levels for different categories of users. This technique can serve as a new practical solution to the design of next generation satellite navigation signals.

References

1. Pratt AR, Owen JI (2003) BOC modulation waveforms. In: Proceedings of the 16th international technical meeting of the satellite division of the institute of navigation (ION-GPS/GNSS-2003), pp 1044–1057

2. Zhang X, Yao Z, Zhang X, Lu M (2011) A method to optimize the spreading code chip waveform in sense of Gabor bandwidth. In: Proceedings of the 24th international technical meeting of the satellite division of the institute of navigation (ION GNSS 2011), 2011, pp 1299–1304
3. Yao Z, Ma J, Zhang J, Lu M (2017) Multicarrier constant envelope composite signal - a solution to the next generation satellite navigation signals. In: Proceedings of the international technical meeting of institute of navigation satellite division (ION GNSS+ 2017), 2017, pp 1520–1533
4. Thevenon P et al (2009) Pseudo-range measurements using OFDM channel estimation. In: Proceedings of the 22nd international technical meeting of the satellite division of the institute of navigation (ION GNSS 2009), pp 481–493
5. Mateu I, et al (2010) A search for spectrum: GNSS signals in S-band, part 2. Inside GNSS 2010, October, pp 46–53
6. Emmanuele A et al (2011) Evaluation of filtered multitone (FMT) technology for future satellite navigation use. In: Proceedings of the 24th international technical meeting of the satellite division of the institute of navigation (ION GNSS 2011), pp 3743–3755
7. Yao Z, Guo F, Ma J, Lu M (2017) Orthogonality-based generalized multicarrier constant envelope multiplexing for DSSS signals. IEEE Trans Aerosp Electron Syst 53(4):2017

Chapter 9
Conclusion

Everything is deemed possible except that which is impossible in the nature of things.

— California Civil Code

This book has provided a systematic discussion of satellite navigation signal structure and a series of key design elements, as well as a detailed description of the theory and technology of signal design. The text has also briefly summarized the evolution of signal design in the development of satellite navigation systems over the course of several decades. In this conclusion, we will attempt to discuss two issues that what type of signal can be regarded as a good satellite navigation signal, and how future satellite navigation signals will develop.

9.1 Performance Evaluation of Satellite Navigation Signals

First, we will discuss what type of satellite navigation signal is deemed to be "good". This is equivalent to the question of quantifying the performance of satellite navigation signals and their evaluation criteria. This is not only an unavoidable issue for developers of satellite navigation systems, but also a key concern of receiver developers. The question of the performance and evaluation of satellite navigation signals is important, and the answers can be extremely complex. One needs to establish a scientific performance index system for satellite navigation signals and determine reasonable performance evaluation methods and evaluation criteria.

In addition to these parameters, the question involves the constraints that a specific satellite navigation system imposes on the signal design technique, the role that a certain satellite navigation signal plays in the satellite navigation system, the various requirements on the receiver's performance imposed by the environment and application purposes, and the relationships among different satellite navigation systems. Thus conducting a comprehensive assessment of the performance of a navigation sig-

Z. Yao and M. Lu, *Next-Generation GNSS Signal Design*, Navigation: Science and Technology 6, https://doi.org/10.1007/978-981-15-5799-6_9

nal can be very complicated. In fact, so far there has not yet been systematic research on this issue. Existing literature provides some basic evaluation of one aspect or one specific indicator of the navigation signal. Research in this field must be improved in the future. In view of this reality, we can only offer a brief introduction of the current scattered research, combined with our own experience in signal design research and engineering practice.

We must point out that the satellite navigation signal is an entity composed of multiple components. The quality of the signal does not depend entirely on the specific performance index of one or even several components. In general, to evaluate the performance of a satellite navigation signal according to its merits and disadvantages it is important to consider the following aspects.

1. Forward-looking design concept

Since the development of a satellite navigation system can take a long time, and the system's life cycle after its completion is also quite long, we must consider the user requirements and application methods in the next 10, 20, or even 30 years. The GPS L1 C/A code signal designed in the 1970s is an example. Currently, even though its successor, the L1C signal, has been deployed, the C/A code signal continues to be used for civilian services in GPS applications, and so far there is no indication that the C/A signal will be discontinued in the near future. The BDS signal, which has been designed in recent years, will be officially put into use in 2020. After the BDS system is completed, it will be used for at least 10 to 20 years. Therefore, to evaluate the merits of a navigation signal, we must see whether the signal design is sufficiently forward-looking. Furthermore, designers should take user requirements and scenarios for the next few decades into consideration. For example, a significant trend is that in the future, most users will use GNSS service in the information network environment. Therefore, signal designers must consider this application feature.

2. Superior tracking and ranging performance

In general, the advanced nature of satellite navigation signals is reflected in two aspects: the scientific and innovative signal design technique, and the improved performance of the signal. In particular, the tracking sensitivity and ranging accuracy of the signal should be the focus of the evaluation of signal performance.

The performance evaluation of a specific signal can be carried out in two aspects, namely theoretical calculation and actual testing. The theoretical calculation aspect has the advantages of objectivity, comprehensiveness, and precision. It can independently, completely, and accurately calculate the various properties of the signal itself, while avoiding the influence on the performance evaluation from the non-ideal characteristics of the signal transmitting device and the receiving device. It is the most basic and credible assessment method and is irreplaceable. Of course, actual performance testing is also a means of evaluating the signal performance and verifying the correctness of the theoretical calculation, but is limited to the conditions that only portions of performance indicators can be accomplished through testing, even in the laboratory. Moreover, in the test results, the signal performance is coupled with the implementation of the transmitting device and the receiving device and is particularly affected by the undesired performance of the receiving device. Therefore, it does not necessarily accurately reflect the actual performance of the signal. Hence, the evaluation of signal performance should focus on the results of theoretical analysis, while referring to the results of actual tests as corroborative data.

We would like to point out that the evaluation on a satellite navigation signal is an entity composed of multiple indicators. For a comprehensive evaluation of signal performance, one should focus on the signal performance parameters which show the highest payoff for enhancement and avoid fruitless efforts to enhance those parameters which are already sufficient. When determining the weight of various performance indicators, we need to focus on the ranging performance of the signal, especially the tracking sensitivity and ranging accuracy that are specific to the satellite navigation system. These two indicators are the most significant, because once tracking is disabled, the receiver will not function. Other indicators, such as demodulation thresholds, are expected to meet basic requirements rather than be improved as much as possible. This is because under special conditions, even if the error rate of demodulation is high, due to the slow rate of change of the message information, the receiver can still work normally. Moreover, the message information can also be obtained from other sources.

3. Adaptability of complex environments

The complete linkage of satellite navigation signals includes signal generation, launch, transmission, reception, and application, in which the transmission environment constitutes an important part of the entire chain. Specifically, the general transmission environment includes the ionosphere and troposphere that the signal travels through, the multipath effect introduced by the environment in which the receiver is located, and possible unintentional and malicious electromagnetic interference. These factors can cause distortion of the signal and can seriously affect the overall performance of the system. Meanwhile, as the signal transmission environment is completely open, it is difficult to improve or control it by artificial means. In order to overcome the influence of environmental factors as much as possible, the traditional approach is to establish a better environmental error model and develop better receiver signal processing algorithms. In fact, a well-designed signal can provide greater environmental adaptability. Then, through appropriate receiver signal processing algorithms, the signal can greatly improve the positioning and navigation performance. In particular, for multipath and interference, designing better navigation signals is an important aspect of research in the field of signal design. Environmental adaptability is also a crucial element of signal design.

4. Compatibility and interoperability with other GNSS

In the course of the development of international GNSS, it is inevitable that there will be a coexistence of the four major GNSSs in the near future. Therefore, in addition to the signal performance in the system, the inter-system performance, namely the signal compatibility and interoperability, should also be taken into consideration. This is the direction of GNSS development that the international organizations have been advocating.

5. The implementation complexity of the technology

The complexity of implementing the signal design technique is due to the on-board payload and the user terminal. For a broadcast system, the number of transmitting devices is limited, while the number of user devices is large. Therefore, we mainly consider the implementation complexity of the user equipment rather than the transmitting device. There are two things to consider in terms of the complexity of user equipment. One is the complexity of the hardware of the user equipment and the implementation complexity of the software algorithm, and the other is the reusability of the receiving channel and software algorithms with other GNSSs. Considering that most of the future GNSS signals use some common frequency bands, different baseband processing channels can share the RF front-end analog hardware, so we mainly consider the complexity and reusability of the baseband channel and software algorithm. Reusable capability is actually the reflection of signal interoperability. For civil signals, where GPS has obvious first mover advantage and high market share, the future direction of development path for receivers must be multi-system joint receiving. Therefore, the interoperability of newly designed signals in multi-system receivers, especially for GPS signals, is also an important part of the evaluation of signal design.

9.2 Possible Future Developments of Satellite Navigation Signals

In the past few decades, and especially in the past 10 years with the modernization of GPS and the construction of Galileo system and BDS, satellite navigation signals have been in a state of constant evolution. Satellite navigation receiver technology has also been developed and become relatively mature. However, some basic issues of satellite navigation signal design and receiver development have not been completely solved. The core of these problems is mainly reflected in the contradiction between limited spectrum resources, limited transmission power, and the increasing demand for PNT services. These contradictions are becoming more prominent with the development of GNSSs. Therefore, in the future development of satellite navigation signals and receivers, we will continue to face the problems that exist in the current design of signals and receivers.

Firstly, both in-band and out-of-band RF interference in satellite navigation will only become more serious in the future. Therefore, in an environment with various interference sources, maintaining reliable and continuous operation of the navigation system is still a major issue faced by system construction and application.

Secondly, as the demand for positioning accuracy and service coverage continues growing, there will be increasing demand for the availability of navigation systems in harsh environments, such as indoor spaces, tunnels, and urban canyons. The joint design of signals and receivers is necessary to deal with problems, such as power attenuation caused by severe occlusion of satellites and severe multipath interference.

Thirdly, the upgrade of navigation signals can not only be driven by the navigation system itself, but also by the development of a national PNT architecture. With the navigation system as its core, a national PNT architecture uses the satellite navigation signal as a link to combine various enhancements, supplements, and backup systems.

All of these aspects have placed new and higher requirements on the design of satellite navigation signals. Although we cannot pin all our hopes of solving these problems on the improvement of signal design, good signal design is essential to finding a solution, and is also the way with the lowest cost and the fewest technical requirements. In summary, it is these requirements that drive the continuous advancement of satellite navigation signal technology and become the main basis for us to predict the development direction for satellite navigation signals.

A basic assumption for predicting the development trend of satellite navigation signals is that, in the foreseeable future, the basic concept of satellite navigation systems will not undergo subversive changes. That is, the navigation signal is still the carrier of the satellite navigation system's transmission ranging code and satellite ephemeris. Therefore the development of satellite navigation signals should be on a gradual rather than a wholesale change. Advanced spreading modulation techniques enable the signals of multiple GNSSs to share spectrum resources more effectively, and significantly improve the ranging performance of the signals in noise, multipath, and interference environments. The optimization of navigation message and coding techniques enables a portion of the transmitted signal power to be allocated to a pilot channel under the condition that the specified demodulation capability is satisfied. In this way, the performance of acquisition, tracking, and ranging can be improved. Advanced multiplexing technology allows satellites to simultaneously broadcast multiple different service signals at one frequency point, and each signal can be optimized for different users. Over the past few decades, the development of navigation signals has mainly been reflected in the aspects of spread-spectrum modulation, message structure and coding, and on-board multiplexing technology. Future development will also be mainly focused on these aspects. Therefore, we can speculate that the possible future trends in the evolution of navigation signals are as follows:

1. In terms of spreading modulation, will future navigation systems use signals with larger bandwidth?

The bandwidth of the existing navigation signals ranges from a few megahertz to tens of megahertz. It is generally considered that a wider signal bandwidth results in improved ranging performance and a better ability to suppress interference and multipath. However, the frequency resources available for satellite navigation are very limited. The corresponding cost of larger bandwidth receivers will increase greatly, and the possibility of being disturbed by narrowband interference will also increase. The advantages of large bandwidth signals in satellite navigation have not been fully realized so far. Therefore, the prospect of using larger bandwidth spread-spectrum modulation, at least in the L-band, for satellite navigation is not optimistic.

2. What will happen to the basic structure of future satellite navigation signals?

Looking back at the development of satellite navigation systems, we see that the first-generation satellite navigation systems, such as Transit in the USA, adopted a pulse signal structure similar to radar signals. Early second-generation satellite navigation systems, including legacy GPS and GLONASS built in the 1990s, used DSSS/BPSK-like signals similar to satellite communications. The recent generation of satellite navigation systems, including modernized GPS, Galileo system, and BDS-3, have undergone major changes in signal structure.

One of the main features is the use of signal structures with orthogonal pilot and data components. Without regard to the pulse signal structure of the first-generation satellite navigation system, which has been completely phased out, the signals of the second-generation satellite navigation systems also experience the evolution of signal structures from the ranging data all-in-one signal structure in the early stage to the recent pilot-plus-data composite signal structure.

According to this evolution and considering the possible employment of small satellites in future satellite navigation systems, it is very likely that the pilot and data components will be further separated in the next generation satellite navigation signals. That is, for a complete satellite navigation signal, its pilot and data components will be transmitted by the MEO satellite and the LEO satellite respectively and will be simultaneously received at the receiver to complete the positioning function. This possible signal structure might be called a spatially separated navigation signal.

The advantage is that the design of the pilot component signal can be specifically optimized for the ranging function. Meanwhile, the data component signal is optimized for the message transmission, without any mutual interference. It can further utilize the characteristics of high-orbit MEO satellites and low-orbit LEO satellites to achieve higher tracking sensitivity and better ranging precision. In addition to the improvement of outdoor positioning accuracy using this constellation structure, it is also possible to achieve GNSS indoor positioning, thereby expanding the GNSS service coverage from outdoors to indoors. This may become the most major breakthrough since the birth of satellite navigation technology.

3. Does the future navigation system broadcast multiple dedicated signals for different applications at the same time, or transmit a small number of multi-purpose signals?

Due to the wide range of applications of GNSS, various practical applications require the system to provide a wide range of services. To meet this need, a variety of targeted dedicated signals can be transmitted, each of which is optimized and customized for different applications, such as aerospace, high-precision surveying, and consumer electronics. Another concept is that the satellite navigation system only transmits a small number of multi-purpose signals and different users obtain their required services through the receiver's flexible signal processing strategies. To this end, in the current signal design, several existing major GNSS systems have totally different design ideas, and have not formed a common direction of development. However, for emerging navigation systems that do not face such concerns, it is obvious that the latter is an ideal choice, but it poses more stringent requirements for signal designers.

4. Will future satellite navigation systems continue to use the existing L-band, or will they continue to develop new bands, such as for mobile communications?

With the construction of new navigation systems and the modernization of existing ones, the number of navigation signals is increasing, and the L-band is becoming crowded due to limited bandwidth. From this perspective, the expansion of satellite navigation bands is absolutely necessary. The ITU has allocated some resources in the S-band and C-band for satellite navigation. However, in the current situation, there are still many uncertainties surrounding how to develop and when to start using these frequency resources. On the one hand, user demand has not yet posed higher requirements on new bands, as the current L-band navigation can basically meet today's user needs. On the other hand, the transmission characteristics of the S-band and C-band as navigation bands have yet to be thoroughly investigated, and it is still difficult to develop low-cost multi-band navigation receivers with existing technology. Therefore, GPS and GLONASS which are undergoing modernization, have not yet proposed a formal plan to explore new frequency bands, and neither Galileo nor BDS have any clear plan for developing new frequency bands.

5. Will future satellite navigation systems be able to achieve deep integration with information networks such as mobile communications?

Currently, it is possible to broadcast enhanced GNSS information over mobile communication networks, enabling mobile communication users with GNSS receiver chips to obtain better GNSS services. This technology has promoted the development of the location-based service industry. It can be considered that GNSS and mobile communication have achieved partial integration. Undoubtedly, with the accelerating pace of informatization, mobile Internet, and the Internet of Things (IoT), GNSS-enhanced networks will soon be widely used. Most GNSS users in the future, including industry users and consumers, will use services in the network environment. Completely independent navigation and positioning devices will mostly disappear. Therefore, in terms of application, the demand for further integration of GNSS and mobile communications is increasing day by day, and the integration of information networks such as GNSS and mobile communication will be strengthened at the information level.

Furthermore, a bolder speculation is that future GNSS and mobile communications can even be integrated at the signal level. Referring to the signal structure in which the pilot component and the data component of the GNSS signal are separated as described above, it is conceivable that by combining the design of the GNSS and the mobile communication signal, it is possible to allocate all the signal power of the GNSS to the pilot component, so that the receiver can utilize all GNSS available signal power for acquisition, tracking, and ranging. Meanwhile, data such as ephemeris would be acquired via means such as mobile communication, the Internet, and the GNSS-enhanced network, enabling network users to obtain better GNSS services. This trend also deserves serious consideration.

In summary, due to the rapid development of electronic information and related technologies, and the growing demands of satellite navigation users, the future navigation systems will develop and change rapidly. Satellite navigation signals, which are important components of satellite navigation systems, will also undergo constant change. This will allow GNSS to offer diversified services with better performance, and will even bring revolutionary progress to the satellite navigation system itself.